THE CONTRACTOR'S FIELD GUIDE

Second Edition

THE
CONTRACTOR'S
FIELD GUIDE

Second Edition

Paul I. Thomas

Library of Congress Cataloging-in-Publication Data

Thomas, Paul I.
    The contractor's field guide / Paul I. Thomas. — Rev. ed.
        p. cm.
    Includes bibliographical references and index.
    ISBN 0-13-012416-8 (lexotone)
    1. Contractors' operations—Handbooks, manuals, etc.   I. Title.
TA210.T56   1999
690'.068–dc21   99–40952
                                                                         CIP

Printed in the United States of America

10   9   8   7   6   5   4   3

PRENTICE HALL
Paramus, NJ 07652

On the World Wide Web at http://www.phdirect.com

*To Ruth D. Thomas, my wife.
Her patience and understanding have
made the planning and writing of this
book possible.*

ABOUT THE AUTHOR

Paul I. Thomas was a Professional Engineer licensed by the State of New Jersey to practice Valuation and Construction Engineering. He was a graduate of the University of New Mexico with a BS in Civil Engineering. His experience in construction estimating spans half a century. He was the author of the standard reference used by property loss adjusters, *How to Estimate Building Losses and Construction Costs*.

HOW TO USE THIS GUIDE

Although this Field Guide is designed for the general contractor, it provides important checkpoints for the subcontractor, owner, architect, and others involved in construction projects. The Guide will also prove helpful to realtors, appraisers, property claim personnel, and to lawyers.

Section I, Legal and Practical Relationships, reviews the important aspects of contracts and documents, such as:

- specifications
- building codes, permits, and inspections
- construction financing
- insurance and bonds
- labor-management relations and legislation

Contract documents and jurisdictional disputes are examined; resolution options are offered along with "helpful hints" for solutions. Project safety and health programs are outlined with a comprehensive checklist for guidance.

Section II, Construction Management, offers a guide for planning and scheduling the construction project and includes the Critical Path Method (CPM). Progress estimating and cost control are discussed and appropriate measures suggested.

Section III, Construction Materials and Methods, reviews the fundamentals and principles of estimating construction costs. It suggests ways to avoid errors in scoping and estimating materials and labor.

This section contains tables and formulas for rapidly determining material quantities and approximating productivity of workers. It covers major trades from construction site exploration through the complete erection and finishing of a building project. Several figures illustrate construction processes. Numerous examples and checklists to guide the reader are included. A review of mathematics is presented in the last chapter.

CONTENTS

SECTION I

LEGAL AND PRACTICAL RELATIONSHIPS

CHAPTER 1

CONTRACTS AND DOCUMENTS

Construction contracts may be simple oral agreements on small projects where the parties have confidence in each other, but most construction contracts consist of formal written documents that clearly define the rights, obligations, and responsibilities of all parties. Because of the size and complexities of many construction projects, several documents are generally necessary. They include but are not limited to the following:

- Plans and drawings
- Written detailed specifications
- Written agreement between owner and contractor
- Instructions covering changes during construction
- Instructions for the selection of subcontractors
- Copies of bids by subcontractors to general contractor
- Proposal of work by general contractor and subcontractors.

STANDARD FORMS OF CONTRACT AGREEMENTS

General Conditions of the Contract for Construction

> "Portions of this text were derived from AIA Document A201, copyrighted 1997 by the American Institute of Architects, and are used here with the permission of the AIA under license number 90006. *Further reproduction is prohibited* without the express written consent of the American Institute of Architects, 1735 New York Avenue, NW, Washington, D.C. 20006."

The American Institute of Architects (AIA) Document A201, 1997, is used extensively in the construction industry. It is particularly notable for its comprehensive treatment of construction elements such as services, rights, responsibilities, and relationships of the construction parties.

The 1997 edition of the AIA Document A201, General Conditions of the Contract for Construction, contains fourteen Articles, the titles of which are:

1. General Provisions
2. Owner
3. Contractor
4. Administration of the Contract
5. Subcontractors
6. Construction by Owner or by Separate Contractors
7. Changes in the Work
8. Time
9. Payments and Completion
10. Protection of Persons and Property
11. Insurance and Bonds
12. Uncovering and Correction of Work
13. Miscellaneous Provisions
14. Termination or Suspension of the Contract

Article 1. GENERAL PROVISIONS

Defines the Contract Documents, the Contract, the Work and services required, the Project, Drawings and Specifications, Execution of the Contract by the Contractor, Correlation and Intent of the Contract Documents, Ownership and Use of Architect's Drawings, Specifications and Other Documents.

Article 2. OWNER

Defines the Owner, information and services required of the Owner, Owner's right to stop the Work, and right to carry out the Work if the Contractor defaults.

Article 3. CONTRACTOR

Defines the Contractor, the Contractor's responsibility to review the Contract Documents and field conditions, supervise and direct the Work in accordance with the Contract Documents, provide and pay for labor, materials equipment, tools, construction equipment and machinery, water, heat, utilities, transportation, and other facilities.

The Contractor warrants to Owner and Architect materials and equipment to be of good quality and Work will be free of defects. The Contractor will pay taxes, secure and pay permits, fees and notices, and all allowances provided in the Contract Documents. The Contractor will furnish a competent superintendent and necessary assistants. The Contractor will prepare a construction schedule for the Work and will maintain at the site for the Owner and available to the Architect copies of Drawings, Specifications, addenda, Change Orders, approved shop drawings, Product Data, samples, and similar required submittals.

3.13 Use of Site

3.13.1 The Contractor shall confine operations at the site to areas permitted by law, ordinances, permits and the Contract Documents and shall not unreasonably encumber the site with materials or equipment.*

3.14 Cutting and Patching

3.14.1 The Contractor shall be responsible for cutting, fitting or patching required to complete the Work or to make its parts fit together properly.*

(Subparagraph 3.14.2 has been omitted here.)

3.15 Cleaning Up

3.15.1 The Contractor shall keep the premises and surrounding area free from accumulation of waste materials or rubbish caused by operations under the Contract. At completion of the Work the Contractor shall remove from and about the project waste materials, rubbish, the Contractor's tools, construction equipment, machinery and surplus materials.*

3.15.2 If the Contractor fails to clean up as provided in the Contract Documents, the Owner may do so and the cost thereof shall be charged to the Contractor.*

Article 4. ADMINISTRATION OF THE CONTRACT

Defines the Architect and states the duties and responsibilities as set forth in the Contract Documents shall not be restricted, modified, or extended without the written consent of Owner, Contractor, and Architect.

4.2 Architect's Administration of the Contract

4.2.1 The Architect will provide administration of the Contract as described in the Contract Documents, and will be the Owner's representative (1) during construction, (2) until final payment is due and (3) with the Owner's concurrence, from time to time during the correction period described in Paragraph 12.2. The Architect will have authority to act on behalf of the Owner only to the extent provided in the Contract Documents, unless otherwise modified by written instrument in accordance with other provisions in the Contract.*

4.2.2   The Architect, as a representative of the Owner, will visit the site at intervals appropriate to the stage of the Contractor's operations (1) to become generally familiar with and to keep the Owner informed about the progress and quality of the portion of the Work completed, (2) to endeavor to guard the Owner against defects and deficiencies in the Work, and (3) to determine in general if the Work is being performed in a manner indicating that the Work, when fully completed, will be in accordance with the Contract Documents. . . . *

(Subparagraph 4.2.3 has been omitted here.)

4.2.4   Communications Facilitating Contract Administration

Except as otherwise provided in the Contract Documents or when direct communications have been specially authorized, the Owner and Contractor shall endeavor to communicate with each other through the Architect about matters arising out of or relating to the Contract. Communications by and with the Architect's consultants shall be through the Architect. Communications by and with Subcontractors and material suppliers shall be through the Contractor. Communications by and with separate contractors shall be through the Owner.*

4.2.5   . . . The Architect will review and certify the amounts due the Contractor and will issue Certificates for Payment in such amounts.*

4.2.6   The Architect will have authority to reject Work which does not conform to the Contract Documents. . . .*

4.2.7   The Architect will review and approve or take other appropriate action upon the Contractor's submittals such as Shop Drawings, Product Data and Samples, but only for the limited purpose of checking for conformance with information given and the design concept expressed in the Contract Documents. . . .*

4.2.8 The Architect will prepare Change Orders and Construction Change Directives, and may authorize minor changes in the Work. . . .*

4.2.9 The Architect will conduct inspections to determine the date or dates of Substantial Completion and the date of final completion. . . .*

(4.2.10 has been omitted here.)

4.2.11 The Architect will interpret and decide matters concerning performance under and requirements of the Contract Documents on written request of either the Owner or Contractor. . . .*

4.3 Claims and Disputes
Under Paragraphs 4.3 and 4.4 of Article 4, the following provisions are described, among others:

- Time Limits on Claims
- Claims for Concealed or Unknown Conditions
- Claims for Additional Cost
- Claims for Additional Time
- Injury or Damage to Person or Property
- Resolution of Claims and Disputes
- Arbitration

Article 5. SUBCONTRACTORS

This Article, under Paragraphs 5.1.1 and 5.1.2, defines Subcontractor and Subsubcontractor respectively.

5.2 Award of Subcontracts and Other Contracts for Portions of the Work
5.2.1 Unless otherwise stated in the Contract Documents or the bidding requirements, the Contractor, as soon as practicable after award of the Contract, shall furnish in writing to the Owner through the Architect the names of persons or entities (including those who are to furnish materials or equipment fabricated to a special design) proposed for each principal portion of the Work. The Architect will promptly reply to the Contractor in writing stating

whether or not the Owner or the Architect . . . has reasonable objection to any such proposed person or entity . . .*

Article 6. CONSTRUCTION BY OWNER OR BY SEPARATE CONTRACTORS

6.1 Owner's Right to Perform Construction and Award Separate Contracts

6.1.1 The Owner reserves the right to perform construction or operations related to the Project with the Owner's own forces, and to award separate contracts in connection with other portions of the Project or other construction or operations on the site under Conditions of the Contract identical or substantially similar to these including those portions related to insurance and waiver of subrogation. If the Contractor claims that delay or additional cost is involved because of such action by the Owner, the Contractor shall make such claim as provided elsewhere in Paragraph 4.3.*

Article 7. CHANGES IN THE WORK

7.1 Changes

7.1.1 Changes in the Work may be accomplished after execution of the Contract, and without invalidating the Contract, by Change Order, Construction Change Directive or order for a minor change in the Work, subject to the limitations stated in this Article 7 and elsewhere in the Contract Documents.*

7.1.2 A Change Order shall be based upon agreement among the Owner, Contractor and Architect; a Construction Change Directive requires agreement by the Owner and Architect and may or may not be agreed to by the Contractor; an order for a minor change in the Work may be issued by the Architect alone.*

7.1.3 Changes in the Work shall be performed under applicable provisions of the Contract Documents, and the Contractor shall proceed promptly, unless otherwise provided in the Change Order, Construction Change Directive or order for a minor change in the Work.*

7.2 Change Orders

7.2.1 A Change Order is a written instrument prepared by the Architect and signed by the Owner, Contractor and Architect, stating their agreement upon all of the following:

.1 a change in the Work;

.2 the amount of the adjustment in the Contract Sum, and

.3 the extent of the adjustment in the Contract Time

7.2.2 Methods used in determining adjustments to the Contract Sum may include those listed in Subparagraph 7.3.3.*

7.3 Construction Change Directives

7.3.1 A Construction Change Directive is a written order prepared by the Architect and signed by the Owner and Architect, directing a change in the Work and stating a proposed basis for adjustment, if any, in the Contract Sum or Contract Time, or both. The Owner may by Construction Change Directive, without invalidating the Contract, order changes in the Work within the general scope of the Contract consisting of additions, deletions or other revisions, the Contract Sum and Contract Time being adjusted accordingly.*

7.3.2 A Construction Change Directive shall be used in the absence of total agreement on the terms of a Change Order.*

7.3.3 If the Construction Change Directive provides for an adjustment to the Contract Sum, the adjustment shall be based on one of the following methods:

.1 mutual acceptance of a lump sum properly itemized and supported by sufficient substantiating data to permit evaluation;

.2 unit prices stated in the Contract Documents or subsequently agreed upon;

.3 cost to be determined in a manner agreed upon by the parties and a mutually acceptable fixed or percentage fee; or

.4 as provided in Subparagraph 7.3.6.*

7.3.4 Upon receipt of a Construction Change Directive, the Contractor shall promptly proceed with the change in the Work involved and advise the Architect of the Contractor's agreement or disagreement with the method, if any, provided in the Construction Change Directive for determining the proposed adjustment in the Contract Sum or Contract Time.*

(Subparagraph 7.3.5 has been omitted here.)

7.3.6 If the Contractor does not respond promptly or disagrees with the method for adjustment in the Contract Sum, the method and the adjustment shall be determined by the Architect on the basis of reasonable expenditures and savings of those performing the Work attributable to the change, including, in case of an increase in the Contract Sum, a reasonable allowance for overhead and profit. In such case, and also under Clause 7.3.3.3, the Contractor shall keep and present, in such form as the Architect may prescribe, an itemized accounting together with appropriate supporting data. Unless otherwise provided in the Contract Documents, costs for the purposes of this Subparagraph 7.3.6 shall be limited to the following:

.1 costs of labor, including social security, old age and unemployment insurance, fringe benefits required by agreement or custom, and worker's compensation insurance;

.2 costs of materials, supplies and equipment, including cost of transportation, whether incorporated or consumed;

.3 rental costs of machinery and equipment, exclusive of hand tools, whether rented from the Contractor or others;

.4 costs of premiums for all bonds and insurance, permit fees, and sales, use or similar taxes related to the Work; and

.5 additional costs of supervision and field office personnel directly attributable to the change.*

(Articles 8, 9, and 10 have been omitted.)

Article 11. INSURANCE AND BONDS

Describes requirements of Contractor to carry Liability Insurance, Performance and Payment Bonds; Owner's responsibility to carry Liability, Property, and Boiler and Machinery Insurance. States conditions relating to purchase other insurance protecting Owner or Contractor.

Article 12. UNCOVERING AND PROTECTION OF WORK

This pertains to uncovering Work for Architect's observation; discusses circumstances under which Contractor or Owner is responsible for the cost of uncovering and replacement; also correction of Work rejected by the Architect or Work failing to conform to the Contract Documents.

Article 13. MISCELLANEOUS PROVISIONS

Describes rights and remedies, tests and inspections; commencement of Statutory Limitations Period between Owner and Contractor.

Article 14. TERMINATION OR SUSPENSION OF THE CONTRACT

States reasons for termination of the Contract by the Contractor, and conditions where Owner may suspend, delay, or interrupt the Work for Convenience.

CONTRACTUAL AGREEMENTS BETWEEN OWNER AND CONTRACTOR

There are two widely used AIA Standard Forms of Agreement between Owner and Contractor.

- The Stipulated Sum Agreement, A101, 1997 edition, or an abbreviated form for Construction Projects of Limited Scope, A107.
- Cost of Work Plus a Fee Agreement, with or without a Guaranteed Maximum Price, A111, 1997 edition, or an abbreviated form for Construction Projects of Limited Scope, A117, 1987 edition.

CONSTRUCTION MANAGEMENT

The concept of a construction manager (CM) evolved in the late 1960s and early 1970s, primarily as a result of the rampant inflation in the construction costs led by labor wages and material prices. Contributing factors were the advance in technology and increasingly specialized methods of construction. It became apparent to the owners that the traditional approach to construction contracting, that is, designing the project, advertising for bids, and awarding the contract to the lowest bidder for a lump sum amount, resulted too often in adversarial relationships and financial losses to the owner, architect, contractor, and even subcontractors. The lines of authority, responsibility, and accountability of these entities obviously had to be clearly defined and regulated as in other successful businesses, under a management type of contract that represented the owner's interests. Thus, the construction manager came into being. The CM's role was to act as the sole agent of the owner managing the construction process from the design phase to final delivery of the complete facility. The job profile of a construction manager included expertise and knowledge of construction methods and techniques, materials, labor costs, skill and judgment in scheduling, and coordination and supervision of construction processes from design to completion of the project.

As the construction industry evolved in the 1980s and 1990s, Design Build came into being. It picked up where CM left off and incorporated the designer, the contractor, and the owner into the building team. Design Build firms now are awarded the task of conceptualizing the project and seeing it through to completion with the owner expressing his needs and budget and the construction team implementing the plan. All of the entities share in the risk and success of the project. It seems like a logical approach to the construction process and will be interesting to watch it evolve in the next millennium.

The construction manager may be a consulting architectural and engineering firm, a general contractor, or a professional construction management company. Construction management contract and agreement forms are published by the American Institute of Architects (AIA), The Associated General Contractors of America, and the National Society of Professional Engineers.

SELECTED REFERENCES

AGC Manual of Contract Documents, The Associated General Contractors of America, 1957 E Street, NW, Washington, D.C. 20006, 1999.

AIA Contract Documents, The American Institute of Architects, 1735 New York Avenue, NW, Washington, D.C. 20006.

CHAPTER 2

DRAWINGS AND SPECIFICATIONS

Written specifications and working drawings are two of the most important documents for a construction project; they are inseparable. One without the other would not be complete because constant reference must be made from one to the other.

Caution

- *Specifications* take precedence over plans and drawings.
- *Stipulated dimensions* take precedence over scaled dimensions.
- *Notes on drawings,* such as *"see specifications,"* must be checked carefully.
- *Notes on specifications,* such as *"as shown on drawings,"* must be checked carefully.
- *Misinterpretations* of the specifications will not relieve you, the contractor, from the requirements of the contract.

Most contracts stipulate that the contractor shall perform no portion of the work without contract documents or, where required, approve shop drawings, product data, or samples for such portion of the work. A condition of standard construction contracts is that you, the contractor, shall study the contract documents and report to the architect any errors, inconsistencies, or omissions you discover. If you knowingly fail to make such report, you will have to assume responsibility for the performance and an appropriate amount of the costs for correction.

Specifications and drawings, prepared primarily to protect the owner, can also be your protection, provided you insist on clear, unambiguous language which you and the subcontractors, suppliers, and estimators clearly understand. Generalizations, blanket clauses, vague or erroneous descriptions or exceptions, faulty punctuation, errors in measurements, and any omissions must be corrected or amended *before* the contract is signed.

UNIFORM SYSTEM FOR CONSTRUCTION SPECIFICATIONS

In 1966, Construction Specifications Institute (CSI) published a Uniform System for Construction Specifications. This was the result of a joint effort by construction industry professional and trade organizations in the United States. The organizations that helped develop and endorse the Uniform System are:

American Institute of Architects

American Society of Landscape Architects

Associated General Contractors of America

Associated Specialty Contractors

Construction Products Manufacturing Council

Construction Specifications Institute

National Society of Professional Engineers

In 1978, Construction Specifications Canada (CSC) joined CSI to update the unified numerical reference concept. The result was the MASTERFORMAT™. It was introduced by CSI as Document MP-2-1, and concurrently in Canada by CSC as Document 004E. It includes *Bidding Requirements, Contract Forms and Conditions of the Contract,* and the 16-division digital format for *Specifications.*

In the United States, MASTERFORMAT has been adopted as an Industry Standardization Document by the Department of Defense and is reflected in the development of Federal Construction Guide Specifications prepared under the aegis of the Federal Construction Council.

In Canada, Construction Specifications Canada (CSC) has been supported by the Royal Architectural Institute of Canada, the Association of Consulting Engineers of Canada, the Canadian Construction Association, and the Urban Development Association of Canada.

Over the last 30 years, the 16-division MASTERFORMAT has grown into the numbering system of choice for the building construction industry. Several other formats have gained popularity but none have reached the same level of acceptance.

One alternative format is the UNIFORMAT™, an activity-based numbering system with 12 divisions instead of 16. This system mirrors the construction process as it is built while the MASTERFORMAT focuses on the materials themselves. A good example would be cast-in-place concrete; MASTERFORMAT catalogs concrete in only one division, while UNIFORMAT could have it in several places. These differences are a bit confusing but the architectural community seems to be able to function well with either system and it is not uncommon to see both formats used in the same specifications.

The civil engineering profession has adopted neither MASTERFORMAT nor UNIFORMAT and in this type of construction no clear numbering system has emerged. The formats in civil engineering-type projects (roads, site-work, bridges, sewers, etc.) are as unique and varied as the jobs themselves.

It is important for contractors to become familiar with the format of any specification system used in contracts they are bidding on. This will result in fewer overlooked items and a smoother construction process.

Listed below are three types of formats:

UNIFORMAT

STANDARD SPECIFICATIONS FOR PUBLIC WORKS CONSTRUCTION (The "Greenbook")

MASTERFORMAT

UNIFORMAT™

1. FOUNDATIONS

2. SUBSTRUCTURES

3. SUPERSTRUCTURES

4. EXTERIOR CLOSURE

5. ROOFING

6. INTERIOR CONSTRUCTION

7. CONVEYING SYSTEMS

8. MECHANICAL

9. ELECTRICAL

10. GENERAL CONDITIONS

11. SPECIAL CONSTRUCTION

12. SITEWORK

STANDARD SPECIFICATIONS FOR
PUBLIC WORKS CONSTRUCTION™ (The "Greenbook")

PART 1—GENERAL PROVISIONS

SECTION 1—Terms, Definitions, Abbreviations and Symbols
SECTION 2—Scope and Control of Work
SECTION 3—Changes in Work
SECTION 4—Control of Materials
SECTION 5—Utilities
SECTION 6—Prosecution, Progress and Acceptance of the Work
SECTION 7—Responsibilities of the Contractor
SECTION 8—Facilities for Agency Personnel
SECTION 9—Measurement and Payment

PART 2—CONSTRUCTION MATERIALS

SECTION 200—Rock Materials
SECTION 201—Concrete, Mortar and Related Materials
SECTION 202—Masonry Materials
SECTION 203—Bituminous Materials
SECTION 204—Lumber and Treatment with Preservatives
SECTION 205—Piles
SECTION 206—Miscellaneous Metal Items
SECTION 207—Pipes
SECTION 208—Pipe Joint Types and Materials
SECTION 209—Electrical Components
SECTION 210—Paint and Protective Coatings
SECTION 211—Soils and Aggregates Tests
SECTION 212—Landscape and Irrigation Materials
SECTION 213—Engineering Fabrics
SECTION 214—Pavement Markers

PART 3—CONSTRUCTION METHODS

SECTION 300—Earthwork
SECTION 301—Treated Soil, Subgrade Preparation and Placement of Base Materials
SECTION 302—Roadway Surfacing
SECTION 303—Concrete and Masonry Construction
SECTION 304—Metal Fabrication and Construction
SECTION 305—Pile Driving and Timber Construction
SECTION 306—Underground Conduit Construction
SECTION 307—Street Lighting and Traffic Signals

SECTION 308—Landscaping and Irrigation Installation
SECTION 309—Monuments
SECTION 310—Painting
SECTION 311—Special Protective Materials
SECTION 312—Pavement Marker Placement and Removal

MASTERFORMAT™*—Level Two Numbers and Titles

INTRODUCTORY INFORMATION

00001	PROJECT TITLE PAGE
00005	CERTIFICATIONS PAGE
00007	SEALS PAGE
00010	TABLE OF CONTENTS
00015	LIST OF DRAWINGS
00020	LIST OF SCHEDULES

BIDDING REQUIREMENTS

00100	BID SOLICITATION
00200	INSTRUCTIONS TO BIDDERS
00300	INFORMATION AVAILABLE TO BIDDERS
00400	BID FORMS AND SUPPLEMENTS
00490	BIDDING ADDENDA

CONTRACTING REQUIREMENTS

00500	AGREEMENT
00600	BONDS AND CERTIFICATES
00700	GENERAL CONDITIONS
00800	SUPPLEMENTARY CONDITIONS
00900	ADDENDA AND MODIFICATIONS

FACILITIES AND SPACES

FACILITIES AND SPACES

SYSTEMS AND ASSEMBLIES

SYSTEMS AND ASSEMBLIES

---

*These BROADSCOPE Section Titles are reproduced with the permission of The Construction Specifications Institute from *MASTERFORMAT—Master List of Section Titles and Numbers,* 1995 Edition.

CONSTRUCTION PRODUCTS AND ACTIVITIES

DIVISION 1—GENERAL REQUIREMENTS

01100	SUMMARY
01200	PRICE AND PAYMENT PROCEDURES
01300	ADMINISTRATIVE REQUIREMENTS
01400	QUALITY REQUIREMENTS
01500	TEMPORARY FACILITIES AND CONTROLS
01600	PRODUCT REQUIREMENTS
01700	EXECUTION REQUIREMENTS
01800	FACILITY OPERATION
01900	FACILITY DECOMMISSIONING

DIVISION 2—SITE CONSTRUCTION

02050	BASIC SITE MATERIALS AND METHODS
02100	SITE REMEDIATION
02200	SITE PREPARATION
02300	EARTHWORK
02400	TUNNELING, BORING, AND JACKING
02450	FOUNDATION AND LOAD-BEARING ELEMENTS
02500	UTILITY SERVICES
02600	DRAINAGE AND CONTAINMENT
02700	BASES, BALLASTS, PAVEMENTS, AND APPURTE-NANCES
02800	SITE IMPROVEMENTS AND AMENITIES
02900	PLANTING
02950	SITE RESTORATION AND REHABILITATION

DIVISION 3—CONCRETE

03050	BASIC CONCRETE MATERIALS AND METHODS
03100	CONCRETE FORMS AND ACCESSORIES
03200	CONCRETE REINFORCEMENT
03300	CAST-IN-PLACE CONCRETE
03400	PRECAST CONCRETE
03500	CEMENTITIOUS DECKS AND UNDERLAYMENT
03600	GROUTS
03700	MASS CONCRETE
03900	CONCRETE RESTORATION AND CLEANING

DIVISION 4—MASONRY

04050	BASIC MASONRY MATERIALS AND METHODS
04200	MASONRY UNITS

04400	STONE
04500	REFRACTORIES
04600	CORROSION-RESISTANT MASONRY
04700	SIMULATED MASONRY
04800	MASONRY ASSEMBLIES
04900	MASONRY RESTORATION AND CLEANING

DIVISION 5—METALS

05050	BASIC METAL MATERIALS AND METHODS
05100	STRUCTURAL METAL FRAMING
05200	METAL JOISTS
05300	METAL DECK
05400	COLD-FORMED METAL FRAMING
05500	METAL FABRICATIONS
05600	HYDRAULIC FABRICATIONS
05650	RAILROAD TRACK AND ACCESSORIES
05700	ORNAMENTAL METAL
05800	EXPANSION CONTROL
05900	METAL RESTORATION AND CLEANING

DIVISION 6—WOOD AND PLASTICS

06050	BASIC WOOD AND PLASTIC MATERIALS AND METHODS
06100	ROUGH CARPENTRY
06200	FINISH CARPENTRY
06400	ARCHITECTURAL WOODWORK
06500	STRUCTURAL PLASTICS
06600	PLASTIC FABRICATIONS
06900	WOOD AND PLASTIC RESTORATION AND CLEANING

DIVISION 7—THERMAL AND MOISTURE PROTECTION

07050	BASIC THERMAL AND MOISTURE PROTECTION MATERIALS AND METHODS
07100	DAMPPROOFING AND WATERPROOFING
07200	THERMAL PROTECTION
07300	SHINGLES, ROOF TILES, AND ROOF COVERINGS
07400	ROOFING AND SIDING PANELS
07500	MEMBRANE ROOFING
07600	FLASHING AND SHEET METAL
07700	ROOF SPECIALTIES AND ACCESSORIES

07800   FIRE AND SMOKE PROTECTION
07900   JOINT SEALERS

DIVISION 8—DOORS AND WINDOWS

08050   BASIC DOOR AND WINDOW MATERIALS AND METHODS
08100   METAL DOORS AND FRAMES
08200   WOOD AND PLASTIC DOORS
08300   SPECIALTY DOORS
08400   ENTRANCES AND STOREFRONTS
08500   WINDOWS
08600   SKYLIGHTS
08700   HARDWARE
08800   GLAZING
08900   GLAZED CURTAIN WALL

DIVISION 9—FINISHES

09050   BASIC FINISH MATERIALS AND METHODS
09100   METAL SUPPORT ASSEMBLIES
09200   PLASTER AND GYPSUM BOARD
09300   TILE
09400   TERRAZZO
09500   CEILINGS
09600   FLOORING
09700   WALL FINISHES
09800   ACOUSTICAL TREATMENT
09900   PAINTS AND COATINGS

DIVISION 10—SPECIALTIES

10100   VISUAL DISPLAY BOARDS
10150   COMPARTMENTS AND CUBICLES
10200   LOUVERS AND VENTS
10240   GRILLES AND SCREENS
10250   SERVICE WALLS
10260   WALL AND CORNER GUARDS
10270   ACCESS FLOORING
10290   PEST CONTROL
10300   FIREPLACES AND STOVES
10340   MANUFACTURED EXTERIOR SPECIALTIES
10350   FLAGPOLES
10400   IDENTIFICATION DEVICES
10450   PEDESTRIAN CONTROL DEVICES

10500	LOCKERS
10520	FIRE PROTECTION SPECIALTIES
10530	PROTECTIVE COVERS
10550	POSTAL SPECIALTIES
10600	PARTITIONS
10670	STORAGE SHELVING
10700	EXTERIOR PROTECTION
10750	TELEPHONE SPECIALTIES
10800	TOILET, BATH, AND LAUNDRY ACCESSORIES
10880	SCALES
10900	WARDROBE AND CLOSET SPECIALTIES

DIVISION 11—EQUIPMENT

11010	MAINTENANCE EQUIPMENT
11020	SECURITY AND VAULT EQUIPMENT
11030	TELLER AND SERVICE EQUIPMENT
11040	ECCLESIASTICAL EQUIPMENT
11050	LIBRARY EQUIPMENT
11060	THEATER AND STAGE EQUIPMENT
11070	INSTRUMENTAL EQUIPMENT
11080	REGISTRATION EQUIPMENT
11090	CHECKROOM EQUIPMENT
11100	MERCANTILE EQUIPMENT
11110	COMMERCIAL LAUNDRY AND DRY CLEANING EQUIPMENT
11120	VENDING EQUIPMENT
11130	AUDIO-VISUAL EQUIPMENT
11140	VEHICLE SERVICE EQUIPMENT
11150	PARKING CONTROL EQUIPMENT
11160	LOADING DOCK EQUIPMENT
11170	SOLID WASTE HANDLING EQUIPMENT
11190	DETENTION EQUIPMENT
11200	WATER SUPPLY AND TREATMENT EQUIPMENT
11280	HYDRAULIC GATES AND VALVES
11300	FLUID WASTE TREATMENT AND DISPOSAL EQUIPMENT
11400	FOOD SERVICE EQUIPMENT
11450	RESIDENTIAL EQUIPMENT
11460	UNIT KITCHENS
11470	DARKROOM EQUIPMENT
11480	ATHLETIC, RECREATIONAL, AND THERAPEUTIC EQUIPMENT

DIVISION 12—FURNISHINGS

DIVISION 13—SPECIAL CONSTRUCTION

13260	SLUDGE CONDITIONING SYSTEMS
13280	HAZARDOUS MATERIAL REMEDIATION
13400	MEASUREMENT AND CONTROL INSTRUMENTA-TION
13500	RECORDING INSTRUMENTATION
13550	TRANSPORTATION CONTROL INSTRUMENTA-TION
13600	SOLAR AND WIND ENERGY EQUIPMENT
13700	SECURITY ACCESS AND SURVEILLANCE
13800	BUILDING AUTOMATION AND CONTROL
13850	DETECTION AND ALARM
13900	FIRE SUPPRESSION

DIVISION 14—CONVEYING SYSTEMS

14100	DUMBWAITERS
14200	ELEVATORS
14300	ESCALATORS AND MOVING WALKS
14400	LIFTS
14500	MATERIAL HANDLING
14600	HOISTS AND CRANES
14700	TURNTABLES
14800	SCAFFOLDING
14900	TRANSPORTATION

DIVISION 15—MECHANICAL

15050	BASIC MECHANICAL MATERIALS AND METHODS
15100	BUILDING SERVICES PIPING
15200	PROCESS PIPING
15300	FIRE-PROTECTION PIPING
15400	PLUMBING FIXTURES AND EQUIPMENT
15500	HEAT-GENERATION EQUIPMENT
15600	REFRIGERATION EQUIPMENT
15700	HEATING, VENTILATING, AND AIR CONDITION-ING EQUIPMENT
15800	AIR DISTRIBUTION
15900	HVAC INSTRUMENTATION AND CONTROLS
15950	TESTING, ADJUSTING, AND BALANCING

DIVISION 16—ELECTRICAL

16050	BASIC ELECTRICAL MATERIALS AND METHODS
16100	WIRING METHODS
16200	ELECTRICAL POWER

16300	TRANSMISSION AND DISTRIBUTION
16400	LOW-VOLTAGE DISTRIBUTION
16500	LIGHTING
16700	COMMUNICATIONS
16800	SOUND AND VIDEO

Each of the BROADSCOPE sixteen divisions is expanded into subheadings that cover almost every type of construction. They are available in the MASTERFORMAT with a BROADSCOPE explanation for each one.

Example

Division 3—Concrete

Section Number	Title
03100	CONCRETE FORMWORK
–110	Structural Cast-in-Place Concrete Forms
–120	Architectural Cast-in-Place Concrete Forms
–130	Permanent Forms
	Permanent Steel Forms
	Prefabricated Stair Forms
03200	CONCRETE REINFORCEMENT
–210	Reinforcing Steel
–220	Welded Wire Fabric
–230	Stressing Tendons
–240	Fibrous Reinforcing

Note that the Sections 03100 and 03200 are all that appear in the BROADSCOPE sixteen divisions.

CHANGE ORDERS

Construction contracts usually contain a provision that any changes or modifications in the specifications, design, materials, or work to be done must be authorized in writing by the owner or the owner's representative. Changes are common on a construction project after the commencement date. Some may be the result of the disclosure of improvements in architectural design of parts, scarcity, shortage or unavailability of materials; substitution of materials for quality and price considerations; error, inconsistencies, or omissions in the contract documents; and changes made necessary by compliance with ordinances when plans are submitted for building permits.

Ten Reasons for Change Orders (Adds and Deducts)*

The causes that form the basis for nearly all changes are relatively few. They include:

1. Design errors
   (a) Contradictions, discrepancies, impossibilities, inconsistencies
2. Changes in market conditions
   (a) Specified product becomes unavailable.
   (b) New products become available, offering price advantages or other benefits.
   (c) New information becomes available, affecting the choice of specified materials.
3. Change in the owner's requirements
   (a) Scope of work
4. The uncovering of undisclosed existing conditions
5. The uncovering of unknown existing (latent) conditions
   (a) Unexpected soil variations
   (b) Conditions uncovered during alterations to an existing structure

---

*Andrew M. Civitello, Jr. *Contractor's Guide to Change Orders*, (Englewood Cliffs, N.J.: Prentice Hall, 1987), pp. 70–71.

6. Suggestions to initiate better, faster, or more economical construction
7. Change in designer preference
8. Discrepancies in the contract documents have described situations contradicting the intent of the project
9. Change in external requirements
   (a) Building codes
   (b) Using agency needs or preferences (public projects)
10. Final coordination with N.I.C. (not in contract) equipment
    (a) Space
    (b) Mechanical and electrical provisions

Change orders may be suggested by the owner, architect, contractor, or subcontractor, although authorization must come from the owner. Change orders are provided for under CHANGES IN THE WORK, Article 7, AIA Document A201, *General Conditions of the Contract for Construction.*

- For the protection of the contractor as well as the owner, the provisions for change orders in this or any contract for construction should be reviewed carefully with particular attention to responsibilities, who is to pay for the changes, and who benefits financially.

- When the change has been decided upon and authorized, it is essential that all parties concerned be notified in writing as to specific work, materials, and specifications.

- The *work order* should include a description and scope of the work covered, the reason for it, the authority, estimated cost, and any extension of time.
- Taking change orders lightly can be costly.

BUILDING CODES

The planning, construction, location, use, and occupancy of building structures are regulated by local ordinances and state and federal laws. These *building codes* provide for standards of workmanship, materials, methods, and systems. They cover structural strength, fire resistance, safety codes, heat, light, ventilation, and other elements in the design, construction, alteration, and demolition of buildings.

Model Building Codes

There are three agencies in the United States that produce "model" building codes.
They are:

- BOCA, The Building Officials and Code Administrators International, Inc.
- ICBO, The International Conference of Building Officials
- SBCCI, The Southern Building Code Congress International, Inc.

The BOCA manual is called the *National Building Code*, and is used east of the Mississippi River in the Midwest and the Northeast.

The ICBO manual is called the *Uniform Building Code*, and is used primarily west of the Mississippi River.

The SBCCI manual is called the *Standard Building Code*, and is used throughout the Southeast.

Each of these codes has specific regional strengths; for example; hurricane resistance in the *Standard Building Code* and earthquake resistance in the *Uniform Building Code*.

In the last several decades, during which the construction industry matured and grew more national in nature, the three model codes became similar in content and philosophy. As the code differences have become less distinct, and as construction and design professionals have expanded their markets into different geographical areas, there arose a demand for one nationwide model building code.

To meet this demand, the three agencies have developed a model code called the *International Building Code*. This code addresses the concerns of the entire country and will replace the three existing model codes after the year 2000. Local building departments will adopt this new national code and change (amend) it as they have done with the three model codes in the past.

Construction Permits

Section 106.1 of the *Uniform Building Code* of ICBO states: "Except as specified in 106.2, no building or structure regulated by this code shall be erected, constructed, enlarged, altered, repaired, moved, improved, removed, converted or demolished unless a separate permit for each building or structure has first been obtained from the building official." Section 106.2 lists a few exemptions such as sheds, playhouses, small retaining walls and fences; also things like painting and decorating or window awnings.

The *Uniform Building Code* of ICBO also states: "Plans, specifications, . . . and other data shall constitute the submittal documents and shall be submitted in one or more sets with each application for a permit. . . . The building official may require plans, computations and specifications to be prepared and designed by an engineer or architect licensed by the state to practice as such even if not required by law."

It is essential to check local building departments and their regulations before you start a project.

Inspections Required

(a) As the work covered by the permit progresses, local inspectors shall make as many inspections thereof as necessary to satisfy them that the work is being done in accordance with this Code, any other applicable State and local laws, and the terms of the permit.

(b) When required, the Inspection Department shall make at least the following inspections of all work being performed under the permit and shall either approve that portion of the construction as completed or shall notify the permit holder or his agent wherein the same fails to comply with the law. The permit holder or his agent shall give timely notice to the Inspection Department when the work for these inspections are ready.

- *Foundation Inspection:* To be made after excavation and forms, if any, are erected and reinforcing steel, if any, is placed and prior to placement of concrete.

- *Frame Inspection:* To be made after the roof, all framing, fire-blocking and bracing is in place and all pipes, chimneys, and vents are complete.

- *Insulation Inspection:* To be made after framing is complete with insulation being installed and prior to finish being applied.
- *Final Inspection:* To be made after the building or structure is completed and ready for occupancy.*

Checklist of Typical Inspections

1. Temporary pole

2. Footing and/or slab (excavations), check of zoning set-backs

3. Underslab, (electrical, plumbing, insulation)

4. Foundations (wall)

5. Framing (rough)

6. Fireplace construction in process

7. Rough-in electric (prior to covering with insulation)

8. Rough-plumbing (prior to covering with insulation)

9. Rough-in energy (prior to covering with wall covering)

10. Rough-in mechanical

11. Final (electrical, plumbing, energy, mechanical)

12. Final building

SYMBOLS AND ABBREVIATIONS

Symbols and abbreviations used on working drawings and in specifications have not been standardized in the

---

*Standard Building Code with North Carolina Amendments.

construction industry, though much progress is being made in that endeavor.

- Check for local variations or customs, and always include a *Legend* to identify symbols and abbreviations in drawings and specifications.

Symbols and abbreviations shown on pages 35–47 are the result of examining many reliable architectural and construction reference texts and similar sources for the purpose of presenting a coalescence and synthesis that will be practical and useful when reading drawings and specifications.

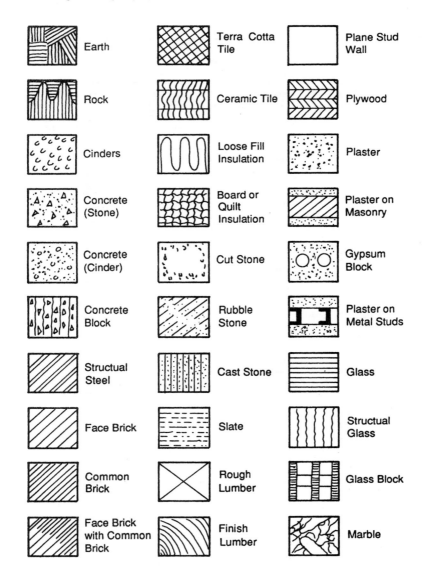

Earth	Terra Cotta Tile	Plane Stud Wall
Rock	Ceramic Tile	Plywood
Cinders	Loose Fill Insulation	Plaster
Concrete (Stone)	Board or Quilt Insulation	Plaster on Masonry
Concrete (Cinder)	Cut Stone	Gypsum Block
Concrete Block	Rubble Stone	Plaster on Metal Studs
Structual Steel	Cast Stone	Glass
Face Brick	Slate	Structual Glass
Common Brick	Rough Lumber	Glass Block
Face Brick with Common Brick	Finish Lumber	Marble

Architectural Symbols—Structural

Electrical Symbols

Symbol	Description	Symbol	Description
⊢◯	Celing Fixture Outlet	S	Single Pole Switch
⊢◯⊣	Celing Fixture Outlet	S$_2$	Double Pole Switch
⊢◯PS	Celing Outlet with Pull Switch	S$_3$	Three Way Pole Switch
⊢⊖	Wall Plug	S$_4$	Four Way Pole Switch
⊢⊖	Duplex Convience Outlet	S$_D$	Automatic Door Switch
⊢⊖$_1$	Single Convience Outlet	S$_P$	Switch and Pilot Light
⊢⊖$_3$	Triple Convience Outlet	S WP	Weatherproof Switch
⊢⊖$_S$	Convience Outlet with Switch	S CB	Circuit Breaker
⊢⊖WP	Weatherproof Convience Outlet	☐	Bell
⊢⊖R	Range Outlet	☐	Buzzer
⊨⊖220	220 Volt Outlet	☐•	Push Button
▲	Special Purpose Outlet	☐D	Electric Door Opener
◉	Floor Outlet	☐TV	Television Outlet
Ⓑ	Blanked Outlet	◀	Telephone
Ⓓ	Drop Cord Outlet	☐	Lighting Panel
Ⓕ	Fan Outlet	▨	Power Panel
Ⓙ	Junction Box	⤛⫤	Fluorescent Celing Fixture
Ⓢ	Pull Switch	⊢Ⓜ	Motor
Ⓧ	Exit Light	⊢Ⓖ	Generator

Architectural Symbols—Electrical

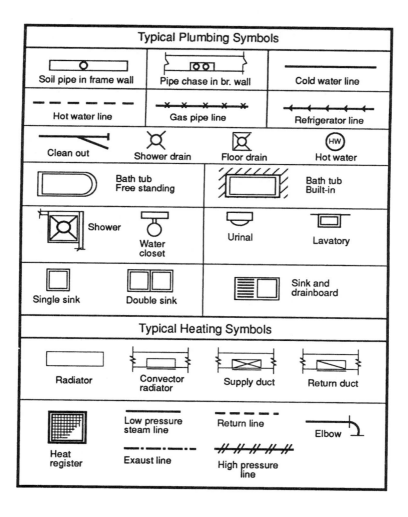

Architectural Symbols—Plumbing & Heating

ABBREVIATIONS

TERM	ABBREVIATION	TERM	ABBREVIATION

A

ABOVE . . . . .ABV
ACCESS DOOR . . . . .AD
ACCESS PANEL . . . . .AP
ACOUSTIC . . . . .ACST
ACOUSTICAL . . . . .ACPL
ACRYLIC . . . . .ACR
ADDITIONAL . . . . .ADDL
ADHESIVE . . . . .ADH
ADJUSTABLE . . . . .ADJ
AGGREGATE . . . . .AGGR
AIR CONDITIONING . . . .A/C
ALTERNATING
  CURRENT . . . . .A/C
ALUMINUM . . . . .AL
ANCHOR . . . . .ANC
ANCHOR BOLT . . . . .AB
APPROVED . . . . .APPD
APPROXIMATE . . . . . . .APX
ASBESTOS . . . . .ASB
ASPHALT . . . . .ASPH
AUTOMATIC . . . . .AUTO

B

BACK-TO-BACK . . . . .B/B
BARREL . . . . .BBL
BASEBOARD . . . . .BB
BASEMENT . . . . .BSMT
BATHROOM . . . . .B
BATHTUB . . . . .BT
BEARING . . . . .BRG

BEARING PLATE . . . .BRG PL
BELOW . . . . .BLW
BENCH MARK . . . . .BM
BETWEEN . . . . .BET
BETWEEN CENTERS . . . .BC
BEVEL . . . . .BEV
BEVELED . . . . .BVLD
BLOCK . . . . .BLK
BLOCKING . . . . .BLKG
BOARD . . . . .BD
BOARD FOOT . . . . .BF
BOILER . . . . .BLR
BOTH SIDES . . . . .BS
BOTTOM . . . . .BOT
BOUNDARY . . . . .BDY
BRACING . . . . .BRCG
BRICK . . . . .BRK
BRIDGING . . . . .BRDG
BUILDING . . . . .BLDG
BUILDING LINE . . . . . . .BL
BULKHEAD . . . . . . .BLKHD
  OR BHD
BUILT-UP ROOFING . . . .BUR

C

CABINET . . . . .CAB
CALK . . . . .CK
CALKING . . . . .CLKG
CAPACITY . . . . .CAP
CARPET . . . . .CARP
CASED OPENING . . . . . . .CO
CASEMENT . . . . .CSMT

ABBREVIATIONS (continued)

TERM	ABBREVIATION	TERM	ABBREVIATION
CASING	CSG	CONCRETE BLOCK	CONC-B
CAST-IN-PLACE CONCRETE	CIPC	CONCRETE FLOOR	CONC-FL
CAST IRON	CI	CONDUIT	CND
CAST STONE	CST	CONNECTION	CX
CATCH BASIN	CB	CONSTRUCTION	CONST
CEILING	CLG	CONTINUOUS	CONT
CEILING HEIGHT	CLG HT	CONTRACT(OR)	CONTR
CELLAR	CEL	COPPER	CPR
CEMENT	CEM	CORRUGATED	CORR
CEMENT PLASTER	CEM PLAS	COUNTER	CNTR
CENTER	CTR	COUNTERSINK	CS
CENTER LINE	CL	COURSE	CRS
CENTER-TO-CENTER	C/C	CUBIC FEET	CU FT
CERAMIC	CER	CUBIC INCH	CU IN
CERAMIC TILE	CER TILE	CUBIC YARD	CU YD
CESSPOOL	CP		
CHALKBOARD	CHBD	**D**	
CHAMFER	CHFR	DAMPER	DMPR
CHANNEL	CHAN	DAMPPROOFING	DMPF
CINDER BLOCK	CIN-BL	DEAD LOAD	DL
CIRCUMFERENCE	CIRC	DEMOLISH (DEMOLITION)	DEMO
CLEAR	CLR	DETAIL	DET
CLOSET	CLO	DIAGONAL	DIAG
CLOSURE	CLS	DIAMETER	DIAM
COLUMN	COL	DIMENSION	DIM
COMBINATION	COMB	DINING ROOM	DRM
COMPARTMENT	COMPT	DISHWASHER	DW
COMPLETE	COMPL	DISTANCE	DIST
COMPOSITION	COMPO	DIVISION	DIV
CONCRETE	CONC		

ABBREVIATIONS (continued)

TERM	ABBREVIATION	TERM	ABBREVIATION
DOOR	DR	ESCALATOR	ESCAL
DOUBLE		ESTIMATE	EST
ACTING	DBL ACT	EXCAVATE	ESC
DOUBLE HUNG	DH	EXHAUST	EXH
DOUBLE STRENGTH		EXISTING	EXG
GLASS	DSG	EXPANSION	
DOWN	DN	JOINT	EXP JT
DOWNSPOUT	DS	EXPOSED	EXP
DRAIN	DR	EXTENSION	EXTN
DRAINBOARD	DB	EXTERIOR	EXT
DRAIN TILE	DT		
DRAWER	DWR		
DRAWING	DWG	**F**	
DRINKING FOUNTAIN	DF		
DRIP CAP	DC	FACE BRICK	FB
DUMBWAITER	DWTR	FACTORY FINISH	FF
		FENCE	FN
		FIBERBOARD	FBD
E		FIBERGLASS	FGL
		FINISH	FIN
EACH	EA	FINISH FLOOR	FIN FL
EAST	E	FINISH GRADE	FIN GR
ELECTRIC PANEL	EP	FIRE ALARM	FA
ELECTRIC WATER		FIRE BRICK	FBRK
COOLER	EWC	FIRE	
ELECTRIC WATER		EXTINGUISHER	FEXT
HEATER	EWH	FIRE HOSE CABINET	FHC
ELEVATION	EL	FIREPLACE	FPL
ELEVATOR	ELEV	FIRE PROOF	FP
EMERGENCY	EMER	FIRE RETARDANT	FRT
ENCLOSE		FIXTURE	FXTR
(ENCLOSURE)	ENCL	FLASHING	FLG
ENGINEER	ENGR	FLEXIBLE	FLX
ENTRANCE	ENTR	FLOOR(ING)	FLR
EQUIPMENT	EQIP		

ABBREVIATIONS (continued)

TERM	ABBREVIATION	TERM	ABBREVIATION
FLOOR DRAIN	FD	GRANITE	GRN
FLUORESCENT	FLUOR	GRAVEL	GVL
FLUSH	FL	GROUND	GND
FOLDING	FLDG	GROUT	GT
FOOT/FEET	FT	GUTTER	GUT
FOOTING	FTG	GYPSUM	GYP
FOUNDATION	FDN	GYPSUM BOARD	GYP BD
FRAME	FR	GYPSUM DRY	
FRESH AIR	FRA	WALL	GYP DW
FULL SIZE	FS	GYPSUM LATH	GYP LATH
FURRED/FURRING	FURR	GYPSUM	
		PLASTER	GYP PL
		GYPSUM TILE	GYP TILE

G

GALLON	GAL		
GALVANIZED	GALV		
GALVANIZED		HALL	H
IRON	GALVI	HAND HOLE	HH
GALVANIZED PIPE	GP	HANDRAIL	HNDRL
GARAGE	GAR	HARDBOARD	HDBD
GAS WATER		HARDWARE	HDWE
HEATER	GWH	HARDWOOD	HDWD
GASKET	GKT	HEAD JOINT	HJT
GAUGE	GA	HEADER	HDR
GENERAL		HEATER	HTR
CONTRACTOR	GC	HEATING	HTG
GLASS	GL	HEATING,	
GLASS BLOCK	GL BLK	VENTILATING, AIR	
GLASS FIBER	GLF	CONDITIONING	HVAC
GLAZING	GLZ	HEAVY DUTY	HD
GOVERNMENT	GOVT	HEXAGONAL	HEX
GRADE (GRADING)	GR	HIGH POINT	HPT
GRADE LINE	GR LN	HOLLOW CORE	HC

H

ABBREVIATIONS (continued)

TERM	ABBREVIATION
HOLLOW METAL	HM
HOLLOW METAL DOOR	HMD
HOOK	HK
HORIZONTAL	HORIZ
HOT AIR	HA
HOT WATER BOILER	HWB
HOT WATER HEATER	HWH
HOT WATER TANK	HWT

I

TERM	ABBREVIATION
INCHES	IN
INCINERATOR	INCIN
INCLUDED	INCL
INSULATION	INSUL
INSULATING CONCRETE	INSC
INSULATING FILL	INSF
INTERIOR	INTR
INTERMEDIATE	INTM
INVERT	INV
IRON	I
IRON PIPE SIZE	IPS

J

TERM	ABBREVIATION
JANITOR'S CLOSET	JC
JOINT	JT
JOINT FILLER	JF
JOIST	JST

K

TERM	ABBREVIATION
KALAMEIN	KAL
KEENE'S CEMENT PLASTER	KCPL
KICK PLATE	KPL
KILN DRIED	KD
KILOWATT	KW
KITCHEN	KIT
KITCHEN CABINET	KIT CA
KITCHEN SINK	KIT SK
KNOCKDOWN	KD
KNOCKOUT	KO

L

TERM	ABBREVIATION
LAUNDRY	LAU
LAVATORY	LAV
LEADER	LDR
LEFT	L
LEFT HAND	LH
LENGTH	LG
LEVEL	LVL
LIBRARY	LIB
LIGHT	LT
LIGHT CONTROL	LC
LIGHTING	LTG
LIGHTPROOF	LP
LIGHTWEIGHT	LT WT
LIGHTWEIGHT CONCRETE	LWC
LIGHTWEIGHT PLASTER	LW PLAS

ABBREVIATIONS (continued)

TERM	ABBREVIATION	TERM	ABBREVIATION
LIMESTONE	LS	METAL ROOF DECK	MRD
LINEAL FOOT	LF	METAL THRESHOLD	MTHR
LINEAR	LIN	METER	M
LINEN CLOSET	L CL	MILLIMETER	MM
LINING	LNG	MILLWORK	MLWK
LINOLEUM	LINO	MINIMUM	MIN
LIVE LOAD	LL	MIRROR	MIR
LIVING ROOM	LR	MISCELLANEOUS	MISC
LOUVER	LVR	MIXTURE	MIX
LUMBER	LBR	MODULAR	MOD
		MOLDING	MLDG
		MOP RECEPTOR	MR

M

TERM	ABBREVIATION	TERM	ABBREVIATION
MACHINE BOLT	MB	MORTAR	MTR
MANHOLE	MH	MOUNTED	MTD
MANUAL	MAN	MOVABLE	MOV
MANUFACTURER	MFR	MULLION	MULL
MARBLE	MARB		
MARK	MK		

N

TERM	ABBREVIATION
MASONRY	MAS
MASTER BEDROOM	MBR
MATERIAL	MATL
MAXIMUM	MAX
MECHANICAL	MECH
MEDICINE CABINET	MC
MEDIUM	MED
MEMBER	MBR
MEMBRANE	MEMB
METAL	MET
METAL FLOOR DECK	MFD
METAL FURRING	MTFR
METAL LATH	ML

TERM	ABBREVIATION
NATURAL	NAT
NICKEL	NI
NOISE REDUCTION	NR
NOMINAL	NOM
NONMETALLIC	NMT
NORTH	N
NOT TO SCALE	NTS
NUMBER	NO

O

TERM	ABBREVIATION
OBSCURE	OBS
OFFICE	OFF

ABBREVIATIONS (continued)

TERM	ABBREVIATION	TERM	ABBREVIATION
ON CENTER	OC	PLATE GLASS	PG
OPAQUE	OP	PLUMBING	PLMB
OPENING	OPNG	PLYWOOD	PLYWD
OPPOSITE	OPP	POINT	PT
OPTIONAL	OPT	PORCELAIN ENAMEL	PE
OUNCE	OZ	PORCH	P
OUTLET	OUT	POUNDS PER	
OUTSIDE DIAMETER	OD	CUBIC FOOT	PCF
OVERALL	OA	POUNDS PER	
OVERHEAD	OVHD	LINEAL FOOT	PLF
		POUNDS PER	
		SQUARE FOOT	PSF
P		POUNDS PER	
		SQUARE INCH	PSI
PAINT	PNT	POWDER ROOM	PR
PAINTED	PTD	PRECAST CONCRETE	PCC
PAIR	PR	PREFABRICATED	PREFAB
PANEL	PNL	PREFINISHED	PREFIN
PANIC BAR	PB	PREFORMED	PRJ
PAPER TOWEL		PRESTRESSED	
DISPENSER	PTD	CONCRETE	PS CONC
PARALLEL	PARR		
PARKING	PRKG		
PARTICLE BOARD	PBD	**Q**	
PARTITION	PTN		
PAVED	PV	QUALITY	QUAL
PAVEMENT	PVMT	QUANTITY	QTY
PAVING	PVG	QUARRY TILE	QT
PEDESTAL	PED		
PERFORATED	PERF		
PERIMETER	PERM	**R**	
PERPENDICULAR	PERP		
PLASTER	PLAS	RABBET	RBT
PLATE	PL	RADIATOR	RADR
		RADIUS	RAD

ABBREVIATIONS (continued)

TERM	ABBREVIATION	TERM	ABBREVIATION
RAILING	RLG	**S**	
RAINWATER CONDUCTOR	RWC	SADDLE	SDL
RAINWATER LEADER	RWL	SAFETY GLASS	SFGL
RECESSED	REC	SCALE	SC
REDWOOD	RW	SCHEDULE	SCHED
REFERENCE	REF	SCREEN	SCRN
REFLECTED	RFL	SEALANT	SNT
REFRIGERATOR/ REFRIGERATION	REFR	SEATING	STG
REGISTER	REG	SECOND	SEC
REINFORCED/ REINFORCING	REINF	SECTION	SECT
REMOVE(ABLE)	REM	SERVICE	SVCE
RESILIENT	RESIL	SERVICE SINK	SSK
RETURN	RET	SEWER	SWR
REVERSE	RVS	SHEATHING	SHTHG
REVISION (REVISED)	REV	SHEET(ING)	SHT
RIGHT HAND	RH	SHEET GLASS	SG
RIGHT-OF-WAY	ROW	SHEET METAL	SM
RISER	R	SHELF	SH
RIVET (RIVETED)	RVT	SHELVES/SHELVING	SHV
ROOF DRAIN	RD	SHIPLAP	SHLP
ROOF HATCH	RFH	SHORING	SHO
ROOFING	RFG	SHOWER	SHR
ROOM	RM	SIDEWALK	SW
ROUGH OPENING	RO	SIDING	SDG
RUBBER	RUB	SIMILAR	SIM
RUBBER TILE	R-TILE	SINGLE STRENGTH GLASS	SSG
RUBBLE STONE	RBL	SINK	SK
RUNWAY	RWY	SKYLIGHT	SKYL
		SLEEVE	SLV
		SLIDE/SLIDING	SL
		SOIL PIPE	SP
		SOLID CORE	SC

ABBREVIATIONS (continued)

TERM	ABBREVIATION	TERM	ABBREVIATION
SOUNDPROOF	.SPRF	SYMBOL	.SYM
SOUTH	.S	SYMMETRICAL	.SYMM
SPACER	.SPC	SYNTHETIC	.SYNTH
SPACING	.SP	SYSTEM	.SYS
SPEAKER	.SPKR		
SPECIAL	.SPCL	**T**	
SPECIFICATION(S)	.SPEC		
SQUARE	.SQ	TACKBOARD	.TK-BD
SQUARE FOOT	.SQ FT	TACKSTRIP	.TKSTR
SQUARE INCH	.SQ IN	TECHNICAL	.TECH
SQUARE YARD	.SQ YD	TELEPHONE	.TEL
STAINLESS STEEL	.SST	TELEVISION	.TV
STAKE	.STK	TEMPERATURE	.TEMP
STANDARD	.STD	TERRA COTTA	.TC
STATION	.STA	TERRAZZO	.TER
STEEL	.STL	THERMOSTAT	.THERMO
STEEL JOIST	.STL JOIST	THICKNESS	.THK
STEEL PLATE	.STL PLATE	THOUSAND BOARD	
STEEL SASH	.STL SASH	FEET	.MBF
STONE	.STN	THOUSAND CUBIC	
STORAGE	.STOR	FEET	.MCF
STORM DRAIN	.SD	THOUSAND CUBIC	
STRAIGHT	.STR	YARDS	.MCY
STRUCTURAL	.STRUCT	THOUSAND SQUARE	
STRUCTURAL		FEET	.MSF
CLAY TILE	.SCT	THRESHOLD	.THR
STRUCTURAL		TOILET PAPER	
STEEL	.STRUCT STL	HOLDER	.TPH
SURFACE	.SURF	TOLERANCE	.TOL
SURVEY	.SURV	TONGUE AND	
SUSPENDED	.SUSP	GROOVE	.T&G
SUSPENDED		TOP OF FINISH	
CEILING	.SUSP CLG	FLOOR	.TFF
SWITCH	.SW	TOP OF FOOTING	.TF
		TOP OF JOIST	.TJ

ABBREVIATIONS (continued)

TERM	ABBREVIATION
TOP OF SLAB	TS
TOP OF WALL	TW
TOTAL	TOT
TOWEL BAR	TB
TOWEL DISPENSER	TD
TREAD	T
TYPICAL	TYP

U

TERM	ABBREVIATION
UNDERGROUND	UGNC
UNEXCAVATED	UNEX
UNFINISHED	UNFIN
URINAL	UR
UTILITY ROOM	UTIL-RM

V

TERM	ABBREVIATION
V-JOINT	VJ
VAPOR BARRIER	VB
VARNISH	VAR
VEHICLE	VEH
VENEER	VNR
VENT	V
VENTILATING (ION)	VENT
VERMICULITE	VRM
VERTICLE	VERT
VERTICLE GRAIN	VG
VINYL	VIN
VINYL COMPOSITION TILE	VCT
VINYL FABRIC	VFABR
VINYL TILE	VT
VOLT	V

TERM	ABBREVIATION
VOLUME	VOL

W

TERM	ABBREVIATION
WAINSCOT	WSCT
WALL CABINET	W-CAB
WALL-TO-WALL	W/W
WAREHOUSE	WHSE
WASHING MACHINE	WM
WATER CLOSET	WC
WATER HEATER	WHTR
WATERPROOF(ING)	WTRPRF
WATER RESISTANT	WR
WEATHERSTRIPPING	WSTR
WEIGHT	WT
WELDED	WLD
WELDED WIRE FABRIC	WWF
WEST	W
WIDTH	WID
WIND LOAD	WL
WINDOW	WDW
WIRE MESH	WM
WIRED GLASS	WGL
WITHOUT	W/O
WOOD	WD
WOOD BASE	WB
WORKING POINT	WPT
WROUGHT IRON	WI

Y

TERM	ABBREVIATION
YARD	YD

SELECTED REFERENCES

Architectural Graphic Standards, Ramsey and Sleeper, John Wiley & Sons, Inc., New York, 1994.

Construction Contracting, Richard H. Clough, John Wiley & Sons, Inc., New York, 1994.

Contractor's Guide to Change Orders, Andrew M. Civitello, Jr., Prentice-Hall, Inc., Englewood Cliffs, NJ, 1988.

CHAPTER 3

CONSTRUCTION FINANCING

Construction loans usually are short-term or interim loans for the period of construction only. While financing of a construction project generally is not the contractor's problem or concern, he should be knowledgeable and understand the processes of financing and how it relates to him and his operation.

SOURCES FOR OBTAINING CONSTRUCTION LOANS

1. Commercial Banks
2. Savings and Loan Associations
3. Mutual Savings Banks
4. Mortgage Banking Companies
5. Life Insurance Companies
6. Real Estate Investment Trusts
7. Government Agencies
8. Other Sources

Commercial Banks

Commercial Banks make single-family short-term and a limited number of long-term construction loans. They are considered the largest construction lenders on multifamily and commercial projects. They also make short-term loans to mortgage banks and to real estate investment trusts (REITs).

Savings and Loan Associations

These are the largest of all lenders of both construction and permanent or long-term loans on single-family housing. They also make a considerable number of construction loans for multifamily residences such as apartment houses and condominiums.

Mutual Savings Banks

These banks are located mainly in the northeastern United States. Their mortgage investments are concentrated for the most part in single-family permanent mortgages. They make only a limited number of construction loans, but do make long-term loans to mortgage bankers and to real estate investment trusts which in turn make construction loans.

Mortgage Banking Companies

These banks make a significant number of loans for construction and land development but are mainly intermediaries between borrowers and lenders.

Life Insurance Companies

Principal commitments of life insurance companies are long-term loans on commercial and multifamily projects. They do a minimum amount of temporary construction lending.

Real Estate Investment Trusts

These trusts provide long-term mortgages on commercial and multifamily projects and a limited amount of construction loans.

Government Agencies

The Federal Housing Administration (FHA) insures mortgage loans made by approved lending institutions. FHA does not lend money. The Veterans Administration (VA) makes construction loans on housing for veterans, their dependents, and other beneficiaries of deceased veterans. Every sixth house built in the United States since World War II has been financed by the GI loan program.

Other Sources for Loans

Other sources for loans of less significance are individuals, syndicates, service organizations, and Community Housing Authorities.

TEMPORARY VS. PERMANENT LOANS

A *temporary* construction loan is a series of short-term notes to be disbursed at agreed intervals during construction.

A *permanent* construction loan is a single long-term note with disbursements at agreed intervals during construction. The note is secured by a mortgage on the property, payments to begin after the building is occupied. Permanent financing carries lower interest rates and eliminates fees, commissions, and other charges associated with obtaining both temporary and permanent loans.

Temporary construction loans and permanent long-term loans often work hand-in-hand with each other. Most construction lenders, as a precondition, require evidence from the borrower that a permanent "take-out" commitment has been obtained. For this reason the borrower usually shops first for the long-term mortgage loan after which the construction financing loan is easier to obtain. The lenders then enter into an agreement whereby the permanent lender will "take-out" the construction lender when the project has been completed. While custom builders of homes and commercial properties work with loan proceeds that have been arranged for by the owner, speculation builders must seek their own financing. A contractor who is owner and builder is interested in both short- and long-term loans.

CONSTRUCTION LOAN PROCEDURE

The different lending institutions may use the same, similar, or different forms to be completed by the borrower. But, all must comply with local, state, and federal laws. Lenders need assurance that the contractor is of good *character* and *reputation,* and that he is *financially responsible* and has the capacity to carry out the work stipulated in the specifications to a conclusion.

This calls for a full financial statement from the contractor stating his assets and liabilities, investments, property owned, life insurance, and other pertinent information, including a credit report. In addition to the financial status of the contractor, the lender will require a complete set of plans, drawings, and specifications of the proposed construction including the names of the subcontractors and the work contemplated by each of them. The contractor submits a detailed cost estimate. The lender appraises the site of the proposed construction from every aspect and submits the application to the loan committee for approval.

CHAPTER 4

INSURANCE AND BONDS

As contractor, your insurance and bond needs are dependent upon the nature and size of the project undertaken. Minor repairs or alterations usually require little more than the run-of-the-mill property and general liability coverages. On private construction of multifamily residential and commercial projects or public works and public buildings, insurance and bond coverages can be complex, especially when you consider today's lawsuit crisis.

> "Product liability, small claims, contract disagreements, personal injury claims—in total some 16 million civil suits were filed in state and federal courts in one year. That's one lawsuit for every 15 Americans."[1]

This chapter briefly outlines the common kinds of insurance coverage and bonds that are available. Some are required by law; others are made a part of the contract. Careful planning is important, and a knowledgeable insurance agent or broker is essential.

---

[1]*The Lawsuit Crisis*, Insurance Information Institute, 110 William St. New York, NY 1986.

BOND CHECKLIST

Bonds guarantee that the contract documents will be complied with and fulfillment of the contract between the principal and the party for whom the work is to be performed. In writing a contract bond, the surety guarantees the character of the principal and also that his or her capital and capacity are adequate for the work to be done. You, the contractor, can expect the surety to check your

- financial statements, present and past
- availability of credit
- equipment
- experience in projects for which the bond is requested
- organization's professional standing
- integrity

Bid Bond—Given by the contractor to the owner, this bond guarantees that, if the contractor is awarded the contract, he will accept its terms, furnish such performance bonds and payment bonds as required, and carry on the work to a satisfactory conclusion in compliance with the contract documents.

Performance Bonds—These bonds provide for the performance of the agreement by the contractor and payment by the surety of the owner's loss if the contractor defaults.

Completion Bonds—These bonds are generally required by the financial institution which has loaned money to the contractor, guaranteeing the contractor will use the money according to the terms of the contract, and will complete all work undertaken.

Supply Contract Bonds—These bonds are given by a manufacturer or supplier guaranteeing delivery of material contracted for as specified.

Special Indemnity Bonds—In some cases special indemnification is required against injury to persons or property. The contractor should always have contractural liability insurance to protect him against any liability assumed under this type of indemnity bond.

Release Indemnity Bonds—These bonds are given in order to obtain payment from the owner when a mechanic's or any other lien has been filed against the premises or the unpaid contract price. They guarantee the owner not to suffer loss from the lien claimed if payments continue to the contractor.

Maintenance Bond—This bond is given by the contractor to the owner guaranteeing workmanship and materials for a specified time following completion.

Labor and Material Payment Bonds—This bond is given by the contractor to the owner guaranteeing to pay all bills for construction labor and materials used in connection with the project.

Subcontractor Bonds—This bond is given by subcontractors to the contractor guaranteeing performance of his or her contract and payment of all labor and material bills.

Subdivision Bonds—Given by a developer to public authorities, these bonds guarantee construction of all necessary improvements and utilities.

License or Permit Bonds—These bonds are given by the contractor/licensee to public authorities or third parties guaranteeing to pay for violation of the terms of licenses or permits, sometimes holding such parties harmless.

PROPERTY AND LIABILITY INSURANCE

Worker's Compensation and Employer's Liability

The Standard Workers' Compensation and Employer's Liability is, in reality, *two policies in one*. Under insuring clause A, the first covers liability imposed upon the insured/employer *under the workers' compensation laws of the state*. Under insuring clause B, the second is an employers' liability policy applicable to accidents and diseases for which the insured/employer may be held legally responsible, but which are not covered under the state workers' compensation act. In other words, the injured employee can still sue the employer under the common law of negligence.

State governments are the administrators of workers' compensation laws, the functions being carried out through *special state commissions*. These commissions work closely with employers and insurers to perform necessary administrative duties. Rules and procedures for filing, reviewing, and investigating claims are embodied in the laws.

The employer pays for the benefits provided in the state laws through the purchase of private insurance contracts, self-insurance, or insurance in a state fund. The cost varies from state to state, among the various occupations, and for the individual employer. Employment benefits also vary widely by state. Therefore it is essential to consult the laws of the state to determine the amount and duration of the benefits. In general they include medical benefits to injured employees, weekly cash payments for certain types of disability, rehabilitation payments, and death benefits that include funeral benefits and income benefits to surviving spouse or surviving spouse and children.

Comprehensive General Liability Insurance

For the reason stated earlier (current lawsuit crisis) it is important for a contractor to have a qualified insurance consultant, agent, or broker analyze his operations and products' liability exposures. Each construction project has its individual liability requirements. The general liability coverages most frequently needed in the construction field are:

1. *Premises Exposure*—Protection for liability for bodily injury and/or property damage arising out of (a) buildings owned or leased by the contractor, and (b) construction operations in progress.

2. *Contractor's Protective*—Protects the liability of the contractor for injuries to the public arising out of the operations of subcontractors.

3. *Owner's Protective*—Protects the liability of the owner for injuries to the public arising out of operations of independent contractors.

4. *Contractual Liability*—Covers the liability of others assumed by the contractor, as in hold-harmless agreements and written equipment leases.

5. *Products Liability (Completed Operations)*—Protects against claims for bodily injury and/or property damage that occur after the contractor's operations have been completed.

6. *Automobile Liability*—Covers liability for bodily injury and property damage for owned, hired, or leased automobiles; and for non-owned, hired, or leased automobiles driven on behalf of the insured.

Contractor's Equipment Floater

These floaters, written under inland marine policies, may cover all risk of physical loss or named perils.

Builder's Risk Insurance

This is a fire and allied lines insurance contract designed for changing values of a property during construction. The *completed value* form is written for the full anticipated completed value of the structure. The interest of the contractor or the owner, or both, may be covered under this contract. The perils of fire and extended coverage are insured against; vandalism and malicious mischief may also be included. All risk of physical loss coverage is also available.

The interest of subcontractors may be included or excluded. The policy usually covers fences, tool houses, builders' machinery, implements, supplies, and materials used in construction when located on the insured or adjacent premises.

A Builders' Risk *reporting* form is available under which the insured is required to furnish the insurer with a monthly statement of values as the construction progresses. The completed value form, however, is preferable.

Installation Floaters

This is a policy written to cover firms, contractors, and subcontractors whose business is installing such products as heating, plumbing, air conditioning, appliances, store fixtures, and specialized equipment. These policies may insure against fire and other perils including the hazards of transportation.

SELECTED REFERENCES

The Basic Bond Book, Associated General Contractors of America, 1957 E Street, NW, Washington, D.C. 20006, 1992.

Guide to Construction Insurance, Staff of the International Risk Management Institute, Inc., Associated General Contractors of America, 1957 E Street, NW, Washington, D.C. 20006, 1992.

CHAPTER 5

LABOR RELATIONS AND LEGISLATION

Most construction workers are unionized, more heavily in some parts of the country than others. Unions provide experienced and skilled workers in construction trades for the contractor to draw on and, in addition, to a large degree maintain discipline among their members.

Before estimating or bidding on any construction project, a wise contractor always investigates the local labor situation as to:

- Union work rules
- Hourly wages and overtime
- Fringe benefits
- Holidays
- Existing contract expiration date

THE UNION BUSINESS AGENT

The local business agent is your contact with the local union. He serves as intermediary between members of the union and the contractor or subcontractor. He can be helpful in negotiating agreements; he assures compliance by all parties and arbitrates disputes. Get to know him well.

PRECONSTRUCTION MEETINGS BEFORE BIDDING

When the construction project is in an unfamiliar or a remote area and the labor situation cannot be readily determined, a preconstruction meeting is recommended. It should be attended by the contractor, concerned subcontractors, and representatives of unions involved. At such meetings work rules, wages, benefits, transportation of workers to and from the job if necessary, and other conditions can be amicably agreed upon to avoid future disputes and misunderstandings.

LABOR LEGISLATION

Federal and state laws effecting employer and employees in the construction industry are numerous and often quite complex. Those laws pertaining to insurance and bond requirements are readily taken care of by a competent agent or broker. Many others, such as the Sherman Antitrust Act, and the Norris–LaGuardia Act, may require legal counsel for interpretation and application to a specific circumstance.

The following brief summary of various federal and state legislation is presented for study as necessary.

Workers' Compensation Laws

All states have workers' compensation acts that are similar in purpose. The main object is payment of benefits to injured employees or to dependents of those killed.

Most states require the employer to carry insurance to cover these situations.

Occupational Safety and Health Act (OSHA)

This Act covers construction workers and applies to general contractors, subcontractors, construction engineers, and architects. Its purpose is to protect the health and safety of workers. (See Chapter 7, Safety and Health Programs.)

Norris–LaGuardia Act

This Act (1932) defines and limits the powers of the federal courts to issue an injunction in labor disputes. The courts may issue an injunction in cases involving a labor dispute after hearing testimony in open court with opportunity to cross-examine or in cases where irreparable property damage may occur if an injunction is not issued and greater injury will result to the complainant.

Labor Management Relations Act (Taft–Hartley)

Under this Act, as amended, unfair labor practices by employer and by a labor organization are described in detail. The following brief digest of provisions that are pertinent to the construction industry should be carefully reviewed.

- *Unfair Labor Practices by Employer*

  It shall be unfair labor practice for an employer:

  1. to interfere with, restrain, or coerce employees in their rights to form, join, or assist labor organizations to bargain collectively through representatives of their own choosing.

2. to dominate or interfere with the formation or administration of any labor organization or contribute financial or other support to it.

3. to discriminate against an employee to encourage or discourage membership in any labor organization. This applies to hiring, tenure, or any other condition of employment.

4. to discharge or otherwise discriminate against an employee because he has filed charges or given testimony under this act.

5. to refuse to bargain collectively with the representatives of the employees about rates of pay, wages, hours of employment, or other conditions.

- *Unfair Labor Practices by Labor Organizations*

It shall be unfair labor practice for a labor organization or its agents:

1A. to restrain or coerce employees in the exercise of their rights under this Act to self-organization, to form or join a union, to assist labor organizations, or to refrain entirely from such activities, except that such right may be affected by an agreement requiring membership in a labor organization as a condition of employment.

1B. to restrain or coerce an employer in the selection of his representatives for the purpose of collective bargaining or adjustment of grievances.

2. to cause or attempt to cause an employer to discriminate against an employee for the purpose of encouraging or discouraging membership in a labor organization.

3. to refuse to bargain collectively with an employer provided it is the representative of his employees.

4. to engage in, or to induce or encourage any individual employed, to engage in a strike, or a refusal in the course of employment to use, manufacture, process, transport, or otherwise handle or work on any goods, articles, materials, or commodities.

   (a) to force an employer or self-employed person to join any labor or employer organization.

   (b) to force or require any person to cease using, selling, handling, transporting, or otherwise dealing in the products of any other producer, processor, or manufacturer or to cease doing business with any other person.

   (c) to force or require any employer to bargain with a particular labor organization as representative of his employees if another labor organization has been certified as the representative of his employees.

   (d) to force or require any employer to assign particular work to employees in a particular labor organization, or in a particular trade, craft, or class rather than to employees in another organization or in another trade, craft, or class, unless such employer is failing to conform to an order or certification of the National Labor Relations Board (NLRB) determining the bargaining representative for employees performing such work. (See Chapter 6, Dispute Resolution Options.)

5. to require employees covered by an agreement authorized under this Act, as a condition precedent to becoming a member, payment of a fee the NLRB finds excessive or discriminating.

6. to cause or attempt to cause an employer to pay or agree to pay for services that are not performed or not to be performed (example: featherbedding).

7. to picket, cause, or threaten to picket any employer with the object of forcing him to recognize or bargain with a labor organization, or forcing his employees to accept such labor organization as their collective bargaining representative

   (a) where the employer has lawfully recognized any other labor organization.

   (b) where a valid election has been conducted within the preceding twelve months under this Act, or

   (c) where picketing has been conducted without NLRB petition being filed within reasonable time not to exceed thirty days from commencement of such picketing.

Labor–Management Reporting and Disclosure Act

This Act (1959) requires unions to file with the Secretary of Labor details on their organization, election of officers, and finances. The purpose of the Act was to remedy some of the questionable scruples of internal activities of some unions. It also amended the original Taft–Hartley Act provisions.

Davis-Bacon Act

This Act (1931, amended 1964) requires every bidder on federal construction projects, in excess of $2,000, to pay the "prevailing wage" for similar projects. It requires the Secretary of Labor to determine prevailing wage rates in the area in advance for inclusion in advertised contract specifications.

National Environmental Policy Act

This Act requires an Environmental Impact Statement be prepared for any construction projects in which federal grants or loans of money or permits of any sort are required. The statement must show the effect the construction will have on such things as land use, zoning, ground water and geology, surface drainage, erosion, wildlife, real estate values, agriculture, and social and psychological effects on the population.

AFL–CIO INTERNATIONAL BUILDING TRADES OFFICIALS

The following list of International Building Trades Officials with addresses and telephone numbers as of April 1999, is provided for ready reference. A Directory of the Trade Councils for each state and Canada may be obtained from the General Office in Washington, D.C.

International Building Trades Officials
AFL-CIO BUILDING AND CONSTRUCTION TRADES DEPARTMENT

General Office
815 16th Street, NW, Suite 603
Washington, D.C. 20006-1461

AFFILIATED INTERNATIONALS

International Association of Heat and Frost Insulators and Asbestos Workers

1300 Connecticut Avenue, NW, #505
Washington, D.C. 20036
(202) 785-2388

International Brotherhood of Boilermakers, Iron Ship Builders, Blacksmiths, Forgers, and Helpers

753 State Avenue, #565
Kansas City, Kansas 66101
(913) 371-2640

International Union of Bricklayers and Allied Craftsmen

Bowen Building
815 15th Street, NW
Washington, D.C. 20005
(202) 783-3788

United Brotherhood of Carpenters and Joiners of America

101 Constitution Avenue, NW
Washington, D.C. 20001
(202) 546-6206

International Brotherhood of Electrical Workers

1125 15th Street, NW
Washington, D.C. 20005
(202) 833-7000

International Union of Elevator Constructors

Clark Building, #530
5565 Sterrett Place
Columbia, Maryland 21044
(301) 997-9000

International Association of Bridge, Structural, and Ornamental Iron Workers

1750 New York Avenue, NW, #400
Washington, D.C. 20006
(202) 383-4800

Laborers' International Union of North America

905 16th Street, NW
Washington, D.C. 20006-1765
(202) 737-8320

International Union of Operating Engineers

1125 17th Street, NW
Washington, D.C. 20036
(202) 429-9100

Operative Plasterers' and Cement Masons' International Association of the United States and Canada

1125 17th Street, NW, 6th Floor
Washington, D.C. 20036
(202) 393-6569

International Brotherhood of Painters and Allied Trades

1750 New York Avenue, NW
Washington, D.C. 20006
(202) 637-0700

United Union of Roofers, Waterproofers, and Allied Workers

1125 17th Street, NW, 5th Floor
Washington, D.C. 20036
(202) 638-3228

Sheet Metal Workers' International Union

1750 New York Avenue, NW
Washington, D.C. 20006
(202) 783-5880

International Brotherhood of Teamsters, Chauffeurs, Warehousemen, and Helpers of America

25 Louisiana Avenue, NW
Washington, D.C. 20001
(202) 624-6800

United Association of Journeymen and Apprentices of the Plumbing and Pipe Fitting Industry of the United States and Canada

901 Massachusetts Avenue, NW
Washington, D.C. 20001
(202) 628-5823

All Mail To:

P.O. Box 37800
Washington, D.C. 20013

SELECTED REFERENCES

Construction Contracting, Richard H. Clough, John Wiley & Sons, Inc., New York, 1994.

Labor and Employment Law Desk Book, Gordon E. Jackson, Prentice-Hall, Inc., Englewood Cliffs, NJ, 1994.

CHAPTER 6

DISPUTE RESOLUTION OPTIONS

Disputes that arise after construction has started are costly in money and time. While many are minor and are resolved by meetings of the parties or in some cases by decision of the architect, engineer, or superintendent, others can be settled only by arbitration or litigation. Most disputes fall under one of two types:

- Contract document disputes over interpretation of contract terms by owner, contractor, subcontractor, or architect.
- Jurisdictional disputes involving union claims to perform certain types of work.

CONTRACT DOCUMENT DISPUTE RESOLUTIONS

Under general conditions of construction contracts, provisions usually are made for the resolution of claims and disputes over interpretation of the contracts. Initially a claim is required to be referred to the architect–engineer for decision. Any claim under the contract documents not resolved to the satisfaction of the parties will be decided by arbitration in accordance with the Construction Industry Arbitration Rules of the American Arbitration Association.

American Arbitration Association (AAA)

The AAA is a nonprofit organization offering a broad range of dispute resolution services available through their headquarters in New York City and regional offices located throughout the United States. It administers arbitrations and maintains panels of competent arbitrators from which the parties can choose. Once designated, the arbitrator decides the issues and renders a final, binding award.

When an agreement to arbitrate is included in the construction contract, it may expedite peaceful settlement without the necessity of going to arbitration.

Standard Arbitration Clause

Arbitration of *future* disputes can be provided for by including the following clause in the contract:

> "Any controversy or claim arising out of or relating to this contract, or the breach thereof, shall be settled by arbitration in accordance with the Construction Industry Arbitration Rules of the American Arbitration Association, and judgment upon the award rendered by the arbitrator(s) may be entered in any court having jurisdiction thereof."[1]

The arbitration of *existing* disputes may be accomplished by using the following:

> "We, the undersigned parties, hereby agree to submit to arbitration under the Construction Industry Arbitration Rules of

---

[1]From the Construction Industry Arbitration Rules of American Arbitration Association, 140 West 51st Street, New York, NY 10020-1203.

the American Arbitration Association the following controversy: (cite briefly). We further agree that the above controversy be submitted to (one)(three) arbitrator(s) selected from the panels of arbitrators of the American Arbitration Association. We further agree that we will faithfully observe this agreement and the rules, and that we will abide by and perform any award rendered by the arbitrator(s) and that a judgment of the court having jurisdiction may be entered upon the award."[2]

Rules of the Construction Industry Arbitration, as amended September 1988, and a list of interpreters' names and addresses may be obtained from the American Arbitration Association, 140 West 51st Street, New York, NY 10020-1203.

JURISDICTIONAL DISPUTE RESOLUTIONS

Jurisdiction disputes occur between trade unions when each claims exclusive right to perform a specific type of work. The parties involved may represent two different unions, different trades in the same union, or nonunion and local union employees.

Examples:

- Carpenters are assigned the erection of scaffolding. A laborer's union claims jurisdiction.
- Laborers are assigned to operate a fork lift. Operating engineers claim jurisdiction.
- A subcontractor is assigned to install insulation in back of brickwork. Bricklayers claim jurisdiction.
- Masons and laborers both claim the right to mix mortar for the brickwork.

---

[2]Ibid.

Threats of strikes or actual strikes, picketing, slowdowns, or other work interferences may accompany jurisdictional demands.

Unfair Labor Practice to Strike

The Taft–Hartley Act (1947), Section 8(b)(4)(D), makes it an unfair labor practice for a union to engage in a jurisdictional strike or other work stoppage for the purpose of

> "(D) forcing or requiring any employer to assign particular work to employees in a particular labor organization in a particular trade, craft, or class rather to employees of another labor organization or another trade, craft, or class unless such employer is failing to conform to an order or certification of the Board (NLRB) determining the bargaining representative for employees performing such work . . ."

Jurisdictional Dispute Resolution by NLRB

The Taft–Hartley Act, Section 10(k), provides for the National Labor Relations Board to:

> "hear and determine any dispute out of which such unfair labor practice (*see above*) shall have arisen unless, within ten days after notice that such charge has been filed, the parties to such dispute submit to the Board satisfactory evidence that they have adjusted or agreed upon methods for voluntary adjustment of the dispute . . ."

Though the Board was empowered and directed to hear disputes, it refused to award the work in dispute until 1961, when the NLRB was ordered by the U.S. Supreme Court to decide and award the party which was to perform the work in disputes.

Studies have shown that the NLRB upholds the employer's original assignment in more than 95 percent of the cases it decides.

Who Can File Charges?

The Taft–Hartley Act provides that any party can file charges under the Section 10(k) proceeding. More often charges are filed by the prime contractor and subcontractor on the job, by the owner, or by one of the labor unions involved.

Procedure for Filing Charges with the NLRB

Initiating a 10(k) proceeding is usually done by the employer–contractor, subcontractors, the owner, or by a labor union involved in the jurisdictional dispute. A NLRB Form 508 must be filed stating:

- The name and address of the labor organization against which the charge is brought.
- The union representative to contact.
- A description of the unfair labor practice within the meaning of Section 8(b) and any subsections of the National Labor Relations Act.
- The specific basis of the charge including names, addresses, location of the jobsite involved, dates, etc.
- The name of the employer and employer's representative to contact.
- The type of project.
- The name and address of the party filing the charge.

NLRB Form 508 is available from the NLRB Regional Office having jurisdiction over the location where the dispute is taking place.

Once the NLRB Form 508 is filed, the matter is turned over to a field examiner who will investigate the facts and decide whether there is reasonable cause to believe Section 8(b)(4)(D) has been violated. If it has, all parties will be informed that the NLRB has accepted the charge for processing.

The charged union will have ten days to prove that all parties have agreed to voluntary resolution of the dispute. Otherwise a "Notice of Hearing" will be issued by the Regional Office. During the hearing, held before the field examiner, all evidence is compiled and forwarded to Washington, D.C. for further action by the NLRB for review. Any party may file a brief with the NLRB within seven days of the conclusion of the hearing. Based on the written record of the field examiner, and any briefs, the NLRB makes its decision.

All parties are required to notify the Regional Director that they will comply with the NLRB decision.

Secondary Options for Jurisdictional Dispute Resolutions

"The Plan for the Settlement of Jurisdictional Disputes in the Construction Industry" was formed in 1948 to decide and award the party which was to perform the work in dispute, prior to the NLRB being ordered by the U.S. Supreme Court. The Plan is a voluntary system stipulated to by several specialty contractor associations, all national and international unions affiliated with the Building and Construction Trades Department, AFL–CIO, and their local constituent bodies. Individual contractors may also stipulate to the Plan.

Parties to the Plan "dedicate their efforts to improving the construction industry by providing machinery for the handling of disputes over work assignments without strikes or work stoppages thus stabilizing employment in the industry . . ."[3]

---

[3]Preamble of Procedural Rules and Regulations of the Plan.

A Joint Administrative Committee (JAC) oversees the operation of the Plan. It is made up of four nominees from the Building and Construction Trades Department and four from the Employers' Association. The JAC appoints a full-time administrator who processes jurisdictional disputes when they are referred to him by a union or by an employer directly effected by the dispute. The parties are given a list of impartial arbitrators and have three days to choose one. If unable to agree on an arbitrator, the administrator will appoint one. The decision of the arbitrator is final and binding.

Rules and Regulations of the Plan are available from Plan for Settlement of Jurisdictional Disputes in the Construction Industry, 815 16th Street, NW, Washington, D.C. 20006.

NOTE: In 1982 the Associated General Contractors of America, an original stipulator of the Plan, withdrew their participation in the Plan. This automatically terminated the stipulation of AGC member–contractors.

HELPFUL HINTS*

The following "helpful hints" are offered for contractors who are faced with jurisdictional disputes and wish to pursue the 10(k) procedure:

- "Preparation and attention to detail are the key words for successfully initiating a jurisdictional charge with the NLRB," according to attorney Charles E. Murphy.

---

*From *Resolving Jurisdictional Disputes in the Construction Industry,* published by the Associated General Contractors of America (AGA), 1957 E Street, NW Washington, D.C. 20006, 1983.

- While economy and efficiency are the key factors in the NLRB decision-making process, employers should not ignore area practice, past employer practice, or specific collective bargaining agreement language when making assignments. In particular, contractors should avoid including specific jurisdictional language in the labor agreement.

- Statements and actions of all parties, be they union officials, contractors, or supervisors, can be used at the 10(k) hearing, as well as at arbitration hearings or future lawsuits for damages, and should be well documented if possible.

- A key element necessary to initiate the 10(k) process is a demand for a reassignment of work. Thus a union demand for a broad jurisdictional clause in its agreement may be considered to be the basis for an 8(b)(4)(D) charge, if it is coupled with a strike threat, because this is a demand for a reassignment coupled with a threat of economic coercion.

- Legal counsel is not required for the 10(k) process, but is recommended. According to Charles E. Murphy, it is estimated that an average of 20 hours of attorney time is generally required to prepare a 10(k) case for hearing. More time will be required if the attorney makes an appearance at the 10(k) hearing itself.

SELECTED REFERENCE

Construction Industry Arbitration Rules, The American Arbitration Association, 140 West 51st Street, New York, 10020-1203, 1999.

CHAPTER 7

SAFETY AND HEALTH PROGRAMS

OCCUPATIONAL SAFETY AND HEALTH ACT (OSHA)

The contractor is responsible for initiating, maintaining, and supervising accident prevention programs in connection with the work. This provision or similar ones are fairly standard in construction contracts. It facilitates compliance with the Occupational Safety and Health Act (OSHA) passed by Congress in 1970. Under this Act, basically, no contractor or subcontractor shall require any laborer or mechanic to work in surroundings that are *unsanitary, hazardous,* or *dangerous to health or safety.* It extends to all fifty states, the District of Columbia, and Puerto Rico. The Act covers a variety of fields in addition to construction.

Inspections

- The employer is required to conduct periodic inspections.
- OSHA Compliance Officers must be permitted to inspect for safety and health hazards at reasonable times and to privately question employees.

Citations

If the Secretary of Labor or an authorized representative believes the employer has violated a requirement of OSHA, he or she shall with reasonable promptness issue a citation to the employer in writing and describe the nature of the violation.

Enforcement

- After issuance of a citation, the Secretary is required to notify the employer by certified mail within a reasonable time of the penalty, and that the employer has fifteen working days to notify the Secretary of his intention to contest the citation or the penalty. Notification must be given to affected employees and unions.

- If the Secretary has reason to believe the employer has not corrected a violation within the period set forth in the citation, he shall so notify the employer and include a proposed penalty. The employer has fifteen working days to notify the Secretary of a contest of the penalty.

Penalties

- The citation for a "serious" violation of the Act carries a mandatory civil penalty for each violation.

- For willful or repeated violations of the Act, an employer may be assessed a civil penalty for each violation.

Adoption of OSHA by States

Most states have adopted an approved State Plan for Occupational Safety and Health that vests the administrative and enforcement authority for all matters relating to occupational safety and health in the State Department of Labor. Under conditions for state plan approval, a state's standards must be as effective as their federal counterparts. To meet this obligation, the states adopt by reference, as provided by law, the federal OSHA Standards for Construction.

The word *shall* appears with great frequency in the OSHA standards, rules or regulations and means *mandatory*. Penalties for noncompliance can be severe. On larger construction projects, compliance with the Act can add materially to the cost of operations.

OSHA as Directed to Heavy Construction

Since the safety, health rules, and regulations of OSHA were first promulgated for the construction industry, it has been recognized that they were directed to and applied to heavy construction rather than to home builders. Investigation and enforcement have diminished substantially to the extent that home builders are almost unaffected.

SUMMARY OF SIGNIFICANT OSHA RULES

OSHA Standards are published in the *Federal Register* which is available in many public libraries, and from the Superintendent of Documents, U.S. Government Printing Office, Washington, D.C. 20402. A summary of some of the more significant rules is given here for your information and further study.

Agreement Between Prime Contractor and Subcontractor

The prime contractor and subcontractor(s) may agree on who shall furnish (for example) toilet facilities, first-aid services, fire protection, or drinking water, but such agreement will not relieve one of legal responsibility.

Machines and Tools

Machines or power tools not in compliance with the regulations must be tagged unsafe, locked, or removed from the premises.

Harmful Animals, Plants, Flammable Liquids, and Gases

Employers must instruct employees in avoidance of unsafe conditions to avoid exposure to illness or injury. If harmful animals or plants (poison ivy, for example) are present, the employee is to be instructed regarding the hazards and in any first aid. The same rules will apply where employees are required to handle any flammable liquids, gases, or toxic materials.

Medical Attention and First-Aid Supplies

- Employers must insure availability of medical personnel for advice and consultation on matters of occupational health.

- Provisions shall be made for prompt medical attention in cases of serious injury.

- If no infirmary, clinic, hospital, or physician is reasonably accessible, a person certified in first aid shall be available at the work site.

- First-aid supplies shall be easily accessible. The first-aid kit must contain approved materials in a waterproof container, and contents must be checked by the employer before being sent to the job and *at least weekly.*

Transportation to Physician or Hospital

- Equipment must be furnished to transport any injured person to a physician or a hospital, or else communication must be available for ambulance service.
- Telephone numbers of physicians, hospitals, and ambulances must be conspicuously posted.

Potable Water Supply

An adequate supply of potable water must be supplied in all places of employment. Containers shall be capable of being tightly closed and supplied with a tap (no dipping of water from containers). Container is to be clearly marked. A common drinking cup is prohibited. Where single-service cups are supplied, a sanitary container for unused and one for used cups are to be provided.

Toilets

Toilets are to be provided in accordance with the regulations.

Protection Against Noise Exposure

Protection against the effects of noise exposure shall be provided when sound levels exceed those set forth by OSHA table "Permissible Noise Exposures." The protective devices inserted in the ears must be fitted individually by competent persons. Cotton is not acceptable.

Lighting of Construction Areas

All construction areas must be lighted to minimum intensities listed in a table designating "foot-candles" for specific areas.

Hardhats and Goggles

Hardhats (usually supplied by employer) are to be worn by employees working in areas where there is a possible danger of head injury from impact, falling objects, or electrical shock or burns. Goggles and face protection are to be provided to employees who are engaged in operations where machines or equipment present possible eye or face injury.

Firefighting Equipment and Alarms

Firefighting equipment shall be furnished, maintained, conspicuously located, and periodically inspected. A fire alarm system (telephone, siren, etc.) shall be established so employees and local fire department can be alerted. Alarm code and reporting instructions shall be conspicuously posted at entrance.

Ladders and Scaffolding

Ladders and scaffolding requirements are extensive. They must be equipped with guardrails and also toeboards on all open sides and ends of platforms over 6 feet above ground or floor. If persons are permitted to work or pass under the scaffold, a 1/2-inch mesh screen must be installed between guardrail and toeboard.

SUBPART T—DEMOLITION

Preparatory Operations to Demolition

(a) Prior to permitting employees to start demolition operations, an engineering survey shall be made, by a competent person, of the structure to determine the condition of the framing, floors, and walls, and possibility of unplanned collapse of any portion of the structure. Any adjacent structure where employees may be exposed shall also be similarly checked. The employer shall have in writing evidence that such a survey has been performed.

(b) When employees are required to work within a structure to be demolished that has been damaged by fire, flood, explosion, or other cause, the walls or floor shall be shored and braced.

(c) All electric, water, gas, steam, sewer, and other service lines shall be shut off, capped, or otherwise controlled, outside the building line before demolition work is started. In each case, any utility company that is involved shall be notified in advance.

(d) Where a hazard exists from fragmentation of glass, such hazard shall be removed.

(e) Where a hazard exists to employees falling through wall openings, the opening shall be protected to a height of approximately 42 inches.

(f) When debris is dropped through holes in the floor without the use of chutes, the area onto which the material is dropped shall be completely enclosed with barricades not less than 42 inches high and not less than 6 feet back from the projected edge of the opening above. Signs warning of the hazard of falling materials shall be posted at each level. Removal shall not be permitted in this lower area until debris handling ceases above.

Stairs, Passageways, and Ladders

(a) All stairs, passageways, ladders, and incidental equipment thereto, which is covered by this section, shall be periodically inspected and maintained in a clean and safe condition.

Chutes

(a) No material shall be dropped to any point lying outside the exterior walls of the structure unless the area is effectively protected.

(b) A substantial gate shall be installed in each chute at or near the discharge end. A competent employee shall be assigned to control the operation of the gate and the backing and loading of trucks.

Removal of Walls, Masonry Sections, and Chimneys

(a) Masonry walls, or other sections of masonry, shall not be permitted to fall upon the floors of the building in such masses as to exceed the safe carrying capacities of the floors.

(b) No wall section that is more than one story in height shall be permitted to stand alone without lateral bracing, unless such wall was originally designed and constructed to stand without such lateral support and is in a condition safe enough to be self-supporting. All walls shall be left in a stable condition at the end of each shift.

INSPECTION CHECKLIST[1]

An inspection checklist should be developed by the safety inspector or safety engineer, with the help of executives, superintendents, and foremen, tailored to the requirements of the individual operations of the firm. General checklists that apply to most of the problems which are likely to be encountered will be easily developed. In addition, special checklists for each individual job will be necessary, as peculiar problems are bound to occur. These will include, but will not be limited to, soil conditions, access roads, storage areas, power and utility lines, special equipment, heating camp housing, and other features.

The following list is offered only as a guide and is not intended to be comprehensive, nor should it be followed in form if job requirements would be better served otherwise.

General Checklist

1. Accident Prevention Organization:
   a. Schedule for posting safety material.
   b. Hardhat requirements.
   c. Safety meetings scheduled and posted.

2. Housekeeping and Sanitation:
   a. General neatness of working areas.
   b. Regular disposal of waste and trash.
   c. Passageways and walkways clear.

---

[1]Inspection Checklist is reprinted by courtesy and with the permission of the Associated General Contractors of America, taken from their *Manual of Accident Prevention in Construction.*

    d. Adequate lighting.

    e. Projecting nails removed.

    f. Oil and grease removed.

    g. Waste containers provided and used.

    h. Sanitary facilities adequate and clean.

    i. Drinking water tested and approved.

    j. Adequate supply of water.

    k. Salt tablets.

    l. Drinking cups or sterilized bubblers.

3. First Aid:

    a. First-aid station.

    b. First-aid supplies.

    c. First-aid instruction on the job.

    d. Telephone numbers and locations of nearby physicians.

    e. Telephone number and location of nearest hospital.

    f. Injuries reported promptly to proper persons.

4. Fire Prevention:

    a. Fire instructions to personnel.

    b. Fire extinguishers identified, checked, lighted.

    c. Phone number of fire department posted.

    d. Hydrants clear; access to public thoroughfare open.

    e. Good housekeeping.

    f. NO SMOKING posted and enforced where needed.

5. Electrical Installations:

    a. Adequate wiring, well insulated.

    b. Fuses provided.

    c. Fire hazards checked.

    d. Electrical dangers posted.

    e. Proper fire extinguishers provided.

6. Hand Tools:
    a. Proper tool being used for each job.
    b. Neat storage, safe carrying.
    c. Inspection and maintenance.
    d. Damaged tools repaired or replaced promptly.

7. Power Tools:
    a. Good housekeeping where tools are used.
    b. Tools and cords in good condition.
    c. Proper grounding.
    d. Proper instruction in use.
    e. All mechanical safeguards in use.
    f. Tools neatly stored when not in use.
    g. Right tool being used for the job at hand.
    h. Wiring properly installed.
    i. Enough people used to handle material.

8. Power-actuated Tools:
    a. Local laws and ordinances complied with.
    b. All operators qualified.
    c. Tools and charges protected from unauthorized use.
    d. Competent instruction and supervision.
    e. Tools checked and in good working order.
    f. Tools not used on any but recommended materials.
    g. Safety goggles or face shields.
    h. Flying hazard checked by backing up, removal of personnel, or use of captive stud tool.

9. Ladders:
    a. Stock ladders inspected and in good condition.
    b. Stock ladders not spliced.
    c. Properly secured, top and bottom.

    d. Siderails on fixed ladders extend above top landing.

    e. Built-up ladders constructed of sound materials.

    f. Rungs not over 12 inches on center.

    g. Stepladders fully open when in use.

    h. Metal ladders not used around electrical hazards.

    i. Proper maintenance and storage.

10. Scaffolding:

    a. Erection under proper supervision.

    b. All structural members adequate for use.

    c. All connections adequate.

    d. Safe tie-in to structure.

    e. Ladders and working areas free of debris, snow, ice, grease.

    f. Proper footings provided.

    g. Passersby protected from falling objects.

    h. Supports plumb; adequate crossbracing provided.

    i. Guardrails and toeboards in place.

    j. Scaffold machines in working order.

    k. Ropes and cables in good condition.

    l. Frequent inspection.

11. Hoists, Cranes, and Derricks:

    a. Inspect cable and sheaves.

    b. Check slings and chains, hooks and eyes.

    c. Equipment firmly supported.

    d. Outriggers used if needed.

    e. Power lines inactivated, removed, or at safe distance.

    f. Proper loading for capacity at lifting radius.

    g. All equipment properly lubricated and maintained.

h. Signalmen where needed.

i. Signals understood and observed.

12. Heavy Equipment:

   a. Regular inspection and maintenance.

   b. Lubrication and repair of moving parts.

   c. Lights, brakes, warning signals operative.

   d. Wheels chocked when necessary.

   e. Haul roads well maintained and laid out properly.

   f. Protection when equipment is not in use.

13. Motor Vehicles:

   a. Regular inspection and maintenance.

   b. Qualified operators.

   c. Local and state vehicle laws and regulations observed.

   d. Brakes, lights, warning devices operative.

   e. Weight limits and load sizes controlled.

   f. Personnel carried in a safe manner.

14. Conveyers and Cableways:

   a. Proper inspection and maintenance.

   b. Good housekeeping.

   c. Screens and other protection where needed.

   d. Adequate inspection, maintenance ladders, and walkways.

   e. Lighting.

15. Marine Equipment:

   a. Waterways rules and regulations observed.

   b. Life jackets.

   c. Life preservers.

   d. Fire protection.

   e. Engine and machinery room safety.

   f. Floating pipelines, walkways, and handrails.
   g. Handrails aboard vessels.
   h. Decks clear, free of oil and grease.
   i. Lights.
   j. Cable.

16. Equipment Maintenance:
   a. Planned maintenance and inspection program.
   b. Adequate equipment records.
   c. Proper oils, fuels, lubrication used.

17. Garages and Repair Shops:
   a. Fire hazards.
   b. Dispensing of fuels and lubricants.
   c. Good housekeeping.
   d. Lighting.
   e. Carbon monoxide dangers.

18. Barricades:
   a. Floor openings planked over or barricaded.
   b. Roadways and sidewalks effectively protected.
   c. Adequate lighting provided.
   d. Traffic controlled.

19. Handling and Storage of Materials:
   a. Neat storage area, clear passageways.
   b. Materials neatly stacked.
   c. Stacks on firm footings, not too high.
   d. Proper number of people for each operation.
   e. People picking up loads correctly.
   f. Materials protected from heat and moisture.
   g. Protection against falling into hoppers and bins.
   h. Dust protection observed.

    i. Extinguishers and other fire protection.

    j. Traffic routing and control.

20. Excavation and Shoring:

    a. Shoring of adjacent structures.

    b. Shoring and sheathing as needed for soil and depth.

    c. Public roads and sidewalks supported and protected.

    d. Materials not too close to edge of excavations.

    e. Lighting at night.

    f. Water controlled.

    g. Equipment at safe distance from edge.

    h. Ladders or stairs provided where needed.

    i. Equipment ramps adequate, slope not too great.

    j. Frequent inspection.

21. Pipelines:

    a. Shoring and bracing as needed.

    b. Equipment in working order.

    c. Proper access to deep trenches.

22. Demolition:

    a. Operations planned ahead.

    b. Shoring of adjacent structures.

    c. Material chutes.

    d. Sidewalk and other public protection.

    e. Clear operating space for trucks and other vehicles.

    f. Adequate access ladders or stairs.

    g. Regular supervision.

23. Pile Driving:

    a. Proper storage procedures.

    b. Unloading only by properly instructed workers.

    c. Steam lines, slings, etc., in operating condition.

    d. Piledriving rigs properly supported.

    e. Ladders on frames.

    f. Care observed by top worker.

    g. Cofferdams maintained and inspected.

    h. Adequate pumping available.

    i. Proper bracing and ties in current.

24. Tunnels:

    a. Adequate ventilation.

    b. Adequate lighting.

    c. Good housekeeping.

    d. Tunnel supports.

    e. Electrical lines.

    f. Operation of hauling equipment.

    g. Proper personal protection.

    h. Gas detectors used.

    i. Proper transportation of personnel.

    j. Control of blasting operations.

    k. Dust protection.

    l. Drilling safety observed.

25. Shafts:

    a. Ladders, stairs adequate and safe.

    b. Top of shaft barricaded, toeboards provided.

    c. Adequate lighting.

    d. Ventilation.

    e. Inspection and maintenance of elevators and hoists.

    f. Signals being used.

    g. Shoring and bracing.

26. Compressed Air Work:

a. Time limits for pressure and exertion observed.

b. Decompression chamber provided.

c. Gauges operating.

d. Exhaust valves.

e. Safety shields.

f. Signals.

g. Ventilation.

h. Sanitary facilities.

27. Boilers:

a. State and local regulations and codes observed.

b. Pressure gauges checked and operating.

c. Equipment in good working order.

28. Pressure Vessels:

a. Gauges operating and checked regularly.

b. Equipment in sound working condition.

c. Connections properly made.

29. Explosives:

a. Qualified operators.

b. Proper transport vehicles.

c. Local laws and regulations observed.

d. Storage magazines constructed per regulations or as recommended.

e. Experienced personnel handling explosives at all times.

f. Cases opened with wooden tools only.

g. NO SMOKING posted and observed where appropriate.

h. Detonaters tested before each shot.

i. All personnel familiar with signals, and signals properly used at all times.

j. Inspection after each shot.

k. Proper protection and accounting for all explosives at all times.

l. Proper disposition of wrappings, waste, and scrap.

m. Advise residents nearby of blasting cap danger.

n. Check radio frequency hazards.

30. Flammable Gases and Liquids:

   a. All containers clearly identified.

   b. Proper storage practices observed.

   c. Fire hazards checked.

   d. Proper storage temperatures and protection.

   e. Proper types and number of extinguishers nearby.

   f. Carts for moving cylinders.

31. Welding and Cutting:

   a. Qualified operators.

   b. Screens and shields.

   c. Goggles, gloves, clothing.

   d. Equipment in operating condition.

   e. Electrical equipment grounded.

   f. Power cables protected and in good repair.

   g. Fire extinguishers of proper type nearby.

   h. Inspection for fire hazards.

   i. Flammable materials protected.

   j. Gas cylinders chained upright.

   k. Gas lines protected and in good condition.

32. Steel Erection:

   a. Safety nets.

   b. Hardhats, safety shoes, gloves.

   c. Taglines for tools.

   d. Fire hazards at rivet forge and welding operations.

   e. Floor opening covered and barricaded.

   f. Ladders, stairs, or other access provided.

   g. Hoisting apparatus checked.

33. Concrete Construction:

   a. Forms properly installed and braced.

   b. Adequate shoring plumbed and crossbraced.

   c. Shoring remains in place until strength is attained.

   d. Proper curing period and procedures.

   e. Heating devices checked.

   f. Mixing and transport equipment supported, and traffic planned and routed.

   g. Adequate runways.

   h. Protection from cement dust.

   i. Hardhats, safety shoes, shirts covering skin.

   j. Nails and stripped form material removed from area.

34. Masonry:

   a. Proper scaffolding.

   b. Masonry saws properly equipped; dust protection provided.

   c. Safe hoisting equipment.

35. Highway Construction:

   a. Laws and ordinances observed.

   b. Competent flagmen properly dressed, instructed, posted.

   c. Adequate warning signs and markers.

   d. Equipment not blocking right-of-way.

   e. Traffic control through construction site.

   f. Adequate marking and maintenance of detours.

   g. Dust control.

   h. Adequate lighting.

36. Quarries and Gravel Pits:

a. Proper signals and safety procedures for blasting.

b. Screening plants operated safely.

c. Traffic routed and controlled.

37. Railroad Construction:

a. Railroad operating regulations observed.

b. Right-of-way maintained and clear.

c. Coupling and uncoupling.

d. Maintenance of equipment.

e. Transportation of personnel.

SELECTED REFERENCES

Construction Planning Equipment and Methods, 5th Ed., R.L. Peurifoy, McGraw-Hill Publishing Company, New York, 1995.

Manual of Accident Prevention in Construction, Associated General Contractors of America, 1957 E Street, NW, Washington, D.C. 20006, 1998.

SECTION II

CONSTRUCTION MANAGEMENT

CHAPTER 8

PLANNING AND SCHEDULING

Orderly management and control of a construction project requires careful planning and scheduling of each construction activity to start when desired, proceed smoothly, coordinate with each other activity, and terminate when expected or specified.

Activities are defined as work units or operations in construction and may involve:

- Labor only, such as excavating by hand
- Labor and materials, such as building concrete forms
- Labor and equipment, such as excavating by machine
- Time only, such as waiting for concrete to cure

The responsibility for planning and scheduling lies largely with the general contractor who, in turn, depends on subcontractors and various trades to cooperate. On large construction projects the recommendations of the owner, architect, engineer, and construction manager, if any, may be sought.

Each activity has its own time schedule or requirement so that planning for any project must take into account:

- Activities that can be performed at the same time as other activities
- Activities that cannot start until the completion of other activities
- Activities that must be completed before other activities can start

BAR CHARTS

Traditional bar charts show the activities of work and a calendar covering the start and desired or estimated completion dates of a project. The estimated duration of each activity is indicated by a line from starting to completion date. Bar charts have been used for many years in scheduling construction, but they are limited in clearly showing the relationships between activities.

CRITICAL PATH METHOD OF SCHEDULING

The critical path method (CPM) is a more logical system for planning and scheduling than using bar charts. It graphically shows each activity and its interrelationship from start to completion with other contiguous or allied activities. The CPM also shows the status of the project at any time, and it enables management to monitor and revise scheduling during construction to meet the desired completion date of the project. In essence the CPM establishes:

- The quantity of work for each activity
- The start-up and sequence or order in which the work of the activity is to be done
- The rate at which the work will be performed to reach completion

An important element of CPM is a diagram consisting of arrows and nodes (or events). An arrow represents the activity for a particular trade. The tail of the arrow is the beginning of the activity; the head of the arrow is the completion of the activity. Each event is a point in time indicated on the diagram by a number enclosed in a circle. It represents the completion of an activity designated by the arrow or arrows leading up to the event. It also represents the beginning of an activity designated by an arrow or arrows leading away from the event. The arrow diagram shows how each activity depends on others during the sequence of construction. Figure 8–1 shows a very simple example of a CPM arrow diagram.

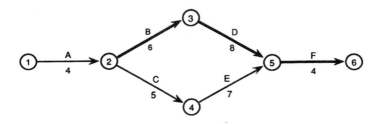

Figure 8–1

Each activity is described or identified by a symbol above the line (in Figure 8–1, the letters *A*, *B*, *C*, *D*, *E*, and *F*). The duration of the activities, in appropriate units (days, weeks, etc.), is shown below the line. In this way the arrow diagram shows the relationship and sequence of the activities needed to complete the project.

In Figure 8–1, *A* is the beginning of the project and must be completed before *B* and *C* can start [concurrently]. *B* must be completed before *D* can start, and *C* must be completed before *E* can start. Both *D* and *E* must be completed before *F* can start.

The *critical path* is the head-to-tail path of activities in an arrow diagram that requires the longest total time for accomplishment. It should be noted that the critical path in Figure 8–1 is *A B D F,* which totals 22 days, against *A C E F,* which totals 20 days. This example of a CPM arrow diagram is very simple. In actual practice a diagram may contain several hundred activities requiring considerable study, drawing, and redrawing.

Computerized Critical Path Method

Most large construction firms use some form of the critical path method. Manual calculations, even using electronic calculators, is time consuming, particularly on large construction projects. CPM concepts coupled with the microcomputer have revolutionized planning and scheduling even for smaller firms. CPM-related software programs are now available for the microcomputer. They greatly reduce the need for elaborate arrow diagramming. Activities and their relationships can be entered into the computer along with other information, such as labor hours, productivity, and quantities. Once entered, the computer can list and summarize interrelationships of activities showing those on the "critical path," and how long other activities can be delayed.

Critical Path Glossary

Activity—The performance of a unit of work or an operation, such as applying roofing.

Activity Duration—The estimated time to start and complete the performance of an activity.

Arrow Diagram or Network—Activity arrows and events drawn to show the relation, dependency, and sequence of the activities.

Critical Path—The path of activities in an arrow diagram that requires the longest total amount of time for completion of the project.

Earliest Finish—The earliest date an activity can be finished.

Earliest Start—The earliest date an activity can be started.

Event (or Node)—A point on an arrow diagram, shown as a circle containing a number, at which all preceding activities leading to the event have been completed.

Free Float—The amount of time an activity can be delayed without delaying or affecting a following activity.

Late Finish—The latest time an activity can be finished without delaying completion of the project.

Late Start—The latest time an activity can be started without delaying subsequent activities.

SELECTED REFERENCES

Construction Planning Equipment and Methods, 5th Ed., R.L. Peurifoy, McGraw-Hill Publishing Company, New York, 1995.

Construction Project Planning and Scheduling Guidelines, Associated General Contractors of America, 1957 E Street, NW, Washington, D.C. 20006, 1986.

CHAPTER 9

PROGRESS ESTIMATES— CONSTRUCTION COST CONTROL

It is seldom that the original bid estimate to perform a particular construction activity, such as brick masonry or poured concrete foundations, is the same as the actual cost to do the work. It will be either higher or lower because of any number of other factors that were unforeseeable or unpredictable. (See Sources of Errors in Estimating in Chapter 10.) Therefore, it is essential, if a contractor is to survive and make a profit, to keep current on a project's costs by maintaining an intelligent and reliable cost-control system. Changes in labor wages or productivity, labor shortages, increased cost of materials, delays in deliveries, accidents, weather conditions, and so forth will all effect the cost and also the original schedule. The probability is that both cost and time will be greater than planned, estimated, or scheduled.

Bar charts or CPM planning systems are ideal for project control by timely updating. The frequency of updating is not subject to any rule. Projects of short duration may require more frequent updating than projects of long duration. Where there are many change orders, updating should be frequent.

Maintaining a cost-control system enables the contractor to analyze the productivity of workers, the performance and efficiency of equipment, and the proper allocation of overhead expenses. In this way, if construction costs are running higher than the bid estimate, or are different than scheduled, corrections can be made while the project is still in progress.

COST CODES

A major component in a project cost-control system is an identifying set of cost codes for the many items of work that comprise a job. There is no standardized system of coding; many contractors establish their own numbering method.

The Uniform System for Construction Specifications, published by the Construction Specifications Institute (CSI), serves as an ideal cost-code system for both bidding and accounting purposes. Major construction industry organizations in the United States and Canada have assisted in developing the Uniform System and endorse it. This system organizes the numerous construction activities into readily understandable categories providing a uniform method for reporting and analyzing costs. The BROADSCOPE Section Titles are reproduced in Chapter 2, pages 19–27. (The more detailed breakdown, including a Key Word Index, is available from the Construction Specifications Institute, 601 Madison Street, Alexandria, VA 22314. An example of the detailed breakdown is shown in Chapter 2.)

COST-CONTROL ACCOUNTING

Some form of bookkeeping is maintained in the office of every contractor, large or small. Its principal function is to record items such as accounts receivable, accounts payable, payroll, taxes, and other financial accounts. Cost-control accounting and progress estimating involve much more than mere bookkeeping. They require the utmost cooperation of every interested party, including the general contractor, owner, architect, subcontractor, purchasing personnel, and suppliers of materials and equipment.

Progress estimating also requires:

- The appointment of a competent and responsible superintendent
- Job meetings held regularly—weekly or biweekly—to discuss existing or new problems and appropriate solutions
- Determination of work quantities produced so that progress estimates can be prepared for a comparison with the scheduled job plan
- A continuous analysis of project costs
- The filing of meaningful daily or weekly progress reports by the superintendent or other appointee

JOB MEETINGS AND REPORTS

Job meetings are usually set up at the start of the construction project. The parties attending should include representatives or key personnel of the following:

- The general contractor
- Each subcontractor
- The owner
- The architect
- Material suppliers if needed
- Equipment operators and renters when necessary

All parties required to attend job meetings should be given prior notification; attendance should be mandatory. Reports or minutes of job meetings usually are prepared by the general contractor's office and should be delivered to each party attending the meeting.

Typical of job meeting reports cover such subjects as:

- Daily activities of the work force
- Number of workers in each trade on the job; where they are working; scope and quantity of work produced; overtime, if any, and reasons for the overtime
- Delays, if any, and the reasons for such delays
- Materials and equipment deliveries
- Material shortages, substitutions, or failures
- Progress estimates
- Decisions that were made at the meeting or instructions received

SELECTED REFERENCES

Builder's Guide to Accounting, Michael C. Thomsett, Craftsman Book Company, Carlsbad, CA, 1996.

Construction Contracting, Richard H. Clough, John Wiley & Sons, Inc., New York, 1994.

Walker's Practical Accounting and Cost Keeping for Contractors, Susan E. Powers, BBA, MBC, Brisbane H. Brown, PhD, P.E., Frank R. Walker Company, Chicago, IL, 1990.

SECTION III

CONSTRUCTION MATERIALS AND METHODS

CHAPTER 10

PRINCIPLES OF ESTIMATING CONSTRUCTION COSTS

By definition, *to estimate* means *to approximate*. Therefore, estimating construction costs is not an exact science. Consequently the highest degree of accuracy and skill is essential for you, the contractor or subcontractor, to obtain a bid and complete a project profitably.

THE QUALIFICATIONS OF A GOOD ESTIMATOR

The following checklist summarizes the special skills and qualifications of a competent construction estimator.

- A background of construction experience
- Ability to visualize a project from plans and specifications
- Ability to scale and read plans and drawings accurately
- Be well grounded in mathematics and geometry; adept in using electronic calculators and, where required, computers
- Understand the importance of visiting the site of a proposed project; visualize step by step its construc-

tion and any unusual conditions that could effect the estimate

- Keep current on construction materials, methods, equipment, and processes
- Be familiar with the production rates of the labor force of his or her own organization as determined from time studies on similar projects
- Understand the difference between overhead chargeable to a specific project and overhead not chargeable to a specific project
- Make certain that quotations of suppliers meet with specifications
- Check all subcontracts for price and for conformity to drawings and specifications
- Be familiar with local and state building codes
- Be familiar with labor regulations and legislation

A Word on Your Subcontractors

The number of construction tasks or trades a general contractor is equipped to estimate and perform varies considerably. On small to medium-size projects, many contractors estimate carpentry, roofing, painting, concrete, and masonry. Many of the large construction companies have their own employees engaged in these trades and have enough projects busy to keep them employed. Most general contractors, however, subcontract work like excavating, built-up roofing, plumbing, heating, air conditioning, electric wiring, elevators, structural steel, and other specialized building trades.

CAUTION: Accepting the lowest bid for a subcontract without checking the financial and professional qualifications of the bidder can be a dangerous practice.

Checking each bid for conformance with drawings and specifications is essential.

ESTIMATING FUNDAMENTALS

Five factors must be considered in estimating a construction project:

- Drawings and Specifications
- Materials
- Labor
- Overhead
- Profit

Drawings and Specifications

Drawings and specifications are discussed in Chapter 2.

Materials

When estimating materials always consider *kind, size, quantity,* and *quality.* Describing wood flooring as 1-by-3-inch pine is meaningless because there is comb-grain pine, flat-grain pine, and each comes in different grades and price. Improper description of such materials as wood, slate, or fiberglass shingles, built-up roofing, lath and plaster, doors and windows, or framing lumber makes it impossible to properly price the material.

CAUTION: When pricing materials, remember to consider:

- local sources when practical
- cash and/or trade discounts
- freight and cartage

- loading and unloading
- price of carload or barge lots

Labor

Many factors influence the judgment of an estimator when considering the hours of labor for a single worker or a crew of workers to install materials or to perform specific operations. Most of these will be found in the following checklist.

1. What unusual or special conditions will the workers encounter?
2. Is the working area confined or cramped, or is it open so the worker or workers can move about freely and unrestrictedly?
3. Can labor-saving machines and equipment be used?
4. Are the workers experienced in the type of work or operation?
5. Is the work quality required high or average?
6. Is the work to be done from ladders or scaffolds?
7. Are the lighting conditions adequate for the workers?
8. Will the workers have to be supervised?
9. What weather conditions are anticipated during the work?
10. Has the productivity of the particular workers been considered?
11. Are the wages paid the workers standard or substandard?
12. Is smoking permitted on the job?
13. Are there union or other regulations limiting productivity?

14. Is the location of the project close to or far from sources of labor?

15. Is the supply of all necessary labor adequate?

16. Will the workers be slowed down to protect the building contents, occupants, or the public?

Sources for Estimating Labor Productivity

There are primarily three sources from which to judge the hours of labor to install materials and perform specific operations.

- Contractors and estimators who have worked as journeymen as carpenters, masons, structural steel fabricators and erectors, and in other specialized trades, rely to a large extent on their own knowledge and experience.
- Contractors, construction superintendents, and time-study engineers develop a knowledge of labor productivity through association with and observing workers during construction.
- The third source for estimating labor is from records of others. Numerous texts are published annually as guides for contractors, subcontractors, and other estimators. (A few of these publications are listed at the end of this chapter.)

In each of the following chapters the estimating of specialized trades is discussed and the subject of labor and labor productivity treated as it relates to the particular trade. Tables are provided *as a guide* for estimating the average hourly labor production rates. The rates shown are, in general, adequate for normal working conditions. They include the time required to start, complete the operation, and clean up. The factors that affect production rates, as outlined and described in this chapter, must be carefully considered when using, increasing, or decreasing the rates of production shown in the tables.

Overhead

Contractors usually consider two types of overhead to be included in their estimates:

- Job overhead, chargeable to each job for which an estimate is prepared
- General overhead, not chargeable to a specific job

The following checklist is provided as a guide.

Overhead Charged to Specific Jobs

Foreman, engineer, superintendent wages
Watchman wages
Timekeeper wages
Taxes
    Social Security
    Federal and state unemployment
    Sales tax on materials
Professional services
    Surveyor
    Engineer
    Architect
    Attorney
Temporary buildings and structures
    Office (trailer, etc.)
    Tool/storage
    Toilet facilities
    Fences, guardrails, ramps, barricades
    Signs and markers

Building permits
  Others, fees, licenses, etc.
Utilities
  Temporary heat, light, and power
  Temporary water
  Temporary telephone
Out-of-town jobs
Travel, maintenance, premium labor
Plant
  Rental of machines and equipment (maintenance)
  Elevators, hoists, chutes
  Depreciation
Insurance
  Public liability
  Fire and extended coverage
  Workers' compensation
Bonds
  Bid, performance, etc.

Overhead Not Charged to Specific Construction Jobs

Contractor's office (building or rental), shop
  Heat, light, power
  Furniture, fixtures, computers, other equipment
Insurance
Depreciation
Office supplies, postage, printing
Telephone, intercom, C.B., cellular phone, pager
Utilities
Employee wages and benefits
  Executives
  Clerks
  Stenographers
  Draftsmen
  Estimators
  Accountants

Payroll taxes on officers and office personnel
Other taxes, licenses, permits,
Outside accountants, legal fees, other professional
services
Dues and subscriptions
Travel and entertainment
Advertising and promotion
Interest on loans and notes
Maintenance

Profit

The net profit a contractor adds to the overall job cost
varies with

- the dollar size of the job
- the complexity of the job
- the competition
- how badly he needs the job

As the dollar amount of the contract increases, the percentage of profit generally decreases. A contractor who
needs work to keep his staff employed and to avoid losing good workers may add 3- or 4-percent profit to the
job cost especially when the competition is strong. At
other times if the job has risks and complexities difficult
to determine, the profit may be justified at as much as 20
percent.

DIFFERENT METHODS OF ESTIMATING

There are different methods used to estimate the cost of
building construction. Each has its particular purpose

dependent on the accuracy desired. The following methods are the most commonly used.

- Cubic foot estimate
- Square foot estimate
- Systems or assembly estimate
- Detailed unit price estimate

Cubic Foot Estimate

When the approximate cost to construct a building is needed, one of the better methods for determining the *replacement* cost is the cubic foot method. If the replacement cost of a particular building is known, that cost divided by the number of cubic feet within the outer walls of the structure equals its cost per cubic foot. This cost per cubic foot then may be used to estimate the replacement cost of buildings of similar size and construction. The cost per cubic foot is applied to the cubic foot volume of the building whose replacement cost is desired.

Example:

$$\frac{\text{Known cost of building}}{\text{Cubic feet in building}} = \frac{\$200,000}{50,000} = \$4.00 \text{ per cubic foot}$$

Cubic feet in building
whose replacement cost
is desired $58,400 \times \$4.00 = \$233,600$

The underlying principle of cubic foot estimating is sound, but it has inherent weaknesses that, unless compensated for, will reduce its reliability.

- Specifications of buildings that appear to be similar on the exterior are not necessarily identical. You have to recognize architectural and construction differences between the building from which the cubic foot cost was developed and the building to which it is to be applied.

- The cubic foot cost of buildings that are identical in design and construction decreases as the cubic foot volume increases, and vice versa.

- The length-to-width ratio of buildings effects their cubic foot cost.

- The number of stories in buildings of like construction affects the cost per cubic foot.

- Special interior finishes, built-in features, porches, patios, and heating and air conditioning systems make it difficult to determine a proper cost per cubic foot.

Cubic foot estimates may be within 10 percent or less of an estimate prepared in detail, but this is considered a coincidence. Their main value is being able to quickly approximate the replacement cost of a building. There are manuals published with tables and charts as guides for selecting a cost per cubic foot. Several are listed at the end of this chapter.

Square Foot Estimate

The cost to construct a building can be approximated by computing the square foot area of each floor and multiplying that by a cost per square foot that has been predetermined from known costs or from published tables. Many of the weaknesses of the square foot method are the same as those discussed under the cubic foot method. A review of those is recommended.

Systems or Assembly Estimate

The areas of exterior walls, interior partitions, floors, and roof of a building are scaled and computed. The estimator, knowing the construction through the walls, partitions, floors, and roof, applies a cost per square foot to the respective areas. Doors and windows are noted and priced installed. Built-in features, plumbing, heating, electric wiring and fixtures, air conditioning, and so forth are added, usually in lump sum amounts. To complete the estimate, overhead and profit are added.

Example:

Assume the exterior walls of a building are constructed as shown below, and the unit costs per square foot are as shown. The total cost per square foot of wall area would be:

	Cost/square foot
Studding, 2" × 4"	$ .80
Sheathing, ½" plywood	.60
Drop siding	1.00
Exterior paint, two coats	.30
Interior lath and plaster	.80
Interior paint, two coats	.35
Total cost per square foot	$3.85

(Prices used are for illustrative purposes)
If the exterior walls of this building measured 140' × 10', the cost of the walls would be:
140' × 10' = 1,400 sq. ft. × $3.85 = $5,390

This method is used to approximate the cost of floors, partitions, roofs, etc. Doors, windows, built-in items, plumbing, heating, electric wiring and fixtures, and other items are added in lump sum amounts. To complete the estimate, overhead and profit are added.

Detailed Unit Price Estimate

The detailed estimate is an itemization of the cost of all materials, labor, and operations required to build a structure from start to its completion in accord with the plans and specifications. It is the most accurate and dependable method of estimating. It has the following advantages over all other methods:

- It identifies rooms, areas, and other parts of the structure.
- Sizes, dimensions, and other measurements are noted appropriately.
- Kind and quality of materials are stated.
- All work to be performed is specified.
- Quantities of materials are shown.
- Hours of labor are noted or can be obtained readily.
- Material prices and wage scales are verifiable.
- Arithmetical calculations can be checked.
- Comparisons with other estimates can be made.
- It provides a basis for negotiating if necessary.

There are four methods commonly used to prepare a detailed estimate.

- Material for each item or operation is priced and the labor is consolidated in a lump sum amount.
- Material and labor are priced individually for each item and operation.
- A unit cost of labor and material is applied to each item.
- The cost of labor and material for each item or operation is processed by a programmed computer.

Method 1—Material Priced, Labor Lumped

Materials are listed for each item of work and operation, quantities are extended and priced. The cost of labor is then determined by approximating the hours or days required to do all the work. Lumping labor as a ballpark guess is not very reliable. Many estimators work out the labor cost for each item of work, then put it in the estimate as a total for labor.

Method 2—Material and Labor Estimated Separately

Material cost is estimated separately for every item and extended to a "Material Cost" column. The labor for every item is computed and extended to a "Labor Cost" column. A third column shows the total cost for both material and labor.

Method 3—Unit Cost Applied to Material and Labor

Considerable estimating on construction is done by the unit cost method. A unit cost is the combined cost of a unit of material and the labor cost to install or apply it. A unit can be any unit of measurement, depending on the type of material and its place in the project. The basic formula for a unit cost would be:

A unit of material	@$ =	$ _____
Hours of labor	@$ =	_____
Unit cost of labor and material		$ _____

EXAMPLE 1. *Framing Lumber*

1,000 MBF lumber	@$	.50	= $	500.00
10 lb. nails	@	.80	=	8.00
25 hr. labor	@	20.00	=	500.00
Unit cost per 1,000 MBF			=	$1,008.00
Unit cost per board foot			= $	1.01

EXAMPLE 2. *One Coat Interior Paint*

(Assume 1 gal. paint covers 450 sq. ft.)

1 gal. of paint	@$ 15.00 =	$15.00
3 hr. painter's labor	@ 20.00 =	60.00
Cost per 450 sq. ft.	=	$75.00
Unit cost per sq. ft.	$\dfrac{\$75.00}{450}$ =	$ 0.167

Method 4—The Computerized Estimate

In this method the work and operations to be done, together with measurements and quantities, are taken from plans and specifications. These are noted on specially coded worksheets, or work lists, designed and furnished by the computerized estimating service being used. The coded worksheets are planned individually for the different construction operations from start to finish. They may consist of a few thousand to tens of thousands of *action* items that are classified by each building trade and by specific operations to be performed. Provisions are made for the individual taking off specifications and measurements to override or bypass the computer by means of specially coded notations on the worksheets.

The *take-off* is fed into a computer that has been previously programmed to local material prices, labor wage rates for various trades, unit costs, and other data. The function of the computer is to extend measurements into areas and quantities and to price out the items onto the coded worksheets. In short, the computer performs all of the mathematical calculations.

Computerized estimating is rapidly gaining in popularity. It is being used by contractors and subcontractors in new construction, remodeling and alterations, and repair work. Its main value is in the speed that an estimate can be made. The efficacy of computerized estimating depends, as with detailed estimating, on

- care and accuracy in complying with the specifications and in examining and scaling plans and drawings, and
- programming the correct material prices and wage rates into the computer. Estimating labor productivity must consider all factors described earlier under estimating labor, page 116. Also, errors made in estimating, listed on page 131, apply equally to computerized estimating.

TABLE 10–1
Units of Measurement Used in Estimating

Material	Unit
Acoustical tile	sq. ft.
Aluminum siding	sq. ft.
Baseboard, moldings, etc.	lin. ft.
Beam, steel	lb. or ton
Beam, wood	b.f. or MBF
Block, concrete, or cinder	ea. or per 100
Boards	b.f. or MBF
Bolts	lb.
Brick (various kinds)	per M (1000)
Brick, glass	ea. or sq. ft.
Building paper	roll, or sq. ft.
Cabinets, kitchen	ea. or lin. ft.
Carpeting	sq. yd.
Casing, door and window	lin. ft. or side
Celotex	sq. ft.
Cement, bagged	bag
Chimney flue lining	ea. or lin. ft.
Cinder block	ea. or per 100
Composition sheathing	per sq. ft.

Units of Measurement Used in Estimating (continued)

Material	Unit
Composition shingles	bdl. or sq.
Concrete block	ea. or per 100
Concrete forms	sq. ft.
Concrete, poured	cu. ft. or cu. yd.
Coping	lin. ft.
Cornice	lin. ft.
Countertop (vinyl, linoleum, Formica)	sq. ft. or lin. ft.
Dimension lumber	b.f. or MBF
Disposal units	ea.
Door hardware	ea. or set
Doors	ea.
Downspouts	lin. ft.
Electrical fixtures	ea.
Electrical wiring	lin. ft. or outlet
Excavating	cu. yd. or cu. ft.
Fixtures, plumbing	ea.
Flagstones	ea. or sq. ft.
Flashing, metal	lin. ft.
Floor covering, carpet and linoleum	sq. yd.
Flooring, wood	b.f., MBF, or sq. ft.
Flue lining	ea. or lin. ft.
Framing, wood	b.f. or MBF
Furring, grounds	per 100 lin. ft.
Girders, steel	lb. or ton
Girders, wood	b.f. or MBF
Glass block	ea.
Glass, plate	sq. in. or sq. ft.
Glass, window	ea. or sq. in.
Gravel or stone	cu. yd. or ton
Hardware, door and window	ea. or set
Hollow tile	ea. or per 100

Units of Measurement Used in Estimating (continued)

Material	*Unit*
Insulation	sq. ft.
Joist, metal	ea. or lb.
Joist, wood	b.f. or MBF
Kitchen cabinets	lin. ft. or unit
Knobs, door and cabinet	ea. or set
Lath, metal or gypsum	sq. yd.
Lath, wood	bdl. or sq. yd.
Linoleum	sq. yd.
Masonite	sq. ft.
Metal lath	sq. yd.
Metal roofing	per sq. or sq. ft.
Mortar, lime	bag or cu. yd.
Moldings, wood	lin. ft.
Nails	lb.
Paint	qt. or gal.
Painting	sq. yd. or sq. ft.
Paper, building	roll or sq. ft.
Pipe, water or gas	lin. ft.
Plaster	bag or lb.
Plastering	sq. yd.
Plate glass	sq. ft.
Plates, wood	b.f. or MBF
Plywood	sq. ft.
Putty	lb.
Rafters, wood	b.f. or MBF
Registers, heat	ea.
Reinforcing rods	lbs. or lin. ft.
Ridge roll	lin. ft.
Roof sheathing, gypsum panels	sq. ft.
Roof sheathing, plywood	sq. ft.
Roof sheathing, wood	b.f. or MBF
Roofing, built-up	sq.

Units of Measurement Used in Estimating (continued)

Material	Unit
Roofing, metal	sq. ft. or sq.
Roofing, roll	roll or sq.
Roofing, shingles (all types)	bdl. or sq.
Sand	cu. yd. or ton
Sanding floors	sq. yd. or sq. ft.
Sash, window	ea.
Screen, window and door	ea.
Screen wire	sq. ft.
Screws	lb., doz., or gross
Sheathing boards	bf. or MBF
Sheathing, plywood	sq. ft.
Sheet rock panels	sq. ft.
Shelving, wood	b.f. or lin. ft.
Shingles (all types)	bdl. or sq.
Shutters, window	ea. or pair
Siding, metal and asbestos	sq. ft.
Siding, wood	b.f. or MBF
Sills, stone or concrete	cu. ft.
Sills, wood	b.f. or MBF
Skylight glass	sq. ft. or unit
Space heaters	ea.
Stairs, metal	assembly
Stairs, wood	assembly, per tread and riser
Steel beams	ea. or ton
Stone, decorative	CWT (100 lb.)
Storm doors and windows	ea.
Structural steel	per lb. or ton
Stucco	sq. yd.
Studding	b.f., per stud, or MBF
Tanks, fuel and water	gal. capacity
Terrazzo floors	sq. ft.
Tile floors and walls	sq. ft.
Tin gutters, ridge and valley	lin. ft.
Toilet bowls and tanks	ea.
Trim, doors and windows	lin. ft. or set

Units of Measurement Used in Estimating (continued)

Material	Unit
Underlay, felt	roll or sq. ft.
Underlay, plywood	sq. ft.
Valley, tin	lin. ft.
Varnish	qt. or gal.
Varnishing	sq. ft. or sq. yd.
Venetian blinds	ea. or sq. ft.
Ventilators, roof	ea. or sq. ft.
Vinyl, roll	sq. ft. or sq. yd.
Vinyl tile	ea. or sq. ft.
Wallpaper	per roll or sq. ft.
Waterproofing	100 sq. ft. or per sq. yd.
Weather stripping	per opening
Whitewashing	sq. ft. or sq. yd.
Wood lath	bdl. or sq. yd.
Wood paneling	per sq. ft.
Wood siding	sq. ft. or MBF

SOURCES OF ERRORS IN ESTIMATING

Few estimates for bidding on construction projects are made that do not contain errors of one kind or another. If the contractor is not awarded the contract, he may never learn what mistakes were made in the estimate. If he is awarded the contract on his estimate, errors that were made, if any, will probably be disclosed before the job is completed. Too many errors on the plus side can be a major reason for a bid not being competitive. Too many errors on the minus side can cause the contractor to lose money if awarded the contract.

Some of the most common sources of errors in estimating are listed and described below. The majority are the result of carelessness or a lack of aptitude. Use it as a CHECKLIST.

1. Errors in Arithmetic. Estimates made on construction projects are arithmetical calculations of quantities and cost of materials and labor costs to install materials or perform various operations. It is recommended that you:

- use an electronic calculator with a recording tape
- have the calculations checked by another person
- attach the tape to the estimate sheet for backup

2. Taking Wrong Measurements from Plans and Specifications. Errors in measurements and dimensions taken from plans, drawings, and specifications result in corresponding mistakes in the cost of construction items based on those measurements.

Examples:
    A concrete slab 6 in. thick mistakenly recorded as 8 in. thick increases the cost by 33.3 percent.
    If 2-by-8-inch floor joists are noted as 2 in. by 10 in., the material cost is increased by 25 percent.

3. Using the Wrong Wage Rates for Labor. Hourly labor wages for construction workers vary countrywide. You should consistently verify current wage rates and fringe benefits for the building trades involved through local union offices, other contractors, supply yards, and other reliable sources. Remember, overtime rates are generally one and one-half to two times regular rates depending on labor agreements and union rules.

4. Insufficient or Excessive Allowances for Labor. A frequent cause of error in estimating is allowing too much or too little for labor to do the job.

5. Materials and Supplies Improperly Priced. Always be sure that building materials and supplies are correctly described as to *kind, quality, size,* and *dimensions.* Also confirm that they are priced competitively.

6. Using Incorrect Units of Measure. Using a wrong unit of measure can result in substantial cost increases or decreases. For example, be careful not to record lineal feet for lineal yards, square feet for square yards, cubic feet for cubic yards, and so forth.

7. Including Poorly Maintained Machinery or Equipment. Machinery or equipment to be used in construction, and included in the estimate or bid, must always be checked for efficient serviceability. Preparing an estimate on a construction project and contemplating the use of poorly maintained machinery or equipment is unwise. Breakdown, repairs, and idle time can be costly, delay completion of the project, and invite penalties.

8. Failure to Visit the Project Site. This source of error might well be number one on the list because of its importance in the early stages of estimating. Visiting the proposed site of the project enables the estimator to inspect the topography, check the soil by boring if necessary, determine if protection of adjacent properties will be needed, and check distances to railroad sidings, supply centers, and the proximity to sources of labor. If existing structures have to be demolished or removed from the premises, the estimator is able to properly determine the probable cost.

9. Overlooking or Miscalculating Haulage Costs. The cost of hauling materials, supplies, machinery, and equipment to a project can be a very expensive item in an estimate. Access to the job site may be difficult because of poor roads or no roads, heavy traffic to and from supply sources, or the requirement to obtain permits, and so forth.

10. Failure to Review Building Codes, Permits, and Inspections. Estimates and bids on construction projects are subject to local, state, and federal building codes, permits, and inspections.

11. Failure to Consider Quality of Workmanship Required. A contractor who is accustomed to working on projects that require high-quality workmanship may not be set up to bid or estimate projects of mediocre, low-grade workmanship. Conversely, a contractor who usually works on cheap structures is frequently handicapped when it comes to bidding on the construction of high-class residences and commercial buildings where only the finest quality of workmanship is acceptable. Failure to give proper consideration to the quality of workmanship a project warrants can lead to overestimating or to underestimating.

12. Omitting Items the Estimator Considers to Be Minor. Sometimes items such as scaffolding, ramps, and guardrails are left out of an estimate on the assumption that their cost is relatively minor and can be absorbed in the overall bid. On small projects a contractor may gamble on his workers handling such items routinely. This can be a costly error.

13. Duplicating the Work of Subcontractors. Subcontractors often prepare their estimates from the plans and specifications without the guidance or supervision of the general contractor. They take off details and include all of the items they assume to fall within their particular trade. As a result there frequently is overlapping with the work of the general contractor or other subcontractors. Examples include:

- A plumbing contractor may include duct work unaware that a sheet metal contractor has included the item in his estimate.
- An electrical contractor may include wiring for air conditioning, not knowing it is also in another subcontractor's estimate.
- Certain subcontractors include cutting through beams, girders, walls, and so forth which is necessary to install their materials, unaware that the general contractor has also included an item "cutting for other trades."
- The cost of scaffolding and hoisting is often included in the estimates of subcontractors as well as in the general contractor's.

14. Failure to Review the Bids of Subcontractors. The owner looks to the general contractor for the completion of the work in compliance with the plans and specifications. The owner does not look to the subcontractors. It is very important that the estimates of subcontractors are carefully reviewed to be sure they comply with the plans and specifications.

15. Overlooking Items. The causes of overlooking items when preparing an estimate or bid are many. The following are representative:

- Lack of attention to details
- In too great a hurry to complete the estimate
- Being interrupted too often

- Too heavy a workload
- Basic lack of experience
- Delegating part of the estimate to others
- Failure to use a reliable checklist

An important safeguard against overlooking items is to have another person independently review and double-check the estimate.

16. Taking Shortcuts in Estimating. Taking shortcuts when making an estimate can be risky. Often there is a temptation to take shortcuts when under pressure because of a time limit in which to complete the estimate or because of a heavy backlog of work. Shortcuts take the form of *guesstimating,* using square foot or cubic foot costs in place of details, and using lump-sum figures picked out of the air, all of which have inherent risks. Guesswork estimates are facetiously called *wag* or *swag* systems. A *wag* is a wild ass guess; a *swag* is a scientific wild ass guess.

17. Not Allowing for Realistic Contingencies. Some construction projects may have inherent and unusual problems that should be recognized when the estimate is being prepared. Failure to make allowances or contingencies may result in not getting the contract or losing money if awarded the contract. These contingencies include severe winter weather conditions, or extremely hot and humid climates. The project may be located in an area of the country subject to heavy rainfall. Justifications for a realistic contingency include anticipated labor troubles, material shortages, or political problems.

18. Including Contingency Items Without Reason. As opposed to making allowances for realistic contingencies, it is poor policy to make flat allowances for contingencies without good reasons, particularly when competition is strong.

19. Inadequate or Excessive Overhead Charges. There are two types of overhead:

- Specific overhead chargeable to a specific job or operation.
- General overhead not chargeable to specific jobs or operations.

The estimator should distinguish between the two types in allocating overhead. Assuming an overhead charge of 10, 15, or 20 percent without analysis may be too high or too low. There are overhead charges that range as high as 30 percent or more of the job cost. But these are usually confined to subcontractors specializing in such things as refrigeration, heating and air conditioning, power installation, and so forth. Whether a general contractor is justified in charging overhead on a subcontractor's estimate is a matter of judgment.

20. Profit: Sufficient or Excessive? Refer to the discussion of profit on page 120.

SELECTED REFERENCES

Construction Costbooks, Building News, BNi Publications, Inc., 1612 S. Clemente St., Anaheim, CA 92802; 1-800-873-6397.

Estimating in Building Construction, Frank R. Dagostino, Reston Publishing Co., Inc., Reston, VA, 1998.

Estimating Tables for Home Building, Paul I. Thomas, Craftsman Book Company, Carlsbad, CA, 1986.

Labor Rates for the Construction Industry, Robert S. Means Company, Inc., 100 Construction Plaza Kingston, MA 02364–0800, 1999.

National Construction Estimator, Martin D. Kiley and Will Moselle (Eds.), published by Craftsman Book Company, Carlsbad, CA, 1999.

Walker's Building Estimator's Reference Book, Frank R. Walker Company, Chicago, IL, 1999.

CHAPTER 11

CONSTRUCTION SITE EXPLORATION AND WORK

Contracts for new construction projects often contain a provision that the contractor will visit the site where the work is to be performed and familiarize himself with the conditions. Frequently too little attention is given to inspecting and examining the site of a construction project before preparing and submitting a bid. Later, after the contract has been awarded and work has begun, unanticipated and costly problems arise.

Examples:

- An adjacent building may require shoring up before excavating.
- Foundations and footings of former structures may be uncovered.
- Subsurface springs, streams, or drainage systems are discovered.
- Needed utilities are not available as stated in the contract.
- Easements, zoning, or environmental requirements are discovered.

Early examination of the site before completing and submitting a bid is also important for learning about the condition of the labor market, including availability, cost, and potential problems. The potential for obtaining materials and supplies should be explored, as should the accessibility to the site for delivery vehicles on all-weather roads.

SITE WORK CHECKLIST

The following checklist of questions to be answered will be helpful when examining and exploring the construction site.

Site Clearing

Tree and Brush Removal
Large track tree and brush removal
Selective tree and shrub removal
Size of area to be cleared
Kind of trees _____ min. diam. _____ max. diam. _____
Equipment required: chain saws _____ dozer _____
    chipper _____ other _____
Saleability of timber as: lumber _____ firewood _____
    chips _____ none _____
Burning on site permitted _____ regulations, permits, etc.
Hauling stumps and debris to dump required?
    distance _____ charges _____

Stripping and Storing Topsoil
Depth of soil to be removed
    Scraped by dozer
    Removed by hand
Storage location: at the site _____ another location _____
    How moved to storage: by dozer _____ by truck _____

Demolition of Existing Structures
Building(s) _____ Number of stories _____
    Construction _____

Old foundations, footings, sidewalks, paving, etc.
  Equipment required:
    Hand tools
    Heavy equipment, cranes, hoists, dozers, compressed
      air tools
    Explosives
  Salvageable materials:
    Doors, windows, metals, wire, appliances
    Brick
    Lumber
  Hauling debris to dump:
    Location of dump _____ Distance _____
    Permit required _____ Charges _____

Structures to Be Relocated
On premises _____ Off premises _____
Type _____ Number _____ Construction _____

Removing/Capping Utilities
Water lines
Electric
Gas
Sewer

Dewatering
Draining subsurface waters
  Springs _____ Creeks _____ Ponds _____ Culverts _____
  Other _____ Use of wellpoints Deep wells _____

Job Access

Nearest R.R. siding
  Freight station
  Airport
  Bus station

Highway routing
  Road surface to job site
  Distance to nearest town
  Roads suitable for heavy trucks and equipment
  Bridge limitations
  Traffic problems
Distance to hospital _____
  Emergency ambulance service _____
  Fire department _____
  Police department: local _____ state _____
  Post office _____
  Motel(s) _____ Name(s) _____ Rates _____

Labor

Skilled
  Organized _____ Unorganized _____
  Supply plentiful _____ Scarce _____
  Wage rates _____
Unskilled
  Organized _____ Unorganized _____
  Supply plentiful _____ Scarce _____
  Wage rates _____

Materials Available

Check kinds required for project.
Available locally _____ Distance to source _____

Taxes

Sales taxes
  City _____ County _____ State _____ Federal _____
Vehicle tax
Fuel tax
Occupancy tax
Other

Utilities

Power
    Name of company _____ Office _____
    Transmission station _____ Distance _____
    Voltage _____ KWH rate _____

Telephone
    Name of company _____ Office _____ Rate _____

Water Supply
    Source _____ Purchased from whom _____
    Distance from site _____
    Supply adequate (GPM) _____ Rates _____
    Temporary line(s) required
        Who to install _____ Cost _____
    Have sample analyzed.

Sanitary sewer lines in proximity of site

Temporary Buildings and Structures

Buildings
    Office(s)
    Tool shed(s)
    Storage shed(s)
    Toilet facilities
    Safety and first-aid building
    Enclosures for equipment

Structures
    Fences, guardrails, ramps
    Barricades
    Protection for adjacent property
    Signs
    Other

SITE CLEARING

Tree and Stump Removal

Clearing and preparing the site is usually the first step taken on a construction project. It may call for the removal of trees and brush; the demolition of old structures and foundations including their removal from the premises; and disconnecting, capping, or removal of existing utilities such as sewer, gas, water, and power lines. The contract plans and specifications should stipulate what work is required; the location of disposal sites for tree branches, stumps, and other debris; and local regulations regarding the burning of trees and brush. An inspection of the site will disclose whether trees can be cut down by hand, chain saw, or bulldozers. Where the burning of limbs is prohibited, a chipping machine may be employed instead of removing them long distances by truck for disposal. Also, chips are often sold to nurseries and private parties for mulching. Consideration should be given to a possible market for wood as either lumber or firewood.

Trees that are to remain on the site should be marked and protected against accidental injury. Trees may be killed by changing the grade of the soil surrounding them, by piling soil around the trunks, or by cutting important roots. Table 11–1 will serve as a guide in estimating labor cutting down trees and removing stumps.

Stripping and Storing Topsoil

Stripping and storing topsoil usually follow the removal of trees and brush. As much topsoil should be scraped as is practical for future use but, if there is an excess, it can be readily and profitably sold to nurseries and landscape gardeners. If possible, the storage location is best located near the construction site to ease redistribution. This will avoid expensive hauling costs, and assure proper protection during the storage period.

TABLE 11-1

Approximate Hours Labor Cutting Down Trees Using Chainsaws and Axes, and Removing Stumps

Diameter of Tree in Inches	Hour Labor	Hours Removing Stumps By Hand	By Dozer
6 to 12	2 to 3	6 to 8	2
13 to 18	3 to 4	8 to 10	3
18 to 24	5 to 6	10 to 12	4
24 to 30	6 to 8	12 to 14	5

The hours shown include cutting branches and trunks into 24-inch lengths. Loading on trucks is not included.

Demolition of Existing Structures

The demolition of existing buildings and structures, including their footings and foundations, can range from tearing down simple shed(s) or the removal of old foundation walls, to the razing of a multi-story steel and concrete building that is located in a congested part of a town. Major demolition jobs are usually contracted out to specialists who have their own heavy equipment and sometimes use explosives. Their work crews are experienced, and they are familiar with all the problems and regulations that are common to that business including permits, licenses, protection of property, safety and health regulations, and so forth.

SOIL TESTING

Construction projects that involve excavating for and building footings and foundations require all available information on the soil's composition and characteristics. If the structure is to be a dwelling in a well-known residential area, the testing of the soil may be a minor performance, or even unnecessary if the information is available from the building department, utility companies or other local sources. But, when the project involves multi-story residential or commercial structures or the soil characteristics are doubtful or unknown, testing is mandatory. It is equally important even though the architect/engineer or the owner accepts in writing responsibility for all subsurface conditions encountered during the construction.

Testing to determine the composition of subsurface soil, uniformity of deposits, thickness of strata for load bearing, and the location of water table is a job for the experienced engineer who is qualified to perform the tests and also to interpret the results. Of the several methods of subsurface exploration, the following are the most frequently employed.

- Test pits
- Sounding rods
- Auger boring
- Core boring
- Dry sample boring

Test Pits

Digging test pits is one of the simplest methods to accurately determine the nature and characteristics of subsurface soils and to obtain samples. The advantage of test pits is that they permit visual examination at the various levels to which the pit is dug, revealing the actual conditions that will be met in excavating and building footings and foundations. Test pits are employed for both shallow foundations and for caissons.

Sounding Rods

The sounding rod is usually a 3/4-in. steel rod in lengths of 5 feet that can be coupled. It is pointed at the lower end. The rod is driven either by hand with a maul or by a mechanical hammer. In the hands of experienced operators, the sounding rod tells much about subsurface composition of soils, and particularly the depth to solid or ledge rock, compacted gravel, and firm bearing soil under landfill. However, its use is rather limited.

Auger Boring

Auger borings are made with 2- to 8-inch augers, similar in design to a wood bit, fastened on sections of pipe encased in a larger one. After a number of turns the auger is raised and samples of soil clinging to the bit are examined. Its use is limited mainly to sandy soil, coarse gravel, and soft clay.

Core Boring

Core boring is primarily used when rock is encountered to determine whether a thin layer of rock has been mistaken for solid or bedrock. The borer is rotated by power-driven equipment, and the coring end may be a diamond drill, a chilled steel shot, or a carbide bit. A core *barrel* that contains samples up to several feet in length is provided. These are removed periodically for examination and testing.

Dry Sample Boring

This method is especially designed for sand or clay soils. A pipe is driven or forced to the sampling level, and a 2-inch O.D. pipe with a sampling spoon at the bottom is driven down by a recorded number of blows of a standard 140-lb. weight, free falling 30 inches to penetrate the ground 12 to 24 inches. The sampling spoon is lifted for laboratory testing.

There are variations of these testing methods depending on the conditions encountered, and on the experience and judgment of the individual directing the work. There also are other acceptable methods of testing, including *boring* and *churn drilling,* some being more reliable than others.

A knowledge of the subsurface characteristics in the particular *geographic region* where the construction project is to be located is essential. The degree to which the area is subject to earthquakes and earth tremors (seismic loads), flash flooding, settlement, subsidence, and mudslides must be given consideration.

Designers and constractors can be held liable for damages resulting from improper or negligently constructed footings and foundations of buildings and other structures. The cause is very often traceable to inadequate attention to subsurface soil conditions and underground waters including springs and streams that could have been disclosed by consulting public records. Liability may also arise for damage to neighboring properties because of poorly designed drainage systems.

Weights and Bearing Capacities of Soils

Published weights and load-bearing capacities of the different types of soils vary depending upon the source of the information. This is understandable when one considers the complexity of individual soils or mixtures, of rock, gravel, sand, silt, and clay, each having different grain sizes and moisture content. A single classification of soils that will meet every situation is, therefore, not practical.

Table 11–2 shows the approximate weights of soil materials, *loose* and *in bank.* The load factors also are shown. To convert the load factors to *swell* of the materials, use Table 11–3. For example, the Load Factor for earth dry-packed, in Table 11–2, is .80. Table 11–3 shows a load factor of .800, which represents a swell of 25 percent.

The approximate bearing capacities of various soils is shown in Table 11–4 as a guideline. For other than routine, light construction, soil testing by experts is recommended.

TABLE 11-2[1]
Weight of Soil Materials and Load Factors

Weight* of Materials	Loose kg/m³	Loose lb/yd³	Bank kg/m³	Bank lb/yd³	Load Factors
Basalt	1960	3300	2970	5000	.67
Bauxite, Kaolin	1420	2400	1900	3200	.75
Caliche	1250	2100	2260	3800	.55
Carnotite, uranium ore . .	1630	2750	2200	3700	.74
Cinders	560	950	860	1450	.66
Clay—Natural bed	1660	2800	2020	3400	.82
Dry	1480	2500	1840	3100	.81
Wet	1660	2800	2080	3500	.80
Clay and gravel—Dry . . .	1420	2400	1660	2800	.85
Wet . . .	1540	2600	1840	3100	.85
Coal—Anthracite, Raw . . .	1190	2000	1600	2700	.74
Washed . .	1100	1850			.74
Ash, Bituminous Coal . .	530–650	900–1100	590–890	1000–1500	.93
Bituminous, Raw . . .	950	1600	1280	2150	.74
Washed . .	830	1400			.74
Decomposed rock—					
75% Rock, 25% Earth . .	1960	3300	2790	4700	.70
50% Rock, 50% Earth . .	1720	2900	2280	3850	.75
25% Rock, 75% Earth . .	1570	2650	1960	3300	.80
Earth—Dry packed . . .	1510	2550	1900	3200	.80
Wet excavated . . .	1600	2700	2020	3400	.79
Loam	1250	2100	1540	2600	.81
Granite — Broken	1660	2800	2730	4600	.61

[1]Reprinted with permission of Caterpillar Inc.

TABLE 11-2 (continued)

Weight* of Materials	Loose		Bank		Load Factors
	kg/m³	lb/yd³	kg/m³	lb/yd³	
Gravel — Pitrun	1930	3250	2170	3650	.89
Dry	1510	2550	1690	2850	.89
Dry 6–50 mm (¼″ – 2″)	1690	2850	1900	3200	.89
Wet 6–50 mm (¼″ – 2″)	2020	3400	2260	3800	.89
Gypsum — Broken	1810	3050	3170	5350	.57
Crushed	1600	2700	2790	4700	.57
Hematite, iron ore, high grade . . .	1810–2450	4000–5400	2130–2900	4700–6400	.85
Limestone — Broken	1540	2600	2610	4400	.59
Crushed	1540	2600	—	—	—
Magnetite, iron ore	2790	4700	3260	5500	.85
Pyrite, iron ore	2580	4350	3030	5100	.85
Sand — Dry, loose	1420	2400	1600	2700	.89
Damp	1690	2850	1900	3200	.89
Wet	1840	3100	2080	3500	.89
Sand & Clay — Loose	1600	2700	2020	3400	.79
Compacted	2400	4050			
Sand & gravel — Dry	1720	2900	1930	3250	.89
Wet	2020	3400	2230	3750	.91
Sandstone	1510	2550	2520	4250	.60
Shale	1250	2100	1660	2800	.75
Slag — broken	1750	2950	2940	4950	.60
Snow — Dry	130	220			
Wet	520	860			
Stone — crushed	1600	2700	2670	4500	.60
Taconite	1630–1900	3600–4200	2360–2700	5200–6100	.58
Top Soil	950	1600	1370	2300	.70
Traprock — broken	1750	2950	2610	4400	.67

*Varies with moisture content, grain size, degree of compaction, etc. Tests must be made to determine exact material characteristics.

TABLE 11-3[2]

Swell—Voids—Load Factors

SWELL (%)	VOIDS (%)	LOAD FACTOR
5	4.8	.952
10	9.1	.909
15	13.0	.870
20	16.7	.833
25	20.0	.800
30	23.1	.769
35	25.9	.741
40	28.6	.714
45	31.0	.690
50	33.3	.667
55	35.5	.645
60	37.5	.625
65	39.4	.606
70	41.2	.588
75	42.9	.571
80	44.4	.556
85	45.9	.541
90	47.4	.526
95	48.7	.513
100	50.0	.500

[2]Ibid.

TABLE 11-4

Approximate Bearing Capacities of Various Soil Materials in Place

Material	Tons Per Square Foot	Material	Tons Per Square Foot
Clay, dry	2 to 5	Mud dry	1 to 1.5
Clay and gravel, dry	2 to 4	Mud, soft, flowing	0 to 1.0
Clay and gravel, wet	2 to 4	Mud, soft, packed	1 to 1.5
Earth, loam, loose and dry	1 to 2	Sand, clean, dry	3 to 5.0
Earth, loam, loose, moist	1 to 2	Sand, wet	2 to 3.0
		Sand and gravel,	
Gravel, dry	4 to 6	dry	4 to 6.0
Gravel, wet	4 to 6	Sandstone	15 to 20.0
Hard rock	40 to 50	Shale	5 to 10.0

ENGINEERING AND DESIGN REQUIREMENTS

Because subsurface soil conditions do vary considerably, building codes require that foundation designs, excavation bracing designs, and lateral earth pressure designs be based on the results of the subsurface exploration and testing at the building site. Such investigation should be planned and supervised by a registered professional engineer, or architect, who shall be responsible for the interpretation of the field and laboratory tests.

Structural Loads

As a guide for designers, two tables are provided. Table 11–5, "Weights of Building Construction Materials" (dead loads), and Table 11–6, "Minimum Uniformly Distributed Live Loads." Both tables are based on minimum unit weights recognized by *The Uniform Building Code (U.B.C.)* and are, of course, approximate averages.

Dead Loads. A dead load is defined in U.B.C. as "the vertical load due to the weight of all permanent structural and nonstructural components of a building, such as walls, floors, roofs, and fixed service equipment."

Live Loads. A live load is defined in the U.B.C. as "the load superimposed by the use and occupancy of the building not including the wind load, snow load, earthquake load, or dead load."

Wind Loads and Snow Loads. Two other important loads that affect buildings are wind and snow loads. Building codes provide for their consideration in the designing of buildings and structures. The requirements range widely according to geographic and local conditions. For example, areas of the country that are exposed to or have a history of tropical storms have stringent code requirements. By the same token, areas of the country with seasonal snowpack exceeding 40 lbs. per square foot also have strict code provisions. In any case, it is recommended that the local building code and building official be consulted.

TABLE 11–5
Weights of Building Construction Materials

Description of Material	Pounds Per Cubic Foot
Masonry	
Brick, unreinforced	120
Brick, reinforced	125
Ashlar	
Granite, Limestone	168
Sandstone	144
Rubble	
Granite	153
Limestone	147
Sandstone	137
Terra cotta, voids filled	120
Terra cotta, voids unfilled	72
Concrete, unreinforced	
Cinder	108
Slag	132
Stone (including gravel)	144
Vermiculite and Perlite aggregate, non–load-bearing	25–50
Other light aggregate, load-bearing	70–105
Concrete, reinforced	
Cinder	111
Slag	138
Stone (including gravel)	150
Concrete block (in inches)	
8 × 8 × 12	34
8 × 8 × 16	45
Cinder block (in inches)	
8 × 8 × 12	22
8 × 8 × 16	30

Description of Material	Pounds Per Square Foot
Concrete slabs	
Concrete, reinforced stone per inch of thickness	12.5
Concrete, reinforced light-weight, per inch of thickness	9

Weights of Building Construction Materials (continued)

Description of Material	Pounds Per Square Foot
Concrete, plain stone, per inch of thickness	12
Concrete, plain lightweight, per inch of thickness	8.5

Wood-joist floors, (double wood floor) Joist sizes in inches	Pounds Per Square Foot	
	12−in. Spacing	16−in. Spacing
2 × 6	6	5
2 × 8	6	6
2 × 10	7	6
2 × 12	8	7
3 × 6	7	6
3 × 8	8	7
3 × 10	9	8
3 × 12	11	9
3 × 14	12	10

Description of Material	Pounds Per Cubic Foot
Timber, seasoned	
Ash, commercial white	41
Cypress, southern	32
Fir, Douglas, coast region	34
Oak, commercial reds and whites	45
Redwood	28
Spruce, red, white, and Sitka	28
Southern pine, short-leaf	39
Southern pine, long-leaf	48
Timber, hemlock	30

Description of Material	Pounds Per Square Foot
Roofing	
Asphalt shingles	2

Weights of Building Construction Materials (continued)

Description of Material	Pounds Per Square Foot
Composition roofing	
Three-ply ready roofing	1
Four-ply felt and gravel	5.5
Five-ply felt and gravel	6
Copper or tin	1
Corrugated asbestos–cement roofing	4
Corrugated iron	2
Skylight, metal frame, ⅜-inch wire glass	8
Slate, ³⁄₁₆-inch	7
Slate, ¼-inch	10
Spanish tile	20
Wood sheathing, per inch thickness	3
Wood shingles	3
Suspended Ceilings	
Gypsum on wood or metal lath	10
Plaster on tile	5
Plaster on wood lath	8
Walls and partitions (unplastered)	
4-in. clay brick, medium absorption	39
8-in. clay brick, medium absorption	79
12-in. common brick	120
12-in. pressed brick	130
4-in. brick, 4-in. load-bearing structural clay tile backing	60
8-in. brick and concrete block	72
8-in. load-bearing structural clay tile	42
12-in. load-bearing structural clay tile	58
8-in. concrete block, heavy aggregate	55
Wood studs, 2 in. × 4 in.	4
Lath and plaster partitions	
2-in. solid cement on metal lath	25
2-in. solid gypsum on metal lath	18
2-in. solid gypsum on gypsum lath	18
2-in. metal studs gypsum and metal lath, both sides	18

Weights of Building Construction Materials (continued)

Description of Material	Pounds Per Square Foot
3-in. metal studs gypsum and metal lath, both sides	19
4-in. metal studs gypsum and metal lath, both sides	20
2-by-4-in. wood studs plastered one side	12
2-by-4-in. wood studs plastered two sides	20
6-in. wood studs plaster and plaster boards, both sides	18
6-in. wood studs plaster and wood lath, both sides	18
6-in. wood studs plaster and metal lath, both sides	18
6-in. wood studs unplastered gypsum board, both sides, (dry wall)	10
Plaster work only	
Gypsum (one side)	5
Cement (one side)	10
Gypsum on wood lath	8
Gypsum on metal lath	8
Gypsum on plaster board or fiber board	8
Cement on wood lath	10
Cement on metal lath	10

TABLE 11-6

Minimum Uniformly Distributed Live Loads*

Occupancy or Use	Live Load (Lbs Per Square Foot)
Apartments (see *Residential*)	
Assembly halls and other places of assembly:	
Fixed seats	60
Moveable seats	100
Platforms (assembly)	100

Minimum Uniformly Distributed Live Loads (continued)

Occupancy or Use	Live Load (Lbs Per Square Foot)
Balcony, exterior	100
Bowling alleys, pool rooms, and similar recreation areas	75
Corridors:	
First floor	100
Other floors, same as occupancy served except as indicated	
Dance halls and ball rooms	100
Dining rooms and restaurants	100
Dwellings (see *Residential*)	
Garages:	
General storage/repair	100
Private pleasure car storage	50
Gymnasiums, main floors and balconies	100
Hospitals:	
Operating rooms, laboratories	60
Private rooms	40
Wards	40
Corridors above first floor	80
Hotels (see *Residential*)	
Libraries	
Reading rooms	60
Stack rooms (books and shelving at 65 pounds per cu ft) but not less than 150 lbs/sq ft Live Load for *Stackrooms*.	150
Manufacturing	
Heavy	150
Light	100
Marquees	75
Morgues	125
Office buildings:	
Offices	50
Lobbies	100
Penal institutions:	
Cell blocks	40
Corridors	100

Minimum Uniformly Distributed Live Loads (continued)

Occupancy or Use	Live Load (Lbs Per Square Foot)
Residential:	
Multifamily houses:	
Private apartments	40
Public rooms	100
Corridors	80
Dwellings:	
First floor	40
Second floor and habitable attics	30
Uninhabitable attics	20
Hotels:	
Guest rooms	40
Public rooms	100
Corridors serving public rooms	100
Corridors	80
Rest rooms	60
Schools:	
Classrooms	40
Corridors	80
Sidewalks, vehicular driveways, and yards, subject to trucking	250
Stairs and exitways, fire escapes	100
Storage warehouse:	
Light	125
Heavy	250
Stores:	
Retail:	
First floor rooms	100
Upper floors	75
Wholesale, use actual weight if greater than	125
Yards and terraces, pedestrians	100

* Standard Building Code

EXCAVATING, BACKFILLING, AND COMPACTING

Excavating is broadly divided into two kinds or types: *general excavating* and *special excavating*. While contractors include stripping and stockpiling topsoil within the term excavating, for purposes of this discussion we consider excavating as the digging of holes for basements, cellars, footings, foundations, substories, trenches, and pits. It includes backfilling and compacting.

General excavating is that which can be done by mechanical equipment, such as power shovels, clamshell cranes, loaders, and bulldozers, including trucks for hauling the excavated material.

Specialized excavating is the type usually done by hand or special equipment and includes trenching for utilities and trimming for footings and foundations.

Factors That Affect Costs

When studying or estimating costs, several factors must be considered to minimize inherent risks—particularly in large construction projects. A review of the following checklist is suggested.

CHECKLIST

1. What type(s) of soils will be involved?

2. Will subsurface water be encountered? How will it be removed? Is continuous pumping required?

3. If the banks are not self-supporting and sloping is not possible because of adjacent property or for any other reason, how will they be braced?

4. If the footings of adjacent structures are above the depth of the excavation, what will it cost to protect and shore up those structures? Who will have that responsibility?

5. Are there any existing underground utilities to be removed or capped?

6. Will all excavated material have to be trucked to a dump site? Will a permit be required?

7. Are the following safety measures required?

   Fencing
   Night lighting
   Night watchman

8. What is the approximate percentage of *swell* over *bank* cubic yardage? (See Table 11–2.)

9. Is rock to be excavated? If so, are there special blasting permits, dynamite and powder storage facilities, blasting mats, and other measures required for the protection of people and adjacent properties?

10. Is there adequate material for any necessary backfilling? Is it in the proximity of the site, or will it have to be trucked in?

Excavating by Hand

Trenches for footings, pits, sewer, water, and some utility lines are frequently dug by hand. Special machines are also used to trench, such as the "ferris wheel" and vertical boring rigs mounted on a bulldozer. A considerable amount of hand excavating is done after general excavating: trimming bottoms and sides of footings, digging pits for tanks, trenches, and so forth. Also, when concrete columns and piers are specified, their footings may require deeper levels for adequate bearing. These deep pits may be dug with power equipment and then trimmed by hand.

When hand labor is used to excavate trenches, piers, and footings on smaller projects, and where machines cannot be used, or are less efficient, Table 11–7 may serve as a guide to estimate labor.

General or Mass Excavating

Today, practically all excavating, backfilling, and grading is performed with power equipment with the exception of small jobs and where pick and shovel will do the work better, more efficiently, and more cheaply. Most of the general excavating is done by *excavating contractors* who have specialized machinery, equipment, and experience. Like other subcontractors, they are generally competitive bidders.

TABLE 11-7

Approximate Hours of Labor to Excavate by Hand

Type of Soil	Depth In Feet	Average Hours Per Cu. Yd.
Earth (loam) or sandy soil	2	1.0
	4	1.3
	6	1.5
Sand and gravel	2	1.2
	4	1.5
	6	1.7
Earth and clay	2	1.2
	4	1.4
	6	1.6
Heavy clay soil	2	1.6
	4	1.8
	6	2.0

The cost of owning power-operated machines, equipment, and trucks requires careful analysis. Some of the major factors to consider are the initial investment, taxes, interest, depreciation, insurance, maintenance, repairs, and storage. In addition there are operating costs that include the wages of the operators of the machines and drivers of the trucks, gasoline, diesel fuel, and lubrication and transmission oil. All of these and other costs must be measured against the probable maximum use, future services, minimum *idle time,* and life expectancy of the units.

An alternative is to lease equipment. This has certain advantages to be weighed against owning the equipment, or subcontracting the work. Some of those benefits include: leasing for a single job, immediate tax reduction for the cost of monthly leases, no large financial commitment, and none of the inherent costs of ownership and its responsibilities.

Excavating with Power Shovels and Draglines

Power shovel and diesel dragline production depends upon type of material, angle of swing, height of bank, and overall job efficiency. Productivity also depends upon the ability of the operator, swell of the material, slope of ground on which the machine is working, and the supply and continuance of hauling units.

Table 11–8 shows the estimated hourly production of dipper-type power shovels operating under the conditions stated in the Table's footnote. Included are correction factors for depth of cut, angle of swing on power shovel output, and bucket fill factor range.

Table 11–9 shows the estimated hourly production of diesel draglines. The quantities are also subject to the conditions in the footnote of the Table.

The figures shown in these two tables are for in-bank (in-ground) quantities, not cubic yards loaded on hauling units which includes swell of the material. As shown in Table 11–2, the swell of excavated material can range from 10 percent to 35 percent for soils, and as much as 60 percent to 70 percent in rock.

TABLE 11–8[3]
Estimated Hourly Production Dipper-Type Power Shovels

Bucket Size:		Diesel Power Shovel Production													Electric Power Shovel Production				
	m³	0.57	0.75	0.94	1.13	1.32	1.53	1.87	2.29	2.62	3.06	3.37	3.82	4.59	4.59	5.35	6.12	6.88	7.65
	yd³	¾	1	1-¼	1-½	1-¾	2	2-½	3	3-½	4	4-½	5	6	6	7	8	9	10
Moist loam or sandy clay	m³	126	157	191	218	245	271	310	356	401	443	485	524	608	696	776	859	929	995
	yd³	165	205	250	285	320	355	405	465	525	580	635	685	795	910	1015	1125	1215	1300
Sand and gravel	m³	119	153	176	206	229	252	298	344	386	424	459	493	566	650	730	807	876	944
	yd³	155	200	230	270	300	330	390	450	505	555	600	645	740	850	955	1050	1145	1235
Common earth	m³	103	134	161	183	206	229	271	310	348	390	428	463	524	604	669	742	813	883
	yd³	135	175	210	240	270	300	355	405	455	510	560	605	685	790	875	970	1065	1155
Clay, hard, tough	m³	84	111	138	161	180	203	237	275	310	344	375	405	463	516	581	650	715	780
	yd³	110	145	180	210	235	265	310	360	405	450	490	530	605	675	760	850	935	1020
Rock, well blasted	m³	73	96	119	138	157	176	210	245	279	313	348	382	440	474	535	596	646	707
	yd³	95	125	155	180	205	230	275	320	365	410	455	500	575	620	700	780	845	925
Common excav. with rocks and roots	m³	61	80	99	119	138	153	187	222	256	291	321	352	413	440	497	558	608	661
	yd³	80	105	130	155	180	200	245	290	335	380	420	460	540	575	650	730	795	865
Clay, wet and sticky	m³	54	73	92	111	126	141	176	206	237	264	294	321	375	394	443	501	543	596
	yd³	70	95	120	145	165	185	230	270	310	345	385	420	490	515	580	655	710	780
Rock, poorly blasted	m³	38	57	73	88	107	122	149	180	206	233	260	287	336	348	398	452	497	547
	yd³	50	75	95	115	140	160	195	235	270	305	340	375	440	455	520	590	650	715

Conditions:
1. Bank m³ (yd³) measure
2. 60 min. hour — 100% efficiency
3. 90° swing
4. Bucket fill factor consideration
5. Optimum digging depth
6. Grade level loading
7. All material loaded into hauling units

³Reprinted with permission of Caterpillar Inc.

TABLE 11–8 (continued)

CORRECTION FACTORS for Depth of Cut and Angle of Swing on Power Shovel Output

Depth of Cut in % of Optimum	Angle of Swing, Degrees						
	45	60	75	90	120	150	180
40	.93	.89	.85	.80	.72	.65	.59
60	1.10	1.03	.96	.91	.81	.73	.66
80	1.22	1.12	1.04	.98	.86	.77	.69
100	1.26	1.16	1.07	1.00	.88	.79	.71
120	1.20	1.11	1.03	.97	.86	.77	.70
140	1.12	1.04	.97	.91	.81	.73	.66
160	1.03	.96	.90	.85	.75	.67	.62

BUCKET FILL FACTOR

Material	Fill Factor Range
Sand & gravel	.90 to 1.00
Common earth	.80 to .90
Hard clay	.65 to .75
Wet clay	.50 to .60
Rock, well blasted	.60 to .75
Rock, poorly blasted	.40 to .50

Note: Depth of cut is vertical distance shovel bucket must travel to obtain load. Optimum distance varies with type of material excavated. Since boom angle must be changed with various types of material, use this rule of thumb: *Optimum depth of cut is equal to vertical distance from dipper stick pivot shaft to ground level.*

It must also be noted that the production figures in both Tables 11–8 and 11–9 are performances under ideal conditions, 100-percent efficiency, 60 min. hours, 90-percent swing, continuous operation, and no interruption of operator or hauling units. In this respect the figures represent

TABLE 11-9[4]
Estimated Hourly Production Diesel Draglines

Bucket Size:	Diesel Dragline — (an attachment for crane-type boom consisting of dragline bucket, hoist cable, drag cable and fairlead)												
m³	0.57	0.75	0.94	1.13	1.32	1.53	1.87	2.29	2.44	3.06	3.37	3.82	4.59
yd³	¾	1	1-½	1-½	1-¾	2	2-½	3	3-¼	4	4-½	5	6
Loam or light, m³	99	122	149	168	187	203	233	268	298	356	386	413	466
moist clay yd³	130	160	195	220	245	265	305	350	390	465	505	540	610
Sand or gravel m³	96	119	141	161	180	195	226	260	291	348	378	405	459
yd³	125	155	185	210	235	255	295	340	380	455	495	530	600
Earth, common m³	80	103	126	145	161	176	203	233	260	287	313	340	390
yd³	105	135	165	190	210	230	265	305	340	375	410	445	510
Clay, hard, tough m³	69	84	103	122	138	149	176	206	233	260	287	313	363
yd³	90	110	135	160	180	195	230	270	305	340	375	410	475
Clay, wet sticky m³	42	57	73	84	99	111	134	161	183	206	229	252	295
yd³	55	75	95	110	130	145	175	210	240	270	300	330	385

Conditions: 1. Bank m³ (yd³) measure
2. 100 % efficiency — 60 min. hr.
3. 90° swing
4. Bucket fill factor consideration

[4]Reprinted with permission of Caterpillar Inc.

perfection seldom achieved and should not be used in estimating or bidding without taking into account every aspect of the particular construction project under consideration. In other words, these tables are starting-point guidelines.

SELECTED REFERENCE

Estimating Earthwork Quantities, Daniel B. Atcheson, Norseman Publishing Company, Lubbock, TX 79493–6617, 1992.

CHAPTER 12

CONCRETE CONSTRUCTION

FORMS FOR CONCRETE

The design of and the material for forms to hold concrete shaped and in place while curing depend largely on the size of the project and the purpose or function of the structural unit. They may range from simple wooden forms for footings, slabs on the ground, and walkways to the more advanced Plyform plywood forms and manufactured forms that can be purchased or rented.

The unit of measurement for forms is the actual square feet of surface area *in contact* with the concrete. An exception is in footings, where the unit of measurement is frequently the lineal feet of the footing, called *ribbon footings*.

Wooden Forms

Footing forms made on the job site and constructed of nominal 1-inch boards and 2-by-4-inch framing are illustrated in Figure 12–1. Wall and sidewalk, or slab-on-fill forms, so constructed, are illustrated in Figure 12–2. Accompanying each illustration is a detail of the material in board feet per square foot of form.

Table 12–1 (on page 170) shows the approximate board feet of lumber and the hours of labor to build and

167

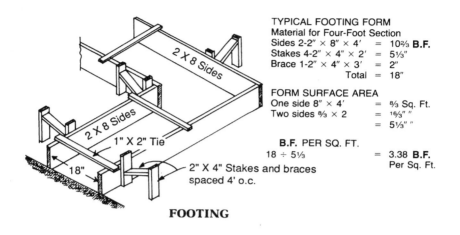

TYPICAL FOOTING FORM
Material for Four-Foot Section

Sides 2-2″ × 8″ × 4′ = 10⅔ **B.F.**
Stakes 4-2″ × 4″ × 2′ = 5⅓″
Brace 1-2″ × 4″ × 3′ = 2″
Total = 18″

FORM SURFACE AREA
One side 8″ × 4′ = ⅔ Sq. Ft.
Two sides ⅔ × 2 = 1⅔″ ″
= 5⅓″ ″

B.F. PER SQ. FT.
18 ÷ 5⅓ = 3.38 **B.F.**
Per Sq. Ft.

2″ X 4″ Stakes and braces
spaced 4′ o.c.

FOOTING

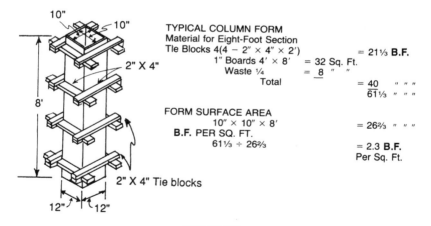

TYPICAL COLUMN FORM
Material for Eight-Foot Section

Tle Blocks 4(4 − 2″ × 4″ × 2′) = 21⅓ **B.F.**
1″ Boards 4′ × 8′ = 32 Sq. Ft.
Waste ¼ = 8 ″ ″
Total
= 40 ″ ″ ″
61⅓ ″ ″ ″

FORM SURFACE AREA
10″ × 10″ × 8′ = 26⅔ ″ ″ ″
B.F. PER SQ. FT.
61⅓ ÷ 26⅔ = 2.3 **B.F.**
Per Sq. Ft.

2″ X 4″ Tie blocks

COLUMN

Figure 12–1
Wood Concrete Forms

168

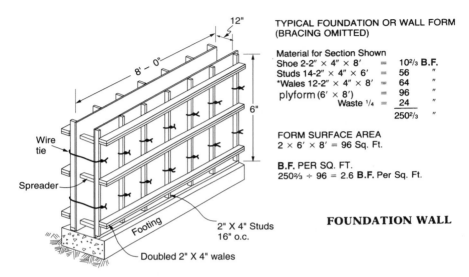

TYPICAL FOUNDATION OR WALL FORM
(BRACING OMITTED)

Material for Section Shown

Shoe 2-2″ × 4″ × 8′	=	10²/₃	B.F.
Studs 14-2″ × 4″ × 6′	=	56	″
*Wales 12-2″ × 4″ × 8′	=	64	″
plyform (6′ × 8′)	=	96	″
Waste ¹/₄	=	24	″
		250²/₃	″

FORM SURFACE AREA
2 × 6′ × 8′ = 96 Sq. Ft.

B.F. PER SQ. FT.
250²/₃ ÷ 96 = 2.6 **B.F.** Per Sq. Ft.

FOUNDATION WALL

*When heavy wales are required, it will be noted that the B.F. per sq. ft. is substantially increased.

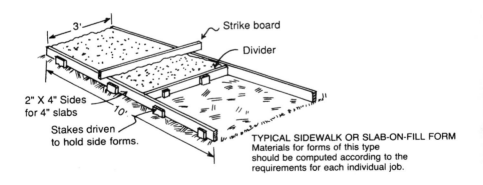

SLAB ON FILL

Figure 12–2
Wood Concrete Forms

169

Concrete Construction

TABLE 12–1
Approximate BF of Lumber and Hours of Labor to Build, Erect, and Strip 100 Sq. Ft. of Wood Forms

Type of Form	BF Lumber* Per 100 Sq. Ft. Contact Area	Hours Labor Per 100 Sq. Ft. Contact Area		
		Build and Erect	Strip	Total
Walls and foundations	260	8	4	12
Slabs above grade	250	10	4	14
Stairs, straight run**	300	16	4	20
Columns	260	8	3	11
Beams and girders	300	12	4	16
Footings for walls	250	6	3	9

*Add for nails, ties, wire, etc.
**Includes stringers, risers, and shoring, 9-ft. to 12-ft. ceilings.

and erect square feet of wood forms of various types. The figures are guidelines, and all factors that affect labor productivity (see Chapter 10) should be taken into consideration when estimating labor.

Plyform

Plyform is a special product produced by the plywood industry designed for concrete forms. The term is proprietory and may be applied only to specific products that bear the APA trademark of the American Plywood Association. Products that bear this specific identification are available in two classes: Plyform Class I and Plyform Class II, with Class I the stronger and stiffer due to species groups required for each class. Either may be ordered with a High Density Overlaid (HDO) surface on each side.

Both Classes I and II of HDO Plyform have a hard, smooth, semi-opaque surface of thermo-setting, resin-impregnated material that forms a durable, continuous bond with the plywood. The abrasion-resistant surface does not require oiling. However, many users wipe the panels lightly with oil or other release agents before each pour to assure easy stripping. With reasonable care, HDO Plyform will normally produce twenty to fifty uses or more.

Textured Plywood

Plywood is manufactured in more than forty surface textures ranging from glass-smooth High Density Overlaid plywood to patterned board-and-batten panels. This special exterior-type textured plywood can provide unusual architectural effects in concrete and is usually applied in one of two ways in formwork design: (1) as a liner requiring plywood backing so that the liner delivers texture, but contributes little to the structure of the formwork; or (2) as the basic forming panel.

Concrete Wall and Slab Form Design

Basic to the designing of forms for concrete walls is that concrete, when placed in the form, is in a fluid-like state causing it to exert a hydrostatic horizontal pressure against the sides of the form. Therefore the initial step in designing forms is to determine pressure of the concrete during placement. The pressure developed will be determined by:

- The rate of pouring, that is, the rate at which the concrete rises in the form
- The height of the form
- The temperature at the time of pouring
- The density of the concrete, that is, the mix and the consistency of the concrete
- The method of vibration

Plyform is manufactured in thicknesses of ½ inch to 1⅛ inch; panels of ⅝ inch and ¾ inch are the most commonly used. Table 12–2 to Table 12–8, for *grade-use guide*

TABLE 12-2

Grade-Use Guide for Concrete Forms*

Use these terms when you specify plywood	DESCRIPTION	TYPICAL TRADEMARKS	VENEER GRADES		
			Faces	Inner Piles	Backs
APA B-B PLYFORM Class I & II**	Specifically manufactured for concrete forms. Many reuses. Smooth, solid surfaces. Mill-oiled unless otherwise specified.	APA PLYFORM B-B CLASS I EXTERIOR 000 PS 1-83	B	C	B
APA High Density Overlaid PLYFORM Class I & II**	Hard, semi-opaque resin-fiber overlay, heat-fused to panel faces. Smooth surface resists abrasion. Up to 200 reuses. Light oiling recommended between pours.	HDO PLYFORM I EXT·APA 000 PS 1-83	B	C-Plugged	B
APA STRUCTURAL I PLYFORM**	Especially designed for engineered applications. All Group 1 species. Stronger and stiffer than PLYFORM Class I and II. Recommended for high pressures where face grain is parallel to supports. Also available with High Density Overlay faces.	APA STRUCTURAL I PLYFORM B-B CLASS I EXTERIOR 000 PS 1-83	B	C or C-Plugged	B

Use these terms when you specify plywood	DESCRIPTION	TYPICAL TRADEMARKS	VENEER GRADES		
			Faces	Inner Piles	Backs
Special Overlays, proprietary panels and Medium Density Overlaid plywood specifically designed for concrete forming.**	Produces a smooth uniform concrete surface. Generally mill treated with form release agent. Check with manufacturer for specifications, proper use, and surface treatment recommendations for greatest number of reuses.				
APA B-C EXT	Sanded panel often used for concrete forming where only one smooth, solid side is required.	APA B-C GROUP I EXTERIOR 000 PS 1-83	B	C	C

*Commonly available in 19/32", 5/8", 23/32" and 3/4" panel thicknesses (4' × 8' size).
**Check dealer for availability in your area.

Reprinted with the permission of the American Plywood Association.

and *form design,* are reprinted with the permission and courtesy of the American Plywood Association of Tacoma, Washington.

Pressures on Column and Wall Forms

Table 12–3 shows the pressure generated by standard weight concrete having a slump of 4 in. or less with normal internal vibration at different pour rates and temperatures. It is based on the recommendations of the American Concrete Institute. For external vibration, double the pressures shown. When concrete is not vibrated, some designers reduce the pressures in the Table by 10 percent.

TABLE 12-3

**Concrete Pressures for Column
and Wall Forms**

| Pour Rate (ft/hr) | Pressures of Vibrated Concrete (psf)[a,b] | | | |
| | 50°F | | 70°F | |
	Columns	Walls	Columns	Walls
1	330	330	280	280
2	510	510	410	410
3	690	690	540	540
4	870	870	660	660
5	1050	1050	790	790
6	1230	1230	920	920
7	1410	1410	1050	1050
8	1590	1470	1180	1090
9	1770	1520	1310	1130
10	1950	1580	1440	1170

[a]Maximum pressure need not exceed 150h, where h is maximum height of pour.
[b]Based on concrete with density of 150 pcf and 4-in. slump.
Reprinted with the permission of the American Plywood Association.

TABLE 12–4

Design Loads for Slab Forms

Slab Thickness (in.)	DESIGN LOAD (psf)	
	Nonmotorized Buggies[a]	Motorized Buggies[b]
4	100[c]	100[c]
5	113	138
6	125	150
7	138	163
8	150	175
9	163	188
10	175	200

[a]Includes 50 psf load for workmen, equipment, impact, etc.
[b]Includes 75 psf load for workmen, equipment, impact, etc.
[c]Minimum design load.
Reprinted with the permission of the American Plywood Association.

Design Loads for Slab Forms

Table 12–4 gives the minimum design loads for slab forms. Forms for concrete slabs must support workers and equipment (live loads) as well as the weight of freshly placed concrete (dead load). Normal weight concrete (150 pcf) will place a load on the forms of 12.5 psf for each inch of slab thickness. The loads shown include the effects of concrete, buggies, and workers.

Recommended Pressures on Plyform

Recommended pressures on the more common thicknesses of Plyform Class I are shown in Tables 12–5 and 12–6. Tables 12–7 and 12–8 show pressures for Structural I Plyform. Tables 12–5 through 12–8 are based on the plywood acting as a continuous beam that spans

TABLE 12-5

**Recommended Maximum Pressures on
Plyform Class I (psf)[a]
Face Grain Across Supports[b],
Architectural Application**

Support Spacing (in.)	Plywood Thickness (in.)						
	$^{15}/_{32}$	$^{1}/_{2}$	$^{19}/_{32}$	$^{5}/_{8}$	$^{23}/_{32}$	$^{3}/_{4}$	$1^{1}/_{8}$
4	2715	2945	3355	3580	4010	4110	5965
8	885	970	1215	1300	1540	1580	2295
12	335	405	540	575	695	730	1370
16	150	175	245	265	345	370	740
20	—	100	145	160	210	225	485
24	—	—	—	—	110	120	275
32	—	—	—	—	—	—	130

[a]Deflection limited to $^{1}/_{360}$ of the span.
[b]Plywood continuous across two or more spans.
Reprinted with the permission of the American Plywood Association.

joists or studs. No blocking is assumed at the unsupported panel edge. However, under conditions of high moisture or sustained load to the panel, edges may have greater deflection than the center of the panel and may exceed the calculated deflection unless panel edges are supported. For this reason, and to minimize differential deflection between adjacent panels, some form designers specify blocking at the unsupported edge, particularly when face grain is parallel to supports.

Prefabricated Concrete Forms

Prefabricated forms and forming systems are much too diverse in design, materials of construction, and accessories to attempt to generalize on their uses and applications. There are a number of outstanding manufacturers of prefabricated forms, and we suggest to those interested

TABLE 12–6

Recommended Maximum Pressures on Plyform Class I (psf)[a] Face Grain Parallel to Supports[b], Architectural Application

Support Spacing (in.)	Plywood Thickness (in.)						
	15/32	1/2	19/32	5/8	23/32	3/4	1 1/8
4	1385	1565	1620	1770	2170	2325	4815
8	390	470	530	635	835	895	1850
12	110	145	165	210	375	460	1145
16	—	—	—	—	160	200	710
20	—	—	—	—	115	145	400
24	—	—	—	—	—	—	255

[a]Deflection limited to 1/360 of the span.
[b]Plywood continuous across two or more spans.
Reprinted with the permission of the American Plywood Association.

TABLE 12–7

Recommended Maximum Pressures on Structural 1 Plyform (psf)[a] Face Grain Across Supports[b]

Spacing Support (in.)	Plywood Thickness (in.)						
	13/32	1/2	19/32	5/8	23/32	3/4	1 1/8
4	3560	3925	4560	4860	5005	5070	7240
8	890	980	1225	1310	1590	1680	2785
12	360	410	545	580	705	745	1540
16	155	175	245	270	350	375	835
20	—	100	145	160	210	230	545
24	—	—	—	—	110	120	310
32	—	—	—	—	—	—	145

[a]Deflection limited to 1/360 of the span.
[b]Plywood continuous across two or more spans.
Reprinted with the permission of the American Plywood Association.

TABLE 12-8

Recommended Maximum Pressures on
Structural 1 Plyform (psf)[a]
Face Grain Parallel to Supports[b]

Support Spacing (in.)	Plywood Thickness (in.)						
	$15/32$	$1/2$	$19/32$	$5/8$	$23/32$	$3/4$	$1\frac{1}{8}$
4	2715	2945	3355	3580	4010	4110	5965
8	885	970	1215	1300	1540	1580	2295
12	335	405	540	575	695	730	1370
16	150	175	245	265	345	370	740
20	—	100	145	160	210	225	485
24	—	—	—	—	110	120	275
32	—	—	—	—	—	—	130

[a]Deflection limited to $1/360$ of the span.
[b]Plywood continuous across two or more spans.
Reprinted with the permission of the American Plywood Association.

that they write directly to one of these manufacturers for their brochures or other literature containing detailed specifications.

Prefabricated forms—built of wood, steel and wood, and aluminum—are available for purchase or rent. Some manufacturers also provide engineering services, supervisory services, or both.

Figures 12-3 and 12-4 illustrate the application of prefabricated forms, courtesy of the Burke Company of San Mateo, California.

CONCRETE REINFORCEMENT

Unreinforced concrete is very high in compressive strength, but very low in tensile strength. For that reason steel bars or welded wire fabric is required to provide the needed tensile strength to resist bending.

Figure 12–3
Prefabricated Column Form
Courtesy of the Burke Company

Figure 12–4
Prefabricated Wall Forms
Courtesy of the Burke Company

For example, a concrete beam loaded in the middle, while supported at each end, will bend downward in the middle causing tensile stress in the bottom and pulling the bottom of the beam apart. Steel reinforcement near the bottom will provide the tensile strength to prevent the beam from bending.

Reinforcement in slabs, beams, walls, and columns of concrete needs to be carefully analyzed and designed by professional and experienced engineers to comply with local building codes, the *ACI Building Code,* and other recognized standards.

> "Safe, reinforced concrete construction depends, to a large extent, on the correct location of all reinforcing steel as shown on the plans. Skilled workmen with a good knowledge

of the best practices are of vital importance to safe construction. . . . Those engaged in placing reinforcing steel must know how to unload, handle, place, and tie reinforcing steel. They should also be able to read placing drawings and have a general knowledge of reinforced concrete design. Knowing why reinforcing steel must be placed a certain way is as important to the skilled workman as knowing how to do it."*

Reinforcing Bars

Reinforcing bars are estimated by the pound or ton, both for material cost and labor placing them. Table 12–9 shows the sizes, weights, and nominal dimensions of deformed reinforcing bars.

The bar size designation denotes the approximate diameter of the bar in eighths of inches. Bar #3, for example, has an approximate diameter of ¾ inch. Bar #10 has an approximate diameter of 1¼ inches (1⁰⁄₈"), and bar #18, the largest, has an approximate diameter of 2¼ inches (1⁸⁄₈")

TABLE 12–9

ASTM STANDARD REINFORCING BARS				
		NOMINAL DIMENSIONS — ROUND SECTIONS		
BAR SIZE DESIGNATION	WEIGHT POUNDS PER FOOT	DIAMETER INCHES	CROSS SECTIONAL AREA SQ. INCHES	PERIMETER INCHES
#3	0.376	0.375	0.11	1.178
#4	0.668	0.500	0.20	1.571
#5	1.043	0.625	0.31	1.963
#6	1.502	0.750	0.44	2.356
#7	2.044	0.875	0.60	2.749
#8	2.670	1.000	0.79	3.142
#9	3.400	1.128	1.00	3.544
#10	4.303	1.270	1.27	3.990
#11	5.313	1.410	1.56	4.430
#14	7.65	1.693	2.25	5.32
#18	13.60	2.257	4.00	7.09

Reprinted by courtesy of Concrete Reinforcing Steel Institute.

---

*PLACING REINFORCING BARS, Concrete Reinforcing Steel Institute, 1997.

ASTM specifications require that each bar producer shall roll onto the bar (a) a letter or symbol to show the producer's mill, (b) a number corresponding to the size number of the bar, (c) a symbol or marking to indicate the type of steel, and (d) a marking, as shown in Figure 12–5, to designate Grade 60.

The grade mark for Grade 60 may be either the number 60 or one (1) single line. When the number 60 is used to designate Grade 60, it is the fourth mark in order. When a single line is used to designate Grade 60, it is smaller and between the two main ribs which are on opposite sides of all U.S. made bars.*

The number 40 identifies 40,000 p.s.i.; the number 50 identifies 50,000 p.s.i.; the number 60 identifies 60,000 p.s.i.; and the number 75 identifies 75,000 p.s.i. yield strength steel.

Labor Placing Reinforcing Bars

The labor to bend and place reinforcing bars depends to a large extent on the weights of the bars and whether they have to be shaped and wired in place. On the average, where the bars are less than ¾ inch, a worker should be able to place them in position, without tying, at the rate of approximately 20 hours per ton. Where the bars have to be tied in place, the rate will run closer to 24 hours per ton. These rates do not include unloading or hoisting to floors above grade. On jobs where the bars are over ¾ inch, the rate for bending, placing, and tying is about 18 hours per ton. Much depends on the experience of the workers and the physical working conditions. Subcontractors and experienced iron workers substantially reduce the labor hours per ton.

---

*For complete identification marks of concrete reinforcing bars produced by all U.S. manufacturers, we refer you to **CRSI MANUAL OF STANDARD PRACTICE**.

Figure 12-5

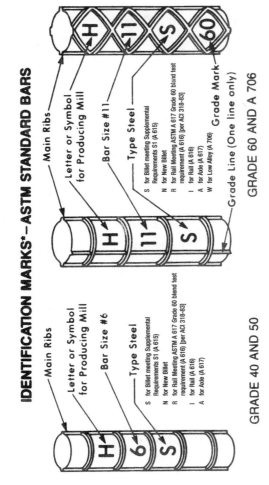

IDENTIFICATION MARKS*–ASTM STANDARD BARS

Main Ribs

Letter or Symbol
for Producing Mill

Bar Size #6

Type Steel

S for Billet meeting Supplemental
 Requirements S1 (A 615)
N for New Billet
R for Rail Meeting ASTM A 617 Grade 60 blend test
 requirement (A 616) [per ACI 318-83]
I for Rail (A 616)
A for Axle (A 617)

GRADE 40 AND 50

Main Ribs

Letter or Symbol
for Producing Mill

Bar Size #11

Type Steel

S for Billet meeting Supplemental
 Requirements S1 (A 615)
N for New Billet
R for Rail Meeting ASTM A 617 Grade 60 blend test
 requirement (A 616) [per ACI 318-83]
I for Rail (A 616)
A for Axle (A 617)
W for Low Alloy (A 706)   Grade Mark

Grade Line (One line only)

GRADE 60 AND A 706

183

Welded Wire Fabric

Table 12–10 (on pages 186–187) shows scheduled unit weights for estimating welded wire fabric with approximate weights in pounds per 100 square feet.

Welded steel wire fabric, commonly but erroneously called "wire mesh," is used to reinforce concrete, especially in driveways and floor slabs. It must conform to ASTM A185 if made of smooth wire, or A497 if made of deformed wire. Welded wire fabric comes in sheets or rolls. Wire sizes smaller than W 1.4 (0.134 in. diam.) are usually manufactured only in rolls. Wire sizes larger than W 4 (0.226 in. diam.) are usually manufactured in sheets.

Smooth wire is denoted by the letter *W* followed by a number indicating cross-sectional area in hundredths of a square inch. Deformed wire is similarly denoted by the letter *D* followed by a number indicating cross-sectional area in hundredths of a square inch.

Welded wire fabric is usually denoted on design drawings as follows: *WWF* followed by spacing of longitudinal wires and then transverse wires and last by sizes of longitudinal and transverse wires.

An example of style designation is *WWF 6 × 12 – W 16 × W 8*. This designation identifies a style of fabric where:

Spacing of longitudinal wires = 6"
Spacing of transverse wires   = 12"
Longitudinal wire size   = W 16
Transverse wire size   = W 8

It is important to note that the terms "longitudinal" and "transverse" are related to the method of manufacture and not to the positions of the wires in a completed concrete structure.

Common widths of welded wire fabric are 60 inches and lengths run 100, 150, and 200 feet. When laid, the wire is lapped a full square.

Figure 12–6
Laborers placing concrete over welded
wire fabric reinforcing.

185

TABLE 12–10

Scheduled Unit Weights for Estimating Welded Wire Fabric*
(Approximate weights in pounds per 100 square feet)

Wire Size Number		Nominal Diameter, Inches	Spacing and Weight of Longitudinal Wires					Spacing and Weight of Transverse Wires					
Smooth	Deformed		2"	3"	4"	6"	12"	3"	4"	6"	8"	10"	12"
W20	D20	0.505	422	286	218	150	82	281	211	141	105	84	70
W18	D18	0.479	379	257	196	135	73	253	190	126	95	76	63
W16	D16	0.451	337	228	174	120	65	225	169	112	84	67	56
W14	D14	0.422	295	200	152	105	57	197	148	98	74	59	49
W12	D12	0.391	253	171	131	90	49	169	126	84	63	51	42
W11	D11	0.374	232	157	120	82	45	155	116	77	58	46	39
W10.5		0.366	221	150	114	79	43	148	111	74	55	44	37
W10	D10	0.357	211	143	109	75	41	141	105	70	53	42	35
W9.5		0.348	200	136	103	71	39	134	100	67	50	40	33
W9	D9	0.338	190	129	98	67	37	126	95	63	47	38	32
W8.5		0.329	179	121	92	64	35	119	90	60	45	36	30
W8	D8	0.319	169	114	87	60	33	112	84	56	42	34	28
W7.5		0.309	158	107	82	56	31	105	79	53	40	32	26
W7	D7	0.299	148	100	76	52	29	98	74	49	37	30	25

186

		Weight											
W6.5		0.288	137	93	71	49	27	91	69	46	34	27	23
W6	D6	0.276	126	86	65	45	24	84	63	42	32	25	21
W5.5		0.265	116	79	60	41	22	77	58	39	29	23	19
W5	D5	0.252	105	71	54	37	20	70	53	35	26	21	18
W4.5		0.239	95	64	49	34	18	63	47	32	24	19	16
W4	D4	0.226	84	57	44	30	16	56	42	28	21	17	14
W3.5		0.211	74	50	38	26	14	49	37	25	18	15	12
W3		0.195	63	43	33	22	12	42	32	21	16	13	11
W2.9		0.192	61	41	32	22	12	41	30	20	15	12	10
W2.5		0.178	53	36	27	19	10	35	26	18	13	11	9
W2.1		0.162	43	29	22	15	8	29	22	15	11	9	7
W2		0.160	42	29	22	15	8	28	21	14	11	8	7
W1.5		0.138	32	21	16	11	6	21	16	11	8	6	5
W1.4		0.134	30	21	16	11	6	20	15	10	7	6	5

*Based on 60″ width, 1″ side overhang each side (62″ overall width), and standard end overhangs.

Note: This table is to be used for estimating purposes only. Exact weights of welded wire fabric will vary from those shown above, depending upon width of rolls or sheets and lengths of overhangs. No allowance is made in this table for the extra weight of fabric required for laps or splices.

Example: Approximate weight of 6 × 6 — W4 × W4

Longitudinal = 30

Transverse $= \frac{28}{58}$ lbs. per 100 sq. ft.

Reprinted by permission of the Concrete Reinforcing Steel Institute from their MANUAL OF STANDARD PRACTICE, 1997.

187

Labor Placing Welded Wire Fabric

Welded steel wire fabric is a commonly used reinforcement for slabs on the ground, driveways, and walks. The square or rectangular styles are sold by the roll, frequently 5 feet wide and 150 feet long. It is also available in flat sheets. When laid, the rolls or sheets are lapped a full square. The labor putting welded wire fabric in floors and driveways varies with the job. On straight slab work like garage floors, a worker should handle about 200 square feet an hour.

CONCRETE

Concrete, freshly mixed, is a semifluid paste composed of portland cement, aggregates, and water. The paste is converted into a hardened mass by the chemical reactions between the cement and the water. Ideal results require *time, favorable temperatures,* and the continued presence of *water.*

For all types of construction, concrete should have four essential physical properties:

- *Workability* for ease of moving to its position of placement for hardening and finishing
- *Strength* to support the design loads
- *Durability* to withstand wear, weathering, and chemical attack
- *Watertightness* to prevent passage of water through the mass

Materials

Cements

There are eight types of portland cement listed by The American Society for Testing Materials (ASTM) in the "Specifications for Portland Cement":

Type I, normal; Type IA, normal, air-entraining
Type II, moderate; Type IIA, moderate, air-entrained
Type III, high early strength; Type IIIA, high early strength, air-entrained
Type IV, low heat of hydration
Type V, sulfate resisting

Type I cement is for use in general concrete construction when the properties of the other types are not specified, especially where the concrete will not be exposed to high sulfate soils or excessive heat during hydration.

Type II cement is for use in general concrete construction exposed to moderate sulfate action, or where moderate heat of hydration is required.

Type III cement is used where high early strength is required in order to remove the forms and put the concrete into use as soon as possible.

Type IV cement is used when it is required to keep the heat of hydration to a minimum.

Type V cement is used where high sulfate resistance is important.

Type I is an all-purpose cement and is more generally used than the other types.

Types IA, IIA, and IIIA have the same composition as their corresponding Types I, II, and III with the addition of air-entraining materials. This improves the resistance of concrete to freezing and thawing and to scaling caused by chloride salts and other chemicals sometimes used for snow and ice removal.

Admixtures

ASTM defines an admixture as "a material other than water, aggregates, and portland cement (including air-entrained portland cement and portland blast-furnace-slag cement), and is added to the batch immediately before or during its mixing." Admixtures are used to modify the properties of concrete to make it more suitable for particular work. The principal kinds of admixtures are, in addition to air-entraining materials:

- *Accelerators,* such as calcium chloride, are used to speed up chemical reactions in the cement and accelerate the setting time. The amount should not exceed 2 percent by weight of the cement and should be introduced into the mix in solution as a part of the water.
- *Retarders* delay the chemical action of the cement. They are used for various reasons such as slowing down the setting time in high temperatures during hot weather.
- *Workability agents* are used to produce smooth plastic mixes.

It is generally admitted that admixtures should be used with great caution. Frequently changing the type of cement, or the amount, and changing the size of the aggregate and the proportions will produce the same results as admixtures with fewer harmful effects.

Water

Water for mixing concrete should be clear and free of such contaminates as oils, acids, or sulfates. Almost any water that is drinkable is satisfactory.

Aggregates

The fine and the coarse aggregates in concrete make up 60 percent to 75 percent of the volume depending on the proportions of the mix and whether it is air-entrained or not. In any prescribed mix for a cubic yard of concrete, the total cubic feet of the cement, sand, and coarse aggregate will exceed 27 cu. ft. But when the materials are mixed, the voids in the coarse aggregate are filled with the fine aggregate, and the voids in the fine aggregate are taken up with the finer particles of the cement. For example, a mix of 1: 2¼: 3 is composed of 6¼ cu. ft. of cement, 14 cu. ft. of sand, and 19 cu. ft. of coarse aggregate for a total of 39¼ cu. ft. After mixing, the mass will occupy 27 cu. ft.

Fine Aggregate. Natural sand is the most commonly used fine aggregate. The ASTM C33 defines sand as "the fine granular material usually less than ¼ inch in diameter resulting from the natural disintegration of rock, or from the crushing of friable sandstone rocks." The sand should be clean and not contain harmful quantities of organic material, clay, coal, loam, twigs, branches, roots, weeds, or other deleterious materials.

Coarse Aggregates. Gravel, crushed stone, and slag are the most commonly used coarse aggregates. It should meet the requirements of ASTM C33 and range in size from ½ inch to a maximum of 1½ inches.

Cinder concrete is a lightweight concrete made by using cinders in place of gravel or crushed stone as a coarse aggregate. It is excellent as a fill between sleepers, for making precast blocks and roof slabs, and for fireproofing.

All coarse aggregates should be free of organic matter, clays, loam, and woody vegetation.

Water–Cement Ratio

The strength and other desired properties of concrete, such as durability and watertightness, are governed largely by the number of gallons of water used with each bag of cement. This water–cement ratio has long been recognized as a fundamental law, and it is applicable as long as the concrete mixture is of a workable plasticity during placement. The smaller the ratio of the volume of water to the volume of cement, the higher the strength of the resulting concrete. Table 12–11 shows the average compressive strength per square inch of concrete for gallons of water per sack of cement and water–cement ratios by weight of water to a sack of cement. Aggregates on the

TABLE 12-11
Maximum Permissible Water–Cement Ratios and Minimum Cement Contents for Concrete (When Strength Data from Trial Batches or Field Experience Are Not Available)

Specified compressive strength f'c psi*	Minimum sks. cement per cu. yd. concrete	Maximum permissible water-cement ratio			
		Non-air-entrained concrete		Air-entrained concrete	
		Absolute ratio by weight	U.S. gal. per 94-lb. bag of cement	Absolute ratio by weight	U.S. gal. per 94-lb. bag of cement
2500	5	0.65	7.3	0.54	6.1
3000	5½	0.58	6.6	0.46	5.2
3500	6	0.51	5.8	0.40	4.5
4000	6½	0.44	5.0	0.35	4.0
4500	7	0.38	4.3	0.30	3.4
5000	7½	0.31	3.5	—	—

*28-day strengths for cements meeting strength limits of ASTM C150 Type I, IA, II or IIA and 7-day strengths for Type III and IIIA.

surface contain some moisture which should be taken into account when computing the water–cement ratios.

Slump Test

The slump test is used as a rough measure of the consistency of a concrete mix of cement, sand, coarse aggregate, and water. A 12-inch high, truncated cone of 16-gauge galvanized metal with a 4-inch open-top diameter and an 8-inch open bottom diameter, is filled with a sample of the concrete mix. The filling of the cone is performed on a flat level surface and in compliance with a specific procedure. The mold is immediately withdrawn vertically, and the amount of subsidence (slump) is measured in inches.

Air-entrained Concrete

Air-entrained concrete contains microscopic air bubbles that tend to keep the aggregate from separating and the water from coming to the top (that is, bleeding). Air-entraining agents are added during the manufacture of the cement or may be added during the mixing of the concrete. This type of concrete was developed, primarily, for road building, as it is more resistant to freezing and the salt used on highways. It is extensively used in most construction today. It is more weather-resistant and has better workability than non–air-entrained concrete.

The compressive strength of air-entrained concrete at 28 days will be slightly less than that of non–air-entrained concrete.

Concrete Mixes

In actual construction the proper proportions, the quality of the materials, and the amount of water used are important and controlled. The Portland Cement Association publishes excellent reference material on the subject of concrete mixes for various kinds of construction work.

TABLE 12-12

Proportions by Weight to Make 1 Cu Ft of Concrete

Maximum-size coarse aggregate, in.	Air-entrained concrete				Concrete without air			
	Cement, lb	Sand, lb	Coarse aggregate, lb*	Water, lb	Cement, lb	Sand, lb	Coarse aggregate, lb	Water, lb
⅜	29	53	46	10	29	59	46	11
½	27	46	55	10	27	53	55	11
¾	25	42	65	10	25	47	65	10
1	24	39	70	9	24	45	70	10
1½	23	38	75	9	23	43	75	9

*If crushed stone is used, decrease coarse aggregate by 3 lb and increase sand by 3 lb.

Metric conversion: 1 lb = 0.454 kg   10 lb = 4.54 kg
1 in. = 25 mm   1 cu ft = 0.028 m³

194

TABLE 12–12 (continued)

Proportions by Volume

Maximum-size coarse aggregate, in.	Air-entrained concrete				Concrete without air			
	Cement	Sand	Coarse aggregate	Water	Cement	Sand	Coarse aggregate	Water
⅜	1	2¼	1½	½	1	2½	1½	½
½	1	2¼	2	½	1	2½	2	½
¾	1	2¼	2½	½	1	2½	2½	½
1	1	2¼	2¾	½	1	2½	2¾	½
1½	1	2¼	3	½	1	2½	3	½

*The combined volume is approximately ⅔ of the sum of the original bulk volumes.
Reprinted by courtesy of the Portland Cement Association.

Table 12–12 (on pages 194 and 195) shows the proportions by *weight* and also by *volume* to make 1 cu. ft. of concrete, reprinted with permission from publication IS174.03T, *Concrete For Small Jobs,* by Portland Cement Association, Skokie, Illinois.

The proportions by *weight* can be used as a starting point to make 1 cu. ft. of strong durable concrete. In the *volumetric* proportions 1 bag of cement is equal to 1 cu. ft. and weighs 94 lb. (42.6 kg). The Canadian bag of cement has a net mass of 40 kg.

These proportions should serve only as a guide and may need adjustments to give a workable mix with locally available aggregates. Publication IS174.03T is recommended for further information.

Estimating Quantities of Concrete

Table 12–13 (on page 198) shows quantities of aggregates for 1 cu. yd. to 29 cu. yd. of concrete when using the four mixes of cement, sand, and stone. Typical concrete shapes are shown in Figure 12–7 on page 200. To estimate quantities of concrete, multiply the area of the cross-section by the length and divide by 27 to obtain the cubic yards. Shortcuts for estimating the cubic yards in footings, walls, slabs, and columns are shown below.

Short Cuts for Estimating Quantities of Concrete

Rather than referring to lengthy tables to determine the number of cubic yards of concrete in footings, walls, slabs, and columns, there is an easier method. It also is simpler than multiplying the cubic feet and dividing by 27 to obtain the cubic yards.

> *Footings*
> Multiply the cross section in square inches by .000257 to get the cubic yards in 1 lineal foot of footing. Multiply that result by the length of the footing.

Example

How many cubic yards of concrete in a footing 24" wide, 10" deep and 40' long?

$$24" \times 10" \times .000257 \times 40' \; 2.47 \text{ cu. yd.}$$

Walls

Multiply the square foot surface area by the wall thickness in inches. Multiply that result by .0031.

Example

How many cubic yards of concrete in a wall 8' high, 50' long and 12" thick?

$$8' \times 50' \times 12" \times .0031 = 14.88 \text{ cu. yd.}$$

Floor Slabs and Concrete Walks

Multiply the floor or walk area in square feet times the thickness in inches times .0031.

Example

How many cubic yards of concrete in a garage floor 24' wide, 25' long and 4" thick?

$$24' \times 25' \times 4" \times .0031 = 7.44 \text{ cu. yd.}$$

Columns

Multiply the cross-section in square inches by .000257. Multiply that result by the height of the column in feet.

Example

How many cubic yards in a column 12" × 12" × 12' high?

$$12" \times 12" \times .000257 \times 12' = .44 \text{ cu. yd.}$$

Table 12–14 (on page 201) shows the cubic yards of concrete per lineal foot of foundation walls from 1 foot to 12 feet high.

Table 12–15 (on page 202) shows the number of square feet of floor or slab of a given thickness that will be covered by one cubic yard of concrete for thicknesses of 2 in. to 12 in.

To estimate the square feet a cubic yard of concrete will cover in slabs, floors and walks, divide the constant 324 by the thickness in inches.

TABLE 12–13

Quantities of Cement, Sand and Stone for 1–29 Cubic Yards of Various Concrete Mixes

Cu Yds Concrete	MIX 1, 1:1:1¾ AGGREGATE			MIX 2, 1:2:2¼ AGGREGATE			MIX 3, 1:2.25:3 AGGREGATE			MIX 4, 1:3:4 AGGREGATE		
	Sacks Cement	Yards Sand	Yards Stone	Sacks Cement	Yards Sand	Yards Stone	Sacks Cement	Yards Sand	Yards Stone	Sacks Cement	Yards Sand	Yards Stone
1	10	.37	.63	7.75	.56	.65	6.25	.52	.70	5.0	.56	.74
2	20	.74	1.26	15.50	1.12	1.30	12.50	1.04	1.40	10.0	1.12	1.48
3	30	1.11	1.89	23.25	1.68	1.95	18.75	1.56	2.10	15.0	1.68	2.22
4	40	1.48	2.52	31.00	2.24	2.60	25.00	2.08	2.80	20.0	2.24	2.96
5	50	1.85	3.15	38.75	2.80	3.25	31.25	2.60	3.50	25.0	2.80	3.70
6	60	2.22	3.78	46.50	3.36	3.90	37.50	3.12	4.20	30.0	3.36	4.44
7	70	2.59	4.41	54.25	3.92	4.55	43.75	3.64	4.90	35.0	3.92	5.18
8	80	2.96	5.04	62.00	4.48	5.20	50.00	4.16	5.60	40.0	4.48	5.92
9	90	3.33	5.67	69.75	5.04	5.85	56.25	4.68	6.30	45.0	5.04	6.66
10	100	3.70	6.30	77.50	5.60	6.50	62.50	5.20	7.00	50.0	5.60	7.40
11	110	4.07	6.93	85.25	6.16	7.15	68.75	5.72	7.70	55.0	6.16	8.14
12	120	4.44	7.56	93.00	6.72	7.80	75.00	6.24	8.40	60.0	6.72	8.88
13	130	4.82	8.20	100.76	7.28	8.46	81.26	6.76	9.10	65.0	7.28	9.62
14	140	5.18	8.82	108.50	7.84	9.10	87.50	7.28	9.80	70.0	7.84	10.36

| Cu Yds Concrete | MIX 1, 1:1:1¾ | | | MIX 2, 1:2:2¼ | | | MIX 3, 1:2.25:3 | | | MIX 4, 1:3:4 | | |
| | AGGREGATE | | | AGGREGATE | | | AGGREGATE | | | AGGREGATE | | |
	Sacks Cement	Yards Sand	Yards Stone	Sacks Cement	Yards Sand	Yards Stone	Sacks Cement	Yards Sand	Yards Stone	Sacks Cement	Yards Sand	Yards Stone
15	150	5.56	9.46	116.26	8.40	9.76	93.76	7.80	10.50	75.0	8.40	11.10
16	160	5.92	10.08	124.00	8.96	10.40	100.00	8.32	11.20	80.0	8.96	11.84
17	170	6.30	10.72	131.76	9.52	11.06	106.26	8.84	11.90	85.0	9.52	12.58
18	180	6.66	11.34	139.50	10.08	11.70	112.50	9.36	12.60	90.0	10.08	13.32
19	190	7.04	11.98	147.26	10.64	12.36	118.76	9.84	13.30	95.0	10.64	14.06
20	200	7.40	12.60	155.00	11.20	13.00	125.00	10.40	14.00	100.0	11.20	14.80
21	210	7.77	13.23	162.75	11.76	13.65	131.25	10.92	14.70	105.0	11.76	15.54
22	220	8.14	13.86	170.05	12.32	14.30	137.50	11.44	15.40	110.0	12.32	16.28
23	230	8.51	14.49	178.25	12.88	14.95	143.75	11.96	16.10	115.0	12.88	17.02
24	240	8.88	15.12	186.00	13.44	15.60	150.00	12.48	16.80	120.0	13.44	17.76
25	250	9.25	15.75	193.75	14.00	16.25	156.25	13.00	17.50	125.0	14.00	18.50
26	260	9.64	16.40	201.52	14.56	16.92	162.52	13.52	18.20	130.0	14.56	19.24
27	270	10.00	17.00	209.26	15.12	17.56	168.76	14.04	18.90	135.0	15.02	20.00
28	280	10.36	17.64	217.00	15.68	18.20	175.00	14.56	19.60	140.0	15.68	20.72
29	290	10.74	18.28	224.76	16.24	18.86	181.26	15.08	20.30	145.0	16.24	21.46

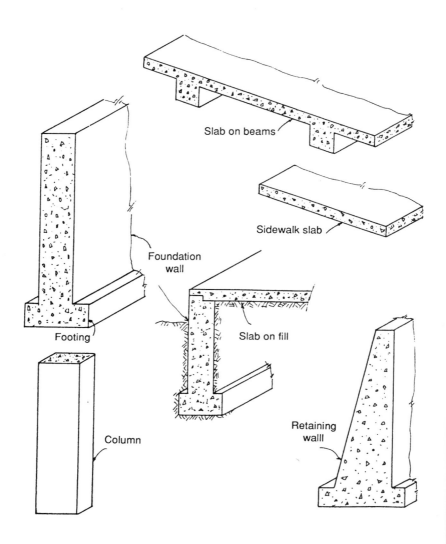

Slab on beams

Sidewalk slab

Foundation
wall

Footing

Slab on fill

Column

Retaining
walll

Figure 12–7
Typical Concrete Shapes

TABLE 12–14

**Cubic Yards of Concrete per Lineal Foot
of Foundation Wall***

Wall Height in Feet	Thickness of Wall in Inches						
	6″	7″	8″	9″	10″	11″	12″
1	.0185	.0217	.0248	.0279	.0310	.0341	.0372
2	.0370	.0434	.0496	.0558	.0620	.0682	.0744
3	.0556	.0651	.0744	.0837	.0930	.1023	.1116
4	.0741	.0868	.0992	.1116	.1240	.1364	.1488
5	.0926	.1085	.1240	.1395	.1550	.1705	.1860
6	.1111	.1302	.1488	.1674	.1860	.2046	.2232
7	.1296	.1519	.1736	.1953	.2170	.2387	.2606
8	.1482	.1736	.1984	.2232	.2480	.2728	.2976
9	.1667	.1953	.2232	.2511	.2790	.3069	.3348
10	.1852	.2170	.2480	.2790	.3100	.3410	.3720
11	.2035	.2387	.2728	.3069	.3410	.3751	.4092
12	.2222	.2604	.2976	.3348	.3720	.4092	.4464

*To deduct window and door openings, multiply the sq. ft. area of the opening
times the wall thickness in inches, by .0031.

Placing Concrete

There are several methods used for moving and placing concrete from mixer to the forms. When choosing the best method, consideration should be given to the one that is least costly and also the quickest in moving the mixture to the forms without the risk of initial setting.

- *Wheelbarrows* with pneumatic tires may be the most efficient for small projects over distances of 100 feet with loads of 3 cubic feet.

- *Hand buggies* or carts with pneumatic tires are superior to wheelbarrows for ease in handling a more balanced load up to 4½ cubic feet in the smaller carts to 8 or 9 cubic feet in the larger ones.

TABLE 12–15

Square Foot Area 1 Cubic Yard Will Cover

Thickness in Inches	Number Square Feet	Thickness in Inches	Number Square Feet
2	162	6	54
3	108	7	46
3½	93	8	40
4	81	9	36
5	65	10	32
5½	59	12	27

- *Motorized carts* can haul up to ½ cubic yard of concrete and are made to walk behind or ride on. They can also move up slight grades and travel 10 mph to 15 mph.

- *Buckets,* which vary in shape, have a capacity of 1 to 12 cubic yards. Buckets with bottom gates that can be opened and regulated to discharge part of the concrete and then closed are preferred. Buckets may be hoisted vertically on multistory projects either by crane or material tower. Each has its advantages.

- *Dropchutes* are used by raising the concrete vertically in a tower and distributing it by gravity through chutes. A uniform slope must be maintained so the flow will be constant.

- *Pumping* through a metal pipeline or flexible hose was used initially as a means to place concrete in hard-to-get-at places. As technology progressed and the pumps became lighter and easier to use, it has become the method of choice for placing concrete on medium to large jobs.

- *Belt conveyors* are portable and of various sizes and capacities. Under certain conditions they perform efficiently with uniform flow of the concrete and high capacity.

Rules or Principles

There are several rules or principles that should be carefully considered when placing concrete. The following checklist will be helpful:

1. When placed directly on earth or in footing forms, moisten the soil or inside of the form to prevent water from being absorbed from the concrete mixture.
2. When placed on concrete that has set, precautions should be taken to obtain a perfect bond. Clean and wet the surface of the older concrete thoroughly, and lightly brush the surface with grout and a thin layer of mortar.
3. Place concrete in forms continuously with minimum lateral flow.
4. In deep forms, free fall of concrete should not exceed 3 or 4 feet.
5. Consolidate fresh concrete immediately to remove entrapped air, by puddling and tamping or using internal vibrators. Move the vibrators up and down; do not drag or leave in one position.
6. Do not add water to the concrete mix when placing because it changes the water–cement ratio, reducing the strength and durability of the concrete.
7. In cold weather, to keep the concrete from freezing, accelerating admixtures may be used; or, use hot water and heated aggregate. The natural heat of curing concrete may be retained by insulating the formwork.

8. Concrete should be placed as near as possible to its final location in the forms.

9. Ideally concrete is best placed when temperatures are between 40°F and 80°F and curing is permitted for at least 7 days.

Table 12–16 shows the cubic yards of concrete, gravel base, and cement topping for concrete slabs, walks, driveways, and walks for large areas of 200 to 3,000 square feet, 1 in. to 12 in. thick. Table 12–17 (on page 206) shows the same data for smaller areas, 20 sq. ft. to 500 sq. ft.

Estimating Labor to Mix and Place Concrete

Mixing concrete by hand and with portable concrete mixers has become relatively limited to small and specialized projects. Also, in rural areas where ready-mixed concrete plants are long distances away, mixing by hand or by concrete mixer is more economical.

Ready-Mixed Concrete

Ready-mixed concrete plants have become the main source of concrete for building construction projects in this country. The plants have developed a high degree of uniformity and, generally, are in compliance with *ASTM Standards C94*, and with the *American Concrete Institute Standards,* regarding specifications for materials, that is, cement, aggregates, water, and admixtures. Their batching plants must measure up to the current National Bureau of Standards Specifications, Tolerances, and Other Technical Requirements for Commercial Weighing and Measuring Devices. Ready-mixed concrete is to be batched, mixed, and transported in accordance with ASTM C94, with few exceptions.

TABLE 12-16

Cubic Yards of Concrete Base and Topping in Slabs, Walks, and Patios with Areas 200 to 3,000 Square Feet, 1" to 12" Thick

Ground Floor Area	CUBIC YARDS OF GRAVEL BASE; CONCRETE OR TOPPING FOR THICKNESS SHOWN											
	1"	2"	3"	4"	5"	6"	7"	8"	9"	10"	11"	12"
200	0.62	1.24	1.86	2.48	3.10	3.72	4.34	4.96	5.58	6.20	6.82	7.44
300	0.93	1.86	2.79	3.72	4.65	5.58	6.51	7.44	8.37	9.30	10.23	11.16
400	1.24	2.48	3.72	4.96	6.20	7.44	8.68	9.92	11.16	12.40	13.64	14.88
500	1.55	3.10	4.65	6.30	7.75	9.30	10.85	12.40	13.95	15.50	17.05	18.60
600	1.86	3.72	5.58	7.44	9.30	11.16	13.02	14.88	16.74	18.60	20.46	22.32
700	2.17	4.34	6.51	8.68	10.85	13.02	15.19	17.36	19.53	21.70	23.87	26.04
800	2.48	4.96	7.44	9.92	12.40	14.88	17.36	19.84	22.32	24.80	27.28	29.76
900	2.79	5.58	8.37	11.16	13.95	16.74	19.53	22.32	25.11	27.90	30.69	33.48
1000	3.10	6.20	9.30	12.40	15.50	18.60	21.70	24.80	27.90	31.00	34.10	37.20
1100	3.41	6.82	10.23	13.64	17.05	20.46	23.87	27.28	30.69	34.10	37.51	40.92
1200	3.72	7.44	11.16	14.88	18.60	22.32	26.04	29.76	33.48	37.20	40.92	44.64
1300	4.03	8.06	12.09	16.12	20.15	24.18	28.21	32.24	36.27	40.30	44.33	48.36
1400	4.34	8.68	13.02	17.36	21.70	26.04	30.38	34.72	39.06	43.40	47.74	52.08
1500	4.65	9.30	13.95	18.60	23.25	27.90	32.55	37.20	41.85	46.50	51.15	55.80
1600	4.96	9.92	14.88	19.84	24.80	29.76	34.72	39.68	44.64	49.60	54.56	59.52
1700	5.27	10.54	15.81	21.08	26.35	31.62	36.89	42.16	47.43	52.70	57.97	63.24
1800	5.58	11.16	16.74	22.32	27.90	33.88	39.06	44.64	50.22	55.80	61.38	66.96
1900	5.89	11.78	17.67	23.56	29.45	35.34	41.23	47.12	53.01	58.90	64.79	70.68
2000	6.20	12.40	18.60	24.80	31.00	37.20	43.40	49.60	55.80	62.00	68.20	74.40
2100	6.51	13.02	19.53	26.04	32.55	39.06	45.57	52.08	58.59	65.10	71.61	78.12
2200	6.82	13.64	20.46	27.28	34.10	40.92	47.74	54.56	61.38	68.20	75.02	81.84
2300	7.13	14.26	21.39	28.52	35.65	42.78	49.91	57.04	64.17	71.30	78.43	85.56
2400	7.44	14.88	22.32	29.76	37.20	44.64	52.08	59.52	66.96	74.40	81.84	89.28
2500	7.75	15.50	23.25	31.00	38.75	46.50	54.26	62.00	69.75	77.50	85.25	93.00
2600	8.06	16.12	24.18	32.24	40.30	48.36	56.42	64.48	72.54	80.60	88.66	96.72
2700	8.37	16.74	25.11	33.48	41.85	50.22	58.59	66.96	75.33	83.70	92.07	100.44
2800	8.68	17.36	26.04	34.72	43.40	52.08	60.76	69.44	78.12	86.80	95.48	104.16
2900	8.99	17.98	26.97	35.96	44.95	53.94	62.93	71.92	80.91	89.90	98.89	107.88
3000	9.30	18.60	27.90	37.20	46.50	55.80	65.10	74.40	83.70	93.00	102.30	111.60

TABLE 12–17

Cubic Yards of Concrete Base and Topping in Slabs, Walks, and Patios with Areas 20 to 500 Square Feet, 1" to 12" Thick

Ground Floor Area	CUBIC YARDS OF GRAVEL BASE; CONCRETE OR TOPPING FOR THICKNESS SHOWN											
	1"	2"	3"	4"	5"	6"	7"	8"	9"	10"	11"	12"
20	.06	.12	.19	.25	.31	.37	.43	.50	.56	.62	.68	.74
30	.09	.18	.28	.37	.47	.56	.65	.75	.84	.93	1.02	1.12
40	.12	.24	.37	.50	.62	.74	.87	.99	1.12	1.24	1.36	1.49
50	.16	.32	.47	.62	.78	.93	1.09	1.24	1.40	1.55	1.71	1.86
60	.19	.38	.56	.74	.93	1.12	1.30	1.49	1.67	1.86	2.05	2.23
70	.22	.43	.65	.87	1.09	1.30	1.52	1.74	1.95	2.17	2.39	2.60
80	.25	.50	.74	.99	1.24	1.49	1.74	1.98	2.23	2.48	2.73	2.98
90	.28	.56	.84	1.12	1.40	1.67	1.95	2.23	2.51	2.79	3.07	3.35
100	.31	.62	.93	1.24	1.55	1.86	2.17	2.48	2.79	3.10	3.41	3.72
120	.37	.74	1.12	1.49	1.86	2.23	2.60	2.98	3.35	3.72	4.09	4.46
140	.43	.87	1.30	1.74	2.17	2.60	3.04	3.47	3.91	4.34	4.77	5.21
160	.49	.99	1.49	1.98	2.48	2.98	3.47	3.97	4.46	4.96	5.46	5.95
180	.56	1.12	1.67	2.23	2.79	3.35	3.91	4.46	5.02	5.58	6.14	6.70
200	.62	1.24	1.86	2.48	3.10	3.72	4.34	4.96	5.58	6.20	6.82	7.44
220	.68	1.36	2.05	2.73	3.41	4.09	4.77	5.46	6.14	6.82	7.50	8.18
240	.74	1.49	2.23	2.98	3.72	4.46	5.21	5.95	6.70	7.44	8.18	8.93
260	.81	1.61	2.42	3.22	4.03	4.84	5.64	6.45	7.25	8.06	8.87	9.67
280	.87	1.74	2.60	3.47	4.34	5.21	6.08	6.94	7.81	8.68	9.55	10.42
300	.93	1.86	2.79	3.72	4.65	5.58	6.51	7.44	8.37	9.30	10.23	11.16
320	.99	1.98	2.98	3.97	4.96	5.95	6.94	7.94	8.93	9.92	10.91	11.90
340	1.05	2.11	3.16	4.22	5.27	6.32	7.38	8.43	9.49	10.54	11.59	12.65
360	1.12	2.23	3.35	4.46	5.58	6.70	7.81	8.93	10.04	11.16	12.28	13.39
380	1.18	2.36	3.53	4.71	5.89	7.07	8.25	9.42	10.60	11.78	12.96	14.14
400	1.24	2.48	3.72	4.96	6.20	7.44	8.68	9.92	11.16	12.40	13.64	14.88
420	1.30	2.60	3.91	5.21	6.51	7.81	9.11	10.42	11.72	13.02	14.32	15.62
440	1.36	2.73	4.09	5.46	6.82	8.18	9.55	10.91	12.28	13.64	15.00	16.37
460	1.43	2.85	4.28	5.70	7.13	8.56	9.98	11.41	12.83	14.26	15.69	17.11
480	1.45	2.98	4.46	5.95	7.44	8.93	10.42	11.90	13.39	14.48	16.37	17.86
500	1.55	3.10	4.65	6.20	7.75	9.30	10.85	12.40	13.95	15.50	17.05	18.60

A simple guide for ordering ready-mixed concrete for smaller jobs is shown in Table 12–18. The Table gives the requirements for the concrete mixture. The ready-mix producer will know how much sand and stone to include so that the mixture will have the correct slump for workability. Also, to conform to standard good practice, the ready-mix supplier will use portland cement, sand, stone, water, and air-entraining agents that meet nationally recognized standards, and will mix and transport the concrete in approved fashion.

TABLE 12–18

Guide for Ordering Ready Mixed Concrete

Exposure	Coarse aggregate, nominal maximum size, in.	Portland cement, minimum lb per cu yd	Water, maximum lb per cu yd	Air entrainment, % by volume
Severe	1½	510	230	5 to 7
Many freeze-thaw	1	564	254	6 to 8
cycles per year;	¾	586	264	6½ to 8½
deicer chemicals	½	640	288	7 to 9
used	⅜	660	300	7½ to 9½
Moderate	1½	470	235	4 to 6
Few freeze-thaw	1	520	260	5 to 7
cycles per year;	¾	540	270	6 to 8
deicer chemicals	½	590	295	6½ to 8½
not used	⅜	610	305	7 to 9
Mild	1½	470	*	**
No freeze-thaw	1	520	*	**
cycles per year;	¾	540	*	**
deicer chemicals	½	590	*	**
not used	⅜	610	*	**

Slump: 5-in. maximum for hand methods of strikeoff and consolidation.
3-in. maximum for mechanical strikeoff and consolidation.

*Amount of water limited by compressive strength needed for service.
**2% to 3% to improve cohesiveness and reduce bleeding.

Reprinted with permission of the Portland Cement Association from Publication IS209HC, Building Concrete Walks, Driveways, Patios, and Steps.

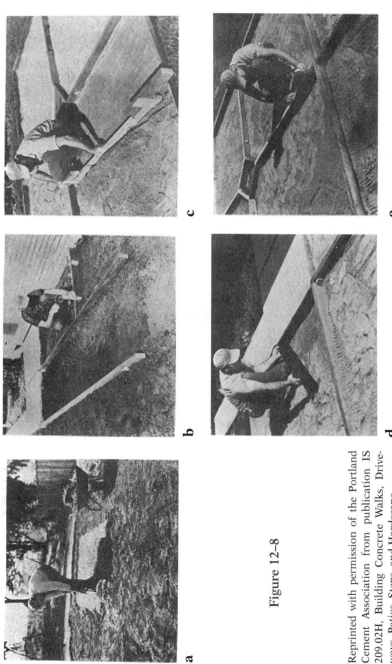

Figure 12–8

Reprinted with permission of the Portland Cement Association from publication IS 209.02H, Building Concrete Walks, Driveways, Patios, Steps, and Hardscape.

Steps in the construction of concrete hardscape: **a.** removing the topsoil, **b.** driving stakes to hold the forms, **c.** leveling off the subgrade, **d.** placing isolation joint material against adjacent construction, **e.** covering divider strips with tape for protection during concreting, **f.** receiving ready-mixed concrete with a wheelbarrow, **g.** striking off the concrete after it is deposited in the forms, **h.** covering the finished concrete with polyethylene to achieve a moist cure.

h

g

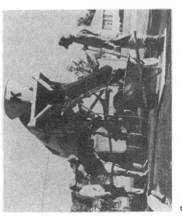

f

Figure 12–8 (on pages 208 and 209) demonstrates, pictorially, the several steps in the construction of a concrete patio, from removing the topsoil to completing the entire project, using ready-mixed concrete.

Ready-mixed concrete is generally more economical than job-mixing because it eliminates the ownership and use of mixers, and the labor of handling materials to be mixed. In many job situations, ready-mixed concrete reduces the labor cost of wheeling and placing when the ready-mixed concrete transport can back in close to the point where the concrete is to be placed. The concrete can also be distributed to the proper spot with chutes attached to the rear of the truck.

Before estimating the labor costs to have the concrete mixed and placed, the project should be studied carefully to determine the best procedure. The following checklist will serve as a guide:

1. Kind of work
   Footings or foundations
   Walls, beams, girders, columns
   Floor slabs, slabs on the ground
2. Method of mixing the concrete
   By hand
   By machine, on the job
   Ready-mixed delivered
3. How is concrete to be placed?
   Directly from ready-mix truck
   By wheelbarrows or buggies
4. Are plank runways to be built?
5. Is hoisting necessary?
   By hand
   By power: crane/elevator

6. Will weather conditions affect labor productivity?
   Winter, or heat of summer
   Protection against rain or wind

While ready-mix concrete does away with the labor of mixing on the job, the cost should be checked against on-the-job mixing. The labor required to wheel concrete long distances, or hoist it up or down to the forms should be estimated in hours per cubic yard over and above the actual mixing. Also, the labor cost to set up and shift runways should be treated as a separate item rather than included in the unit cost per yard.

Concreting in Cold Weather

Mixing and placing concrete in winter months requires great care. The aggregate has to be heated, or some type of admixture or special and costly cements may be necessary. Sometimes using richer mixtures and a minimum of water will provide more heat from hydration and obtain early strength sooner. These considerations, plus protective equipment such as tarps, commercial curing or electric heating blankets, straw, and sawdust, increase the cost of cold weather concreting.

Table 12–19 shows the approximate average hours of labor for job-mixing and placing, also ready-mix and placing different kinds of concrete work. Mixing and placing concrete is usually done by common labor with an allowance for supervision. Cement finishing is primarily done by masons and, in some situations, by laborers who are skilled in the art.

As with most building trades, the hours of labor to perform various operations cannot be cast in stone. There are far too many conditions and factors to be considered. Table 12–19 is presented solely as a guideline for the estimator.

TABLE 12–19

**Approximate Hours of Labor to Mix
and Place Concrete**

Kind of Work	Hours Per Cubic Yard*
Footings and Foundations	
Job machine-mixed and placed	2.5–3.0
Ready-mix and placed by truck	.7– .9
Ready-mix and wheeled	1.0–1.5
Columns piers and pilasters	
Job machine-mixed and placed	2.7–3.5
Ready-mix and placed	1.0–1.3
Walls above and below grade	
Job machine-mixed and placed	2.0–2.5
Ready-mix and placed	.9–1.2
Floors and walks on grade	
Job machine-mixed and placed	1.5–2.5
Ready-mix and placed	.6–1.0
Stairs and landings	
Job machine-mixed and placed	2.0–3.0
Ready-mix and placed	1.0–1.5

Finishing Concrete	Hours Per 100 Sq. Ft.
Wood float finish on slabs	2.0–2.5
Machine finish over large areas	1.3–1.6
Steel trowel finish on slabs	2.5–3.0
Cement wash on walls	2.0–3.0

*Add for hoisting, and for foreman
Deduct for power concrete buggies

FINISHING CONCRETE

Specifications usually require some type of finish to any concrete surface exposed to view. After floor slabs, sidewalks, and similar concrete units have been placed and screed, a smoother surface is usually desired or specified. This may consist of a float finish or a trowel finish.

Floating can begin shortly after screeding while the concrete is still plastic, and while the surface will support the finisher's weight on wooden knee boards. On large slabs, walks, and driveways, long handle floats are used. Floating imbeds the coarser aggregate below the surface; it tends to produce a more level surface and compacts the mortar at the surface for later troweling if needed. Aluminum floats are being used more in preference to wood floats, because the wood float tends to drag or stick to the surface of the concrete.

Troweling generally follows floating and is done with a steel trowel to produce a smooth hard surface.

Machine finishing over large areas of floor slabs, walks, and driveways is becoming increasingly popular with the improvement of the machines. Screeding, floating, and troweling by machine is usually begun when the concrete is firm enough to walk on. Cement finishing is generally done by masons.

SELECTED REFERENCES

Concrete & Formwork–Construction Manual, T. W. Love, Craftsman Book Company, Carlsbad, CA, 1973.

Concrete Construction & Estimating, Ed. by Craig Avery, Craftsman Book Company, Carlsbad, CA, 1980.

Concrete Construction Handbook, Dobrowolski, McGraw-Hill Publishing Company, New York, 1998.

Construction Principles, Materials & Methods, 5th Edition, Harold B. Olin, John L. Schmidt, and Walter H. Lewis, John Wiley and Sons, New York, 1995.

Manual of Standard Practice, The Concrete Reinforcing Steel Institute, 933 North Plum Grove Road, Schaumburg, IL 60173, 1998.

Placing Reinforcing Bars, 7th Edition, The Concrete Reinforcing Steel Institute, 933 North Plum Grove Road, Schaumburg, IL 60173, 1997.

Specifications for Structural Concrete for Buildings, American Concrete Institute, P.O. Box 19150, Redford Station, Detroit, MI 48219, 1996.

CHAPTER 13

UNIT MASONRY

This chapter on *unit* masonry, as distinguished from *concrete* masonry, covers the work of the mason whose principal materials are mortar and masonry units. These units include, but are not limited to, stone, brick, tile, concrete block, and glass block.

MORTAR FOR MASONRY

The most important property of mortar is its ability to provide a strong bond with the masonry unit. Other desirable properties are workability, strength, durability, and water retentivity. There is no ideal combination of ingredients that will produce a mortar that satisfies all of these desirable properties. For example, air-entrained cements may contribute to the workability and water retentivity of the mortar but may reduce its bonding quality. A high-strength mortar may lose some of its durability.

Cements for Unit Masonry Mortar

Portland cements, conforming to the ASTM specifications, both air- and non-entrained, are listed and described in Chapter 12, Concrete Construction. Types I and III, non–air-entrained, are the most commonly used in mortar for unit masonry. Type I is a general-purpose cement and is customarily used in ready-mix plants unless otherwise specified. Type III is a high, early strength cement used where fast setting is needed, as in low temperatures.

Masonry cements are mixtures prepared by manufacturers who seldom disclose their precise formulae. Type II masonry cements weigh 70 lbs. per bag and contain equal parts by weight of portland cement and ground limestone. Additives provide workability and water retentivity.

Hydrated Lime

Hydrated lime is introduced into the mortar mixture as a plasticizer to produce smoothness and workability. The lime also increases the water retentivity which in turn reduces bleeding.

Sand

Sand is the cheapest ingredient in mortar and is commercially available almost everywhere. For a good strong mortar the sand must be clean, free from impurities, and be graded in accord with ASTM Tentative Specifications for Aggregate for Masonry Mortar, C144-52T, which provides:

> "Aggregate for use in masonry mortar shall consist of natural sand or manufactured sand. Manufactured sand is the product obtained by crushing stone, gravel, or air-cooled iron blast furnace slag."

The ASTM specifications fixing grading limits for both natural and manufactured sands are shown in Table 13–1. In addition, the ASTM specifications provide that "the aggregate shall have not more than 50 percent retained between any two consecutive sieves listed [in Table 13–1] nor more than 25 percent between the No. 50 and No. 100 sieves."

TABLE 13–1

Sand Grading Limits (ASTM Specifications)

Sieve Size	Size of Opening in Inches	Percentage Sand Passing	
		Natural	Manufactured
No. 4	0.1940	100	100
No. 8	0.0970	95 to 100	95 to 100
No. 16	0.0490	60 to 100	60 to 100
No. 30	0.0232	35 to 70	35 to 70
No. 50	0.0117	15 to 35	20 to 40
No. 100	0.0059	2 to 15	10 to 25

Mortar Types

Most building codes and building specifications require ASTM mortar types M, S, N, or O. Table 13–2 lists the mortar proportions for each type acceptable under ASTM Specifications C270.

Type M Mortar is a high-strength mortar suitable for general use. It is recommended for below-grade masonry and work in contact with the earth, such as foundations, walls, catch basins, sewers, and manholes.

Type S Mortar is a general-purpose mortar. Its tensile bonding strength on brick approaches the maximum obtainable with cement–lime mortars. It is recommended where the masonry is exposed to severe weathering. It is also recommended where high lateral strength is desired.

Type N Mortar is a medium strength mortar with excellent workability because of its high lime content. It is recommended for exterior walls with light loads and mild exposure to the elements.

Type O Mortar is a low-strength mortar suitable for non–load-bearing partitions. It is the least resistant of the mortars to moisture and freezing.

TABLE 13–2

Mortar Proportions by Volume*

Mortar Type	Portland Cement Cu. Ft.	Masonry Cement Cu. Ft.	Hydrated Lime or Lime Putty Cu. Ft.	Aggregate Measured in Damp Loose Condition Cu. Ft.
M	1	None	¼	
	1	1	None	
S	1	None	Over ¼ to ½	Not less than 2¼ and not more than 3 times the sum of the volumes of cement and lime used.
	½	1	None	
N	1	None	Over ½ to 1¼	
	None	1	None	
O	1	None	Over 1¼ to 2½	
	None	1	None	

Courtesy of North Carolina Building Code.

*For the purpose of these specifications, the weight of one cubic foot of the respective materials used shall be considered to be as follows:

Portland Cement	94 pounds
Masonry Cement	weight printed on bag
Hydrated Lime	40 pounds
Lime Putty (Quicklime)	80 pounds
Sand, damp and loose	80 pounds of dry sand

The following volume proportions of cement, lime, and sand conform to the ASTM specifications for these four mortars.

Type M   1:¼:3
Type S   1:½:4½
Type N   1:1:6
Type O   1:2:9

Based on these proportions, Table 13–3 gives the quantities of cement, lime, and sand for 1 cubic foot of mortar by volume and by weight.

TABLE 13–3

Material Quantities Per Cubic Foot of Mortar

Ma-terial	Quantities by Volume				Quantities by Weight			
	Mortar Type and Proportions by Volume				Mortar Type and Proportions by Volume			
	M 1:¼:3	S 1:½:4½	N 1:1:6	O 1:2:9	M 1:¼:3	S 1:½:4½	N 1:1:6	O 1:2:9
Cement	0.333	0.222	0.167	0.111	31.33	20.89	15.67	10.44
Lime	0.083	0.056	0.167	0.222	3.33	2.22	6.67	8.89
Sand	1.000	1.000	1.000	1.000	80.00	80.00	80.00	80.00

Table 13–4 shows the average compressive strength of mortars M, S, N, and O.

Table 13–5 shows types of mortar permitted in various kinds of masonry in accord with ASTM specifications.

TABLE 13–4

Types of Mortar

Type	Average Compressive Strength 2 inch cubes at 28 days, psi
M	2500
S	1800
N	750
O	350

Coloring Mortar

Mortar may be colored when being mixed by adding pigments. The practice of mixing in a mortar box is not recommended because of the difficulty of producing batches of uniform color. It is better to obtain colored mortar from manufacturers who measure and grind the color pigments uniformly at the plant.

TABLE 13-5

Types of Mortar Required

Type of Masonry	Types of Mortar Permitted
Foundations: (Below grade masonry)	
Walls of Solid Units	M. S or N
Walls of Hollow Units	M or S
Hollow Walls	M or S
Masonry Other Than Foundation Masonry	
Piers of Solid Masonry	M. S or N
Piers of Hollow Units	M or S
Walls of Solid Masonry	M. S. N or O
Walls of Hollow Masonry	M. S. or N
Hollow Walls and Cavity Walls	
(a) Design Wind Pressure Exceeds 20 psf.	M or S
(b) Design Wind Pressure 20 psf or less.	M. S or N
Glass Block Masonry	M. S or N
Non-Bearing Partition and Fireproofing	M. S. N. O or Gypsum
Gypsum Partition Tile or Block	Gypsum
Fire Brick	Refractory Air Setting Mortar
Masonry Other Than Above	M. S or N

Courtesy of North Carolina Building Code.

Mixing Mortar

Mortar is usually mixed by masons' helpers while they are tending the masons by carrying materials to place them within their reach and shifting scaffolding, planks, and masons' horses. On small jobs the mixing is done by hand in a mortar box; on larger projects where many masons are engaged, the mortar is mixed in portable motorized mixers. In most instances the labor of mixing is included in the overall labor allowance made for masons' helpers.

CONCRETE BLOCK MASONRY

Concrete block units, the most widely used material for foundations and for backup for brick walls, are manufactured with heavyweight aggregate and lightweight aggregate. The ASTM specifications classify concrete masonry units into two grades, N and S, and into two types, Type I and Type II.

Grade N units are intended for use below and above ground where they are directly exposed to moisture penetration, freezing, and thawing. Minimum allowable compressive strength per unit is 800 lbs. p.s.i.

Grade S units are limited to above-grade use in walls with weather-protective coatings and in walls not exposed to the weather. Minimum allowable compressive strength per unit is 600 lbs. p.s.i.

Masonry units retain some moisture that, in low humidity areas of the country, may dry out and cause in-place shrinkage. In the manufacturing process, load-bearing and non–load-bearing units can be produced with specified moisture content. ASTM specifications have established moisture limits for Type I units depending upon the average relative humidity at the source of manufacture.

Type II units have no specified moisture content limit and are used where the average relative humidity is moderate to high. Types I and II masonry units are available in grades N and S, and are designated N-I, N-II, S-I, and S-II.

Shapes and Sizes

Typical shapes and sizes of concrete block units are shown in Figures 13–1, 13–2, and 13–3 (on pages 222–227). Both heavy- and lightweight units are available in these shapes and sizes. In addition to the shapes and sizes shown, there

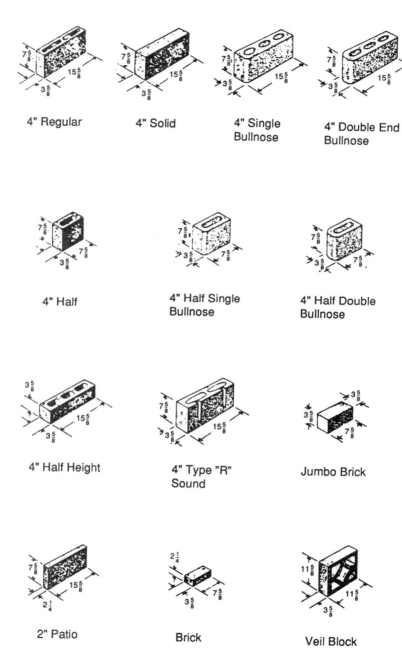

Figure 13–1
Common shapes and sizes of concrete block.

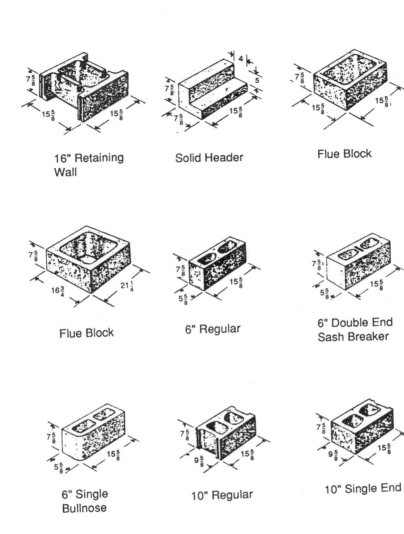

16" Retaining Wall

Solid Header

Flue Block

Flue Block

6" Regular

6" Double End Sash Breaker

6" Single Bullnose

10" Regular

10" Single End

10" Bond Beam

10" Single Bullnose

10" Half Sash

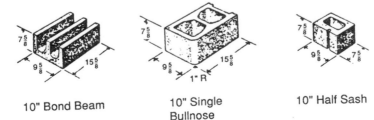

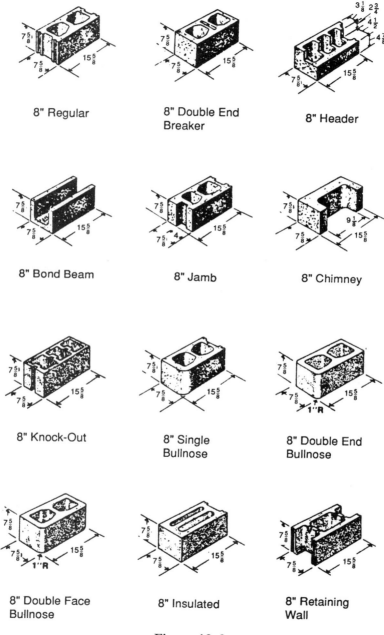

Figure 13-2
Common shapes and sizes of concrete block.

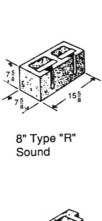

8" Type "R"
Sound

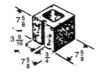

Pilaster

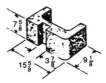

Pilaster

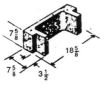

Pilaster

8" Half Sash

8" Half Jamb

8" Half Single
Bullnose

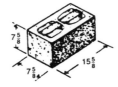

8" Half Double
Bullnose

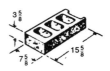

8" Half Height

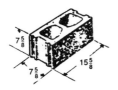

8" 75% Solid

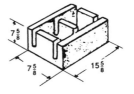

8" Warren
Insulated

8" Bottomless
(Knock-Out)
Bond Beam

Courtesy of Metromont, Spartanburg, South Carolina.

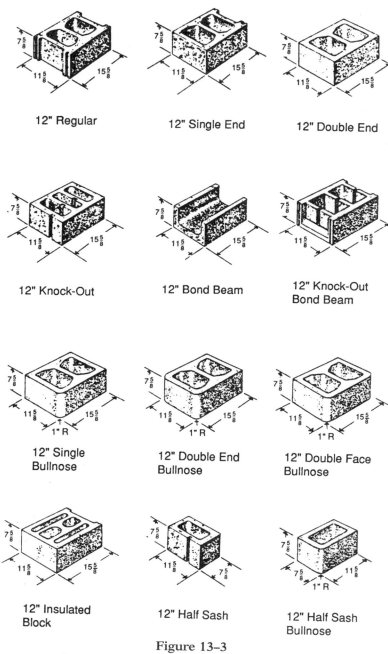

Figure 13–3
Common shapes and sizes of concrete block.

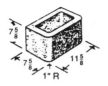

12" Half Double Bullnose

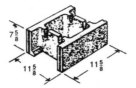

12" Retaining Wall

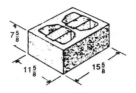

12" Warren Insulated

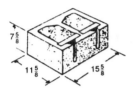

12" Sound Block

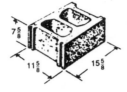

12" 75% Solid

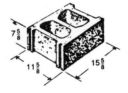

4 Hour

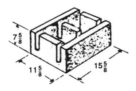

12" Bottomless (Knock-Out) Bond Beam

Workers applying cut stone veneer to
horizontally reinforced concrete block.

Figure 13–4
Concrete blocks and horizontal reinforcing
bars ready for use.

are special and decorative concrete blocks manufactured. They are in stock or on order depending upon the individual manufacturer.

Scored Block is a regular stretcher block on which the manufacturer scores the face one or more times creating an artificial joint, giving the impression of smaller units in the wall.

Textured Block is a regular sized block with molded patterns on the face.

Split Blocks are solid or hollow units that are split lengthwise. The units are laid in the wall with the fractured faces exposed producing a rough surface texture. These units are available in random sizes.

Slump Block is manufactured of special mixtures so that when released from the mold the unit sags or slumps. Standard heights vary from $1\frac{5}{8}$ in. to $3\frac{5}{8}$ in. and are laid in coursed or ashlar patterns creating an artistic effect for both interior and exterior walls.

Glazed Blocks have one or more surfaces glazed on most standard-sized concrete block. The glazing is of hard, smooth material like ceramic or stone in various colors and patterns.

Reinforced Concrete Block Masonry

Building codes and also local conditions may require both vertical and horizontal reinforcement of concrete block walls. For example, areas of the country that are subject to high winds, tornadoes, hurricanes, or seismic disturbances have greater than normal lateral stresses. Thus, states like Florida and California have stringent building code requirements for masonry reinforcement.

The size and spacing of vertical reinforcing bars is usually specified in the local building code. The bars are placed in the footing or foundation and run through the hollow block core in each course. The cores are then filled with grout as each course is laid. This type of reinforcement is used at corners, window and door jambs, and at specified spaces between openings and corners.

Bond Beams

Reinforced concrete bond beams are desirable and in some states are required to strengthen concrete block masonry walls that are subject to abnormal lateral stresses. They are customarily placed at each story height, but again, local building codes may require that they be placed at various courses. Bond beam concrete blocks are available. These are troughed to allow for the reinforcing rods and concrete fill. Aesthetically these are preferable because they look like the regular concrete block.

Concrete Masonry Cavity Walls

A cavity wall consists of two walls separated by an air space not less than 2 in. or more than 3 in. wide. A 10-inch concrete block cavity wall is usually constructed of two 3⅝ in. (nominally 4 in.) blocks with the specified air space between. A 12-inch concrete block cavity wall has nominally 4-inch outside walls and 6-inch inside walls with a 2-inch air space between. Most building codes limit 10-inch cavity walls to 25 ft. in height. The cavity walls are bonded together with metal ties.

Mortar Required to Lay Concrete Block

The amount of mortar required to lay masonry units varies with the thickness of the mortar joints, and with the individual workers, some of whom are much more efficient in controlling waste. Also, where hollow units are to be filled with mortar (grout), as in vertical reinforcement, additional amounts of mortar are required.

Estimating Concrete Block Materials

The most common face dimensions (height and length) of concrete block used in construction are 7⅝ in. × 15⅝ in. and, when laid with ⅜-inch joints, the nominal face dimension is 8 in. × 16 in. The widths vary, but the most common are 4 in., 6 in., 8 in., and 12 in.

The number of blocks per square foot of wall is determined by multiplying the nominal length by the height of the block and dividing that into 144 square inches. For example, an 8 in. × 8 in. × 16 in. block has a face area of (8 × 16) 128 square inches.

$$\frac{144}{128} = 1.125 \text{ blocks per sq. ft. of wall}$$

Thus it requires 112.5 concrete blocks (8 × 8 × 16) per 100 sq. ft. of wall.

Table 13–6 shows the number of concrete blocks and approximate cubic feet of mortar required for 100 sq. ft. of wall. For areas greater or less than 100 sq. ft., the quantities in the Table can be interpolated. Please note that the quantity of mortar needed to lay masonry units does not lend itself to exact or fixed amounts that will apply to all job situations in all regions of the country in all seasons. Approximate averages provide the best approach as guidelines for individual judgment.

Table 13–7 shows the number of 8 × 8 × 16-in. concrete block per lineal feet of foundation wall, one course 8 in., to fifteen courses 10 ft. high. For wall heights greater than 10 ft. high, multiply the desired wall height by 1.125 to obtain the number of 8 × 8 × 16-in. block per lineal foot. For example, a wall 16 ft. high would have 18 blocks per lineal foot of wall.

To obtain the number of courses in a wall, divide the height of the wall by .667. A wall 16 ft. high would have 24 courses.

TABLE 13–6
Number of Concrete Blocks and Amount
of Mortar Required per 100 Sq. Ft.

Nominal Size of Block in Inches	Wall Thickness in Inches	No. Blocks 100 Sq. Ft.	Cu. Ft. Mortar Per 100 Sq. Ft.
4 × 8 × 12	4	150	5 to 6
6 × 8 × 12	6	150	6 to 7
8 × 8 × 12	8	150	7 to 8
4 × 8 × 16	4	112.5	5 to 6
6 × 8 × 16	6	112.5	5 to 6
8 × 8 × 16	8	112.5	6 to 7
12 × 8 × 16	12	112.5	6 to 7

TABLE 13–7
Number of 8 × 8 × 16-in. Concrete Blocks per
Lineal Foot of Wall, Mortar Joints ⅜ in.

Wall Height	Number of Courses	Blocks Per Lineal Feet	Cubic Feet of Mortar Required
0 ft. 8 in.	1	.750	.04
1 ft. 4 in.	2	1.500	.08
2 ft. 0 in.	3	2.250	.12
2 ft. 8 in.	4	3.000	.16
3 ft. 4 in.	5	3.750	.20
4 ft. 0 in.	6	4.500	.24
4 ft. 8 in.	7	5.250	.28
5 ft. 4 in.	8	6.000	.32
6 ft. 0 in.	9	6.750	.36
6 ft. 8 in.	10	7.500	.40
7 ft. 4 in.	11	8.250	.44
8 ft. 0 in.	12	9.000	.48
8 ft. 8 in.	13	9.750	.52
9 ft. 4 in.	14	10.500	.56
10 ft. 0 in.	15	11.250	.60

Estimating Labor Laying Concrete Block

A mason and helper work together laying concrete block. The laborer mixes and carries mortar and blocks to the mason; he shifts or helps in shifting scaffolding when the mason moves to new areas. The hours to lay 100 blocks varies with the kind of work and the size and weight of block. The rate of laying decreases where there are many window or door openings, piers, or other breaks and corners.

Table 13–8 shows the approximate hours for a mason and a laborer to lay heavyweight and lightweight concrete blocks of different sizes. Judgment is recommended in considering specific job conditions, regional and seasonal factors, the productivity of individual workers, and the class of work. (See Estimating Labor in Chapter 10.)

TABLE 13–8
Approximate Hours of Labor Laying 100
Blocks of Different Sizes

Nominal Size of Units in Inches		Hours of Labor	
Heavyweight	**Type of Work**	**Mason**	**Laborer**
8 × 8 × 16	Foundations	6.0 to 7.0	6.0 to 7.0
8 × 12 × 16	Foundations	7.0 to 7.5	7.0 to 7.5
4 × 8 × 12	Above Grade	3.8 to 4.4	3.8 to 4.4
4 × 8 × 16	Above Grade	4.4 to 5.0	4.4 to 5.0
8 × 8 × 12	Above Grade	5.0 to 5.5	5.0 to 5.5
8 × 8 × 16	Above Grade	5.5 to 6.0	5.5 to 6.0
Lightweight	**Type of Work**	**Mason**	**Laborer**
8 × 8 × 16	Foundations	5.5 to 6.5	5.5 to 6.5
8 × 12 × 16	Foundations	6.5 to 7.0	6.5 to 7.0
4 × 8 × 12	Above Grade	3.5 to 4.0	3.5 to 4.0
4 × 8 × 16	Above Grade	4.0 to 4.5	4.0 to 4.5
8 × 8 × 12	Above Grade	4.5 to 5.0	4.5 to 5.0
8 × 8 × 16	Above Grade	5.0 to 5.5	5.0 to 5.5

Figure 13–5
Sizes of Modular and Nonmodular Brick

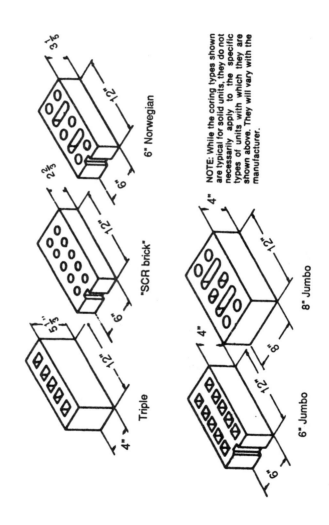

NOTE: While the coring types shown are typical for solid units, they do not necessarily apply to the specific types of units with which they are shown above. They will vary with the manufacturer.

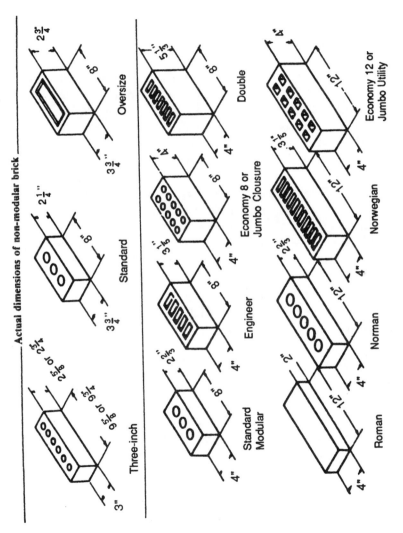

Actual dimensions of non-modular brick

Oversize — $2\frac{3}{4}$, 8", $3\frac{3}{4}$"

Standard — $2\frac{1}{4}$", 8", $3\frac{3}{4}$"

Three-inch — $2\frac{3}{4}$ or $2\frac{13}{16}$, $9\frac{5}{8}$ or $8\frac{5}{8}$, 3"

Double — $5\frac{5}{3}$", 8", 4"

Economy 8 or Jumbo Clousure — 4", 8", 4"

Engineer — $3\frac{3}{4}$", 8", 4"

Standard Modular — $2\frac{3}{4}$", 8", 4"

Economy 12 or Jumbo Utility — 4", 12", 4"

Norwegian — $3\frac{1}{3}$", 12", 4"

Norman — $2\frac{3}{4}$", 12", 4"

Roman — 2", 12", 4"

Courtesy of Brick Institute of America, McLean, Virginia 22101.

BRICK MASONRY

Most of the brick used in construction is one of four types.

Building brick, formerly called common brick, is primarily a construction brick with no special markings. The nominal size is 4 in. × 2⅔ in. × 8 in.

Face brick is manufactured with close control of texture, color, hardness, strength, and uniform size. The nominal size is the same as building brick.

Fire brick is used to line fireplaces, furnaces, and wood stoves. The usual dimensions are the same as building brick, but it is available in many other sizes.

SCR brick, developed by Structural Clay Products Institute Research Division, is wider than other brick. The dimensions are 2⅙ in. × 5½ in. × 11½ in. This brick competes with frame wall construction in dwellings, particularly those of one-story where the walls are 9 feet or less. One thickness of brick with furring inside makes a strong wall.

Nominal dimensions of modular brick and the actual dimensions of nonmodular brick are shown in Figure 13–5 on pages 234–235.

Brick Bonding

The bonding of brick in a wall has two functions. One is structural, which pertains to the interlocking of the brick to produce a strong monolithic unit and to prevent vertical joints between the bricks from being directly above each other. A second function of bonding is its use to create aesthetic patterns on the face of the wall. Eight of the basic and traditional bond patterns are shown in Figure 13–6.

Modern building codes specify that where solid masonry is bonded by means of masonry headers, no less than 4 percent of the wall surface of each face shall be composed of headers extending not less than 3 inches into the backing. The distance between adjacent full-length

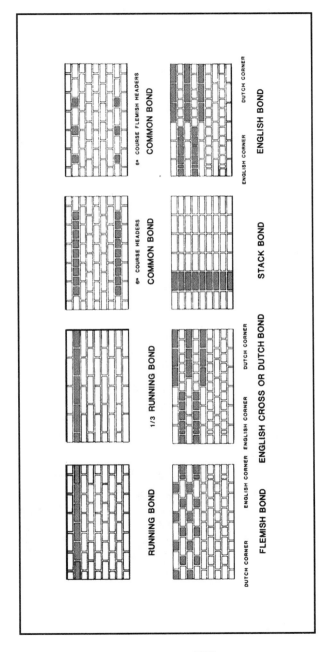

Figure 13–6
Traditional Pattern Bonds
Courtesy of Brick Institute of America

headers shall not exceed 24 inches vertically or horizontally. Also, where solid masonry is bonded with metal ties, the ties shall be corrosive resistant, $\frac{3}{16}$ inches in diameter imbedded in the horizontal mortar joints. There shall be one metal tie for each $4\frac{1}{2}$ square feet of wall area. Ties in alternate courses shall be staggered, the maximum vertical distance between ties not to exceed 24 inches. Additional bonding ties shall be provided at all openings spaced not more than 3 feet apart around the perimeter and within 12 inches of the opening. Cavity walls shall be bonded in accordance with these requirements.

Running Bond uses all stretcher courses with the joints at the center of the brick above and the brick below. There are no bonding masonry units, metal ties being used. Face, building, Roman, or SCR brick is used for this bond.

Common Bond, or American Bond, is similar to the running bond but has header courses every fifth, sixth, or seventh course which ties into the masonry backing material. Face or building brick are generally used for this bond.

Flemish Bond uses alternate headers and stretchers in each course. Headers in one course are centered above and below stretchers in the other course. Where headers are not used for structural bonding, half brick (clipped or snap headers) may be used. Face or building brick are frequently used in this bond.

English Cross or Dutch Bond uses alternate header and stretcher courses. Vertical joints between stretchers are centered on the stretcher in the stretcher courses above and below. Building or face brick is generally used in this bond.

English Bond uses alternate courses of headers and stretchers. The headers are centered on the stretchers, and joints between stretchers in all courses are aligned vertically. Face or building brick are used in this bond.

Stack Bond has all stretchers, and the joints are vertically aligned. This is strictly an aesthetic bond with little structural value. It is usually bonded to the backing with rigid steel ties. The most effective brick for this bond is the Roman.

Estimating Brick Masonry Materials

A widely used estimating procedure for determining the number of brick and amount of mortar is the "wall area" method. The known quantities of brick and mortar per square foot are multiplied by the net wall area being estimated. The net area is the gross area less openings for windows and doors.

Tables 13–9, 13–10, and 13–11, reprinted with permission of the Brick Institute of America, will be helpful in estimating. Table 13–9 shows the number of bricks per 100 square feet of wall for various modular brick sizes, and the cubic feet of mortar per 100 square feet of wall and 1,000 brick. If the facing is built with a backup wall, add the mortar quantities shown in Table 13–10 for the interior vertical, longitudinal joint (collar joint).

TABLE 13–9*
Modular Brick and Mortar Required for Single Wythe Walls in Running Bond (No allowances for breakage or waste)

Nominal Size of Brick in.			Number of Brick per 100 sq ft	Cubic Feet of Mortar			
				Per 100 Sq Ft		Per 1000 Brick	
t	h	l		⅜-in. Joints	½-in. Joints	⅜-in. Joints	½-in. Joints
4 x 2⅔ x		8	675	5.5	7.0	8.1	10.3
4 x 3⅕ x		8	563	4.8	6.1	8.6	10.9
4 x 4 x		8	450	4.2	5.3	9.2	11.7
4 x 5⅓ x		8	338	3.5	4.4	10.2	12.9
4 x 2 x		12	600	6.5	8.2	10.8	13.7
4 x 2⅔ x		12	450	5.1	6.5	11.3	14.4
4 x 3⅕ x		12	375	4.4	5.6	11.7ⁱ	14.9
4 x 4 x		12	300	3.7	4.8	12.3	15.7
4 x 5⅓ x		12	225	3.0	3.9	13.4	17.1
6 x 2⅔ x		12	450	7.9	10.2	17.5	22.6
6 x 3⅕ x		12	375	6.8	8.8	18.1	23.4
6 x 4 x		12	300	5.6	7.4	19.1	24.7

TABLE 13–10*
Cubic Feet of Mortar for Collar Joints

Cubic Feet of Mortar Per 100 Sq Ft of Wall		
¼-in. Joint	⅜-in. Joint	½-in. Joint
2.08	3.13	4.17

Note: Cubic feet per 1000 units $= \dfrac{10 \times \text{cubic feet per 100 sq ft of wall}}{\text{number of units per square foot of wall}}$

Table 13–9 is for *running bonds* with no headers. For bonds with full headers, apply the correction factors shown in Table 13–11. Add quantities obtained, by using these correction factors, to the facing brick quantities and deduct them from the backup. The reason for this is that the bonding face brick, extending into the backing brick, takes more face brick and less backing brick.

TABLE 13–11*
Bond Correction Factors for Walls of Table 13–9
(Add to facing and deduct from backing)

Bond	Correction Factor [1]
Full headers every 5th course only	1/5
Full headers every 6th course only	1/6
Full headers every 7th course only	1/7
English bond (full headers every 2nd course)	1/2
Flemish bond (alternate full headers and stretchers every course)	1/3
Flemish headers every 6th course	1/18
Flemish cross bond (Flemish headers every 2nd course)	1/6
Double-stretcher, garden wall bond	1/5
Triple-stretcher, garden wall bond	1/7

[1] Note: Correction factors are applicable only to those brick which have lengths of twice their bed depths.

*Courtesy of the Brick Institute of America.

Brick Chimneys

Chimneys are built with different sizes and numbers of flues. Nominal flue sizes are 8 × 8 × 12 in. and 12 × 12 × 12 in. Interior chimney walls may be 4 in. thick when lined with terra cotta flue, but exterior lined chimney walls should be 8 in. thick. To estimate the number of bricks in a chimney, draw a sketch, as in Figure 13–7, showing the number of bricks in each course. Multiply the number of bricks *per course* by the number of courses in a lineal foot of chimney. The height of the chimney in feet multiplied by the number of brick per lineal foot will give the total number of bricks required.

STRUCTURAL CLAY TILE

By standard definition, structural clay tiles are *hollow burned clay masonry units with parallel sides.* They are manufactured in various styles, sizes, and shapes for interior and exterior use; they may be load-bearing or non–load-bearing.

Figure 13–8 (on page 243) shows various typical shapes of structural clay *load-bearing tile.* Figure 13–9 (on page 244) shows various typical shapes of structural clay *non–load-bearing tile* used for partitions and for furring.

Load-bearing structural clay tile is produced in two grades, ASTM C34, LBX when exposed to weather or soil, and LB when not so exposed, or where protected by at least 3 inches of masonry facing. Non–load-bearing fireproofing and furring tile are produced in one grade, ASTM C57.

Structural clay *facing tile* is used for interior partitions, for backup for exterior walls and, in combination with other masonry units, as partitions. These tiles are available with modular dimensions in most of the shapes

	WALL THICKNESS	BRICK PER COURSE	BRICK PER LINEAL FOOT
	4"	6	28
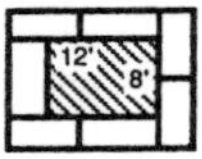	4"	7	32
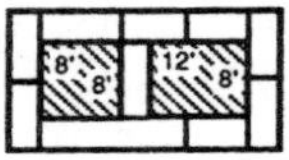	4"	11	50
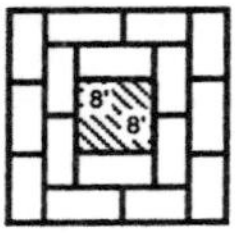	8"	16	73
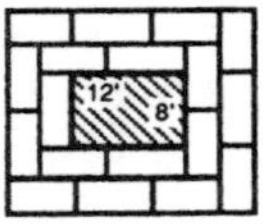	8"	18	82
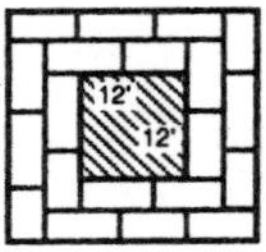	8"	20	91
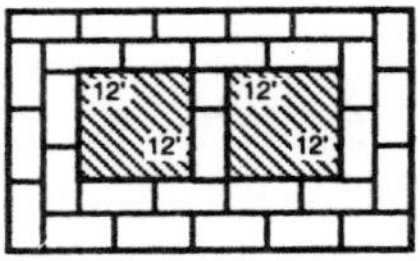	8"	29½	135

(BASED ON ⅜" MORTAR JOINTS.)

Figure 13–7
Number of Bricks per Lineal Foot for
Chimneys with Flue Sizes Shown

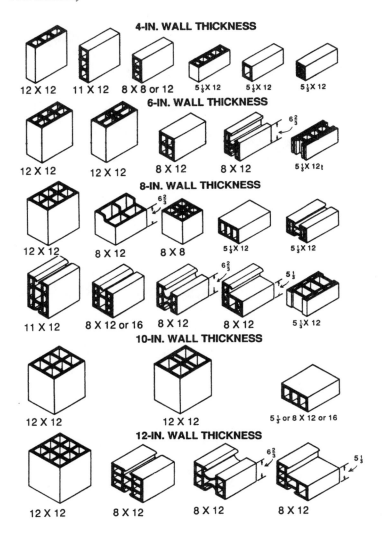

Figure 13–8
Various Standard Shapes of Load-Bearing Structural Clay Tile
Note: All dimensions are nominal. Face dimensions given height by length.

Courtesy of Brick Institute of America

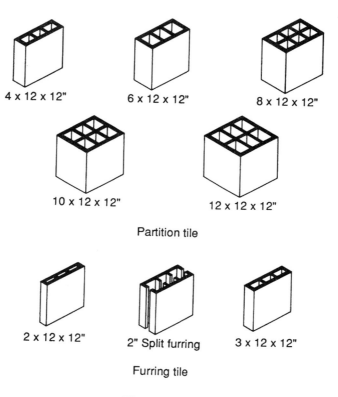

4 x 12 x 12"   6 x 12 x 12"   8 x 12 x 12"

10 x 12 x 12"   12 x 12 x 12"

Partition tile

2 x 12 x 12"   2" Split furring   3 x 12 x 12"

Furring tile

Figure 13–9
Structural Clay Partition Tile and Furring Tile
Courtesy of Brick Institute of America

shown in Figure 13–8. ASTM C212 covers two grades of facing tile: FTX for low water absorption, a high degree of mechanical perfection, and a narrow range of color variation; and FTS for moderate water absorption, a wide range of color variation, and moderate to good mechanical perfection.

Estimating Structural Clay Tile Quantities

The number of structural clay tiles per square foot of wall may be determined from the nominal face size of each unit which will include an allowance for mortar over the actual size of the unit. For example, an 8 × 16 × 4-in. tile has a face area of 128 square inches. This would be 1.125 tiles per square foot of wall.

Estimating Brick Masonry Labor

In laying brick, as in concrete block masonry, masons and masons' helpers work together. Masons are attended by the laborers who mix the mortar, move and set up the scaffolding, and keep the masons supplied with materials. On smaller jobs the labor is figured on the basis of the mason and the laborer working an equal number of hours.

The unit of measurement in brick masonry is 1,000 brick. The number of hours to lay 1,000 brick depends on the kind of work. Straight solid walls without openings require less time per 1,000 brick than walls with numerous openings, or pilasters, where proper plumbing of jambs and corners take extra time. Chimneys, piers, fireplaces, and working in confined areas also affect productivity. Table 13–13 is presented as a broad guide showing approximate hours of labor for a mason and laborer to lay 1,000 brick. Scaffolding is not included.

TABLE 13–12
Estimating Labor to Lay Structural Clay Tile, 12" × 12"

Size of Unit	Hours	
	Mason	**Laborer**
6 × 12 × 12	5	5
8 × 12 × 12	6	6
10 × 12 × 12	7	7
12 × 12 × 12	8	8

TABLE 13–13

**Approximate Average Hours of Labor
Required to Lay 1,000 Bricks**

Kind of Work		Mason	Laborer
Basement foundation walls	8 in.	10	10
	12 in.	9	8
	16 in.	8	8
Common brick walls	8 in.	11	8
	12 in.	10	8
	16 in.	9	8
Common Brick in Flemish or English Cross Bond	8 in.	18	12
	12 in.	14	10
	16 in.	12	8
Face brick veneer, ordinary grade		16	14
Face brick veneer, high grade		20	18
Brick piers and columns		14	14
Ordinary residential fireplaces		10	8
Common brick chimneys		15	15

SELECTED REFERENCES

Masonry Design and Detailing, 4th Ed., Christine Beall, McGraw-Hill Publishing Company, New York, 1997.

Recommended Practice for Engineered Brick Masonry, Harry C. Plummer, Brick Industry Association, Reston, VA 22091, 1977.

Selected ASTM Standards for Brick, compiled by ASTM for the Brick Industry Association, Reston, VA 22091. © ASTM.

CHAPTER 14

ROUGH CARPENTRY

TYPES OF ROUGH CARPENTRY

The carpentry trade is very broadly divided into two classifications: *rough carpentry* and *finish carpentry* or *mill-work*. While the line separating the two cannot always be sharply defined, the following work and materials are generally considered as applying to rough carpentry:

- Sleepers
- Concrete forms
- Wood foundations
- Framing
  —Sills, plates, ridge pieces
  —Studding, fire stops
  —Joists, wood trusses, purlins
- Sheathing, roof boards, and rough flooring
- Furring and grounds
- Window and door frames and bucks
- Temporary and rough basement stairways
- Fences, trellises, arbors, and other yard fixtures
- Rough hardware

- Cutting for other trades, scaffolding
- Runways and barricades, debris chutes, canopies

PERMANENT WOOD FOUNDATIONS

Through the cooperative studies of the U.S. Forest Service, American Wood Preservers Institute, and National Forest Products Association, *all-wood foundations* were developed in the late 1960s.

All-wood foundations consist of pressure-treated lumber and plywood sheathing. A 2-in. plank footing rests on a minimum of 4 in. of gravel, crushed stone, or coarse sand. The foundation studs, 2 in. thick by 4 in., 6 in., or 8 in., depending on the backfill, are erected on the plank footing and spaced 12 in., 16 in., or 24 in. on center. The treated plywood paneling ranges from ½ in. to ¾ in., and is 4- to 5-ply depending upon the height of the wall.

- All-wood foundations may be shop fabricated or built on the job site.
- They are approved for one- and two-story buildings and may be used for full basements or crawl space construction.
- Permanent wood foundations (PWF) are sometimes referred to as all-weather foundations (AWWF).
- Permanent wood foundations are more economical than concrete or masonry.
- Permanent wood foundations can be erected in all kinds of weather—the coldest, hottest, and wettest.
- Permanent wood foundations can be completed faster than other types; utility lines are readily installed through foundation walls; termite and other insect protective measures are unnecessary because the treated wood keeps them out.
- Permanent wood foundations are energy-efficient.

Permanent wood foundations have the acceptance and approval of the Veterans Administration (VA); Department of Housing and Urban Development/Federal Housing Administration (HUD/FHA); most major lending and mortgage insurance institutions; the Home Owners Warranty Corporation (HOW); and most model building codes.

The exterior of the all-wood foundation is covered with a 6 mil polyethylene sheet which acts as a moisture barrier. This construction and the continuously drained gravel pad under the footings and floor make the foundation waterproof and the basement drier.

Figure 14–1 shows details of a typical full basement wall. Figure 14–2 shows the knee wall with brick veneer. Figure 14–3 shows crawl space construction.

For additional information about APA trademarked panel products and the Permanent Wood Foundation System, contact the American Plywood Association, P.O. Box 11700, Tacoma, Washington 98411, or the nearest APA regional office.

For a copy of All-Weather Wood Foundation System, Design Fabrication and Installation Manual, write to National Forest Products Association, 1250 Connecticut Ave., NW (Suite 200), Washington, D.C. 20036.

FRAMING LUMBER

The *conifers* or needle-leaved trees comprise the group that supplies the majority of woods used for structural purposes. These are the *evergreens* such as fir, spruce, hemlock, pine, and cedar. The *broad-leaved* trees usually supply wood for interior floors, trim, and cabinet work. They are the hardwoods such as oak, maple, birch, ash, walnut, and white wood (basswood). In a few sections of the country, depending on choice and local supply, the different types of lumber are used interchangeably.

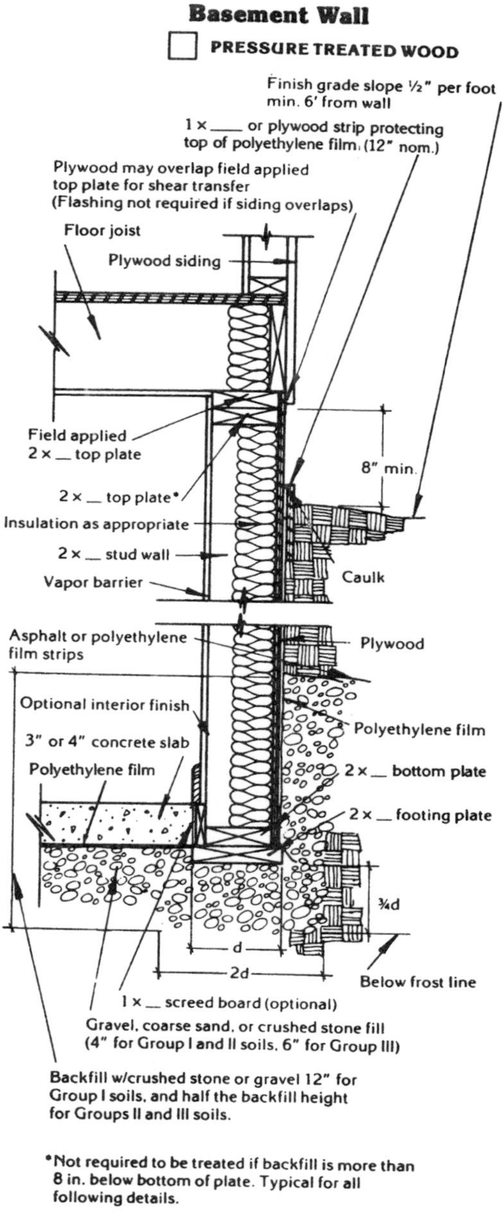

Basement Wall

PRESSURE TREATED WOOD

Finish grade slope ½" per foot
min. 6' from wall

1 × _____ or plywood strip protecting
top of polyethylene film. (12" nom.)

Plywood may overlap field applied
top plate for shear transfer
(Flashing not required if siding overlaps)

Floor joist

Plywood siding

Field applied
2 × __ top plate

8" min.

2 × __ top plate*

Insulation as appropriate

2 × __ stud wall

Vapor barrier

Caulk

Asphalt or polyethylene
film strips

Plywood

Optional interior finish

3" or 4" concrete slab

Polyethylene film

Polyethylene film

2 × __ bottom plate

2 × __ footing plate

¾d

d

2d

Below frost line

1 × __ screed board (optional)

Gravel, coarse sand, or crushed stone fill
(4" for Group I and II soils, 6" for Group III)

Backfill w/crushed stone or gravel 12" for
Group I soils, and half the backfill height
for Groups II and III soils.

*Not required to be treated if backfill is more than
8 in. below bottom of plate. Typical for all
following details.

Figure 14–1
Courtesy of American Plywood Association (APA).

Knee Wall with Brick Veneer

☐ **PRESSURE TREATED WOOD**

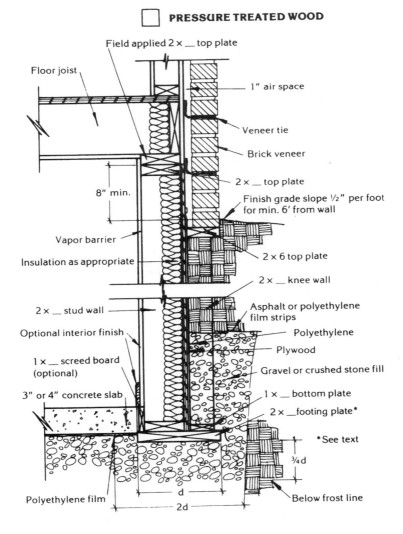

Field applied 2 × __ top plate

Floor joist

1" air space

Veneer tie

Brick veneer

2 × __ top plate

8" min.

Finish grade slope ½" per foot for min. 6' from wall

Vapor barrier

2 × 6 top plate

Insulation as appropriate

2 × __ knee wall

2 × __ stud wall

Asphalt or polyethylene film strips

Optional interior finish

Polyethylene

Plywood

1 × __ screed board (optional)

Gravel or crushed stone fill

3" or 4" concrete slab

1 × __ bottom plate

2 × __ footing plate*

*See text

¾ d

Polyethylene film

Below frost line

d

2d

Figure 14–2
Courtesy of American Plywood Association (APA).

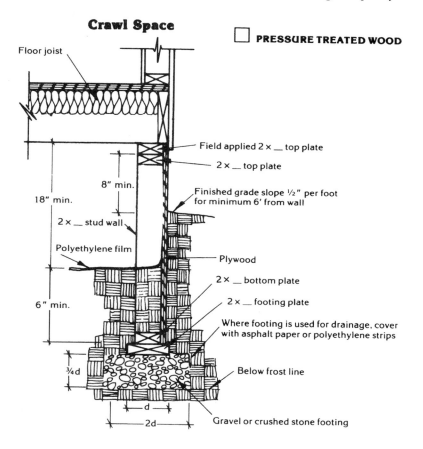

Crawl Space

☐ **PRESSURE TREATED WOOD**

Floor joist

Field applied 2 × __ top plate

2 × __ top plate

8" min.

18" min.

Finished grade slope ½" per foot for minimum 6' from wall

2 × __ stud wall

Polyethylene film

Plywood

2 × __ bottom plate

2 × __ footing plate

6" min.

Where footing is used for drainage, cover with asphalt paper or polyethylene strips

¾d

Below frost line

d

2d

Gravel or crushed stone footing

Figure 14–3
Courtesy of American Plywood Association (APA).

Board Measure

Nearly all lumber, from the time it is surveyed in the forest until it is sawed from logs into timbers and boards, is measured and sold by the unit of a *board foot*. Special millwork like moldings, window and door trim, lattice work, handrails, balusters, shelf cleats, some types of baseboard, molded exterior trim, and wood gutters are sold by the lineal foot. Plywood, particle board, flakeboard, and so forth, as distinguished from lumber, is sold by the square foot.

- A *board foot* is one square foot of wood one inch thick. The symbol for a board foot is BF. See Figure 14–4.

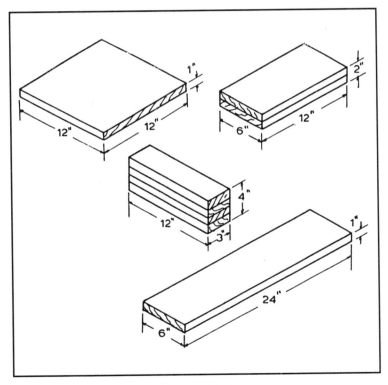

Figure 14–4
(Each unit is one board foot)

- The number of board feet in a piece of wood is the width times the thickness in inches, times its length in feet, divided by 12.
- Table 14–1 shows the number of board feet in various sizes of lumber of various lengths.
- Table 14–2 gives the factor to multiply a piece of lumber to obtain the number of board feet.

TABLE 14–1

**Number of Board Feet in Various
Sizes of Lumber**

Length of Piece

Size Inches	8′	10′	12′	14′	16′	18′	20′	22′
2×3	4	5	6	7	8	9	10	11
2×4	5⅓	6⅔	8	9⅓	10⅔	12	13⅓	14⅔
2×6	8	10	12	14	16	18	20	22
2×8	10⅔	13⅓	16	18⅔	21⅓	24	26⅔	29⅓
2×12	16	20	24	28	32	36	40	44
3×4	8	10	12	14	16	18	20	22
3×6	12	15	18	21	24	27	30	33
3×8	16	20	24	28	32	36	40	44
3×10	20	25	30	35	40	45	50	55
3×12	24	30	36	42	48	54	60	66
4×4	10⅔	13⅓	16	18⅔	21⅓	24	26⅔	29⅓
4×6	16	20	24	28	32	36	40	44
4×8	21⅓	26⅔	32	37⅓	42⅔	48	53⅓	58⅔
6×6	24	30	36	42	48	54	60	66
6×8	32	40	48	56	64	72	80	88
6×10	40	50	60	70	80	90	100	110
8×8	42⅔	53⅓	64	74⅔	85⅓	96	106⅔	117⅓
8×10	53⅓	66⅔	80	93⅓	106⅔	120	133⅓	146⅔
8×12	64	80	96	112	128	144	160	176
10×10	66⅔	83⅓	100	116⅔	133⅓	150	166⅔	183⅓
10×12	80	100	120	140	160	180	200	220
12×14	112	140	168	196	224	252	280	308

TABLE 14–2

Multipliers to Compute the Number of Board Feet in any Length of Dimension Lumber

Nominal Size in Inches	Multiply Length by	Nominal Size in Inches	Multiply Length by
2 × 2	0.333	4 × 4	1.333
2 × 3	0.500	4 × 6	2.000
2 × 4	0.667	4 × 8	2.667
2 × 6	1.000	4 × 10	3.333
2 × 8	1.333	4 × 12	4.000
2 × 10	1.667		
2 × 12	2.000	6 × 6	3.000
		6 × 8	4.000
3 × 3	0.750	6 × 10	5.000
3 × 4	1.000	6 × 12	6.000
3 × 6	1.500		
3 × 8	2.000	8 × 8	5.333
3 × 10	2.500	8 × 10	6.667
3 × 12	3.000	8 × 12	8.000

Kiln-Dried vs. Air-Dried Lumber

Wood is seasoned by exposing it to the air to dry or by kiln-drying. The advantages and objectives of seasoning are:

- The weight of the lumber is reduced by as much as one-third of its green weight. This lowers the cost of handling and shipping.
- Wood that is thoroughly dried will not rot because fungi cannot grow in it.
- Seasoning greatly increases the strength and hardness of the lumber.

Kiln drying, properly done, increases the strength and hardness of lumber the same as air drying and is equal to the best air-dried material in all respects. Air drying is more expensive because it requires as much as ten times as many days. For example, typical hardwoods can take as many as 200 days to reach 20 percent of the green-state moisture, while kiln drying may take 10 to 15 days to reach 6 percent of the original moisture content. Softwoods can take 60 to 70 days to be air-dried, but only 4 to 8 days to be kiln-dried.

Milling Waste

Matched boards are those that have the edges milled in a *tongue and groove* or *lap joint* (shiplap), to join them tightly together. Stock of this kind is used as surface lumber, as distinguished from framing lumber, and consists of such types as rough and finish flooring, exterior sheathing, siding, roof boards, paneling, and ceiling boards.

Most sawed lumber is planed (dressed) at the lumber mill before shipping. As a result the nominal size, for example, a 2-by-4-inch stud, becomes 1½ in. by 3½ in. The difference, termed *milling waste,* is not significant in estimating where planks, timbers, joists, and similar dimension lumber are involved. It does become important, however, when estimating quantities of surface lumber such as common boards, shiplap, tongue and groove boards, and bevel siding. The supplier will figure his charges on the *nominal size,* and if a factor is not added for the so-called milling waste, when placing the order, the quantity delivered will fall short of what is needed.

- Table 14–3 shows the factor by which the area to be covered is multiplied to obtain the exact amount needed. This factor can be calculated for *any piece of surface lumber* by dividing the *nominal* size by the *actual* face size. For example, the factor for a 1-by-

6-inch square edge board (actual size 5½ in.) is 6 divided by 5.5 = 1.09.

- This formula, nominal size/actual face size = factor, is useful when there is no table to refer to. *Normal cutting and fitting waste is to be added.*

- Use Table 14–3 to compute material needed. Multiply the measured square foot area by the factor in Table 14–3 for the particular type of material to be used.

TABLE 14–3

Factor by Which Area to Be Covered Is Multiplied to Determine Exact Amount of Surface Material Needed (From Western Wood Products Association)

Item	Nominal Size	Width Overall	Face	Area Factor
Shiplap	1"×6"	5½"	5 1/8"	1.17
	1×8	7¼	6 7/8	1.16
	1×10	9¼	8 7/8	1.13
	1×12	11¼	10 7/8	1.10
Tongue and Grooved	1×4	3 3/8	3 1/8	1.28
	1×6	5 3/8	5 1/8	1.17
	1×8	7 1/8	6 7/8	1.16
	1×10	9 1/8	8 7/8	1.13
	1×12	11 1/8	10 7/8	1.10
S 4 S	1×4	3½	3½	1.14
	1×6	5½	5½	1.09
	1×8	7¼	7¼	1.10
	1×10	9¼	9¼	1.08
	1×12	11¼	11¼	1.07
Solid Paneling	1×6	5 7/16	5 7/16	1.19
	1×8	7 1/8	6¾	1.19
	1×10	9 1/8	8¾	1.14
	1×12	11 1/8	10¾	1.12
*Bevel Siding	1×4	3½	3½	1.60
	1×6	5½	5½	1.33
	1×8	7¼	7¼	1.28
	1×10	9¼	9¼	1.21
	1×12	11¼	11¼	1.17

Note: This area factor is strictly so-called milling waste. The cutting and fitting waste must be added.
* 1-inch lap

Example

The area to which l-by-8-in. shiplap is to be applied is 100 ft. × 9 ft. or 900 sq. ft. The factor in Table 14–3 for 1-by-8-in. shiplap is 1.16.

900 sq. ft. × 1.16 = 1,044 sq. ft. needed

Cutting Waste

In addition to milling waste, an allowance must be made for material that is wasted in cutting and fitting it into place. This averages 3 percent to 5 percent in framing lumber, and it includes a small amount used on the job for such items as temporary saw-horses, minor scaffolding, bracing, and blocking.

Cutting and fitting waste on surface lumber such as rough flooring, roof boards, and board sheathing usually runs on the average of 5 percent to 10 percent.

ESTIMATING FRAMING MATERIAL

When estimating quantities of framing lumber, one should remember that the lengths needed for plates, sills, posts, joists, rafters, and so forth may have to be cut from the nearest stock length sold at the lumber supply yard. A rafter or joist 11 ft. 6 in. may have to be cut from a 12-foot length. On the other hand, pre-cut studs are often available in lengths of 92⅝ in., 93 in., 94½ in., and 96 in.

Sills and Plates

Sills and plate quantities are taken off by the lineal foot. The 2-by-6-inch sill in a shed 30 feet wide and 40 feet long would be 140 feet, which is equal to the perimeter of the building. If the plate on top of the studding is 4 in. by 4 in., and carries around the building as does the sill, it also is 140 feet long. The material would be listed in this manner.

| Sill | 140 lin. ft. | 2 in. × 6 in. = 140 BF |
| Plate | 140 lin. ft. | 4 in. × 4 in. = 187 BF |

- Figure 14–5 illustrates sill construction in platform framing.

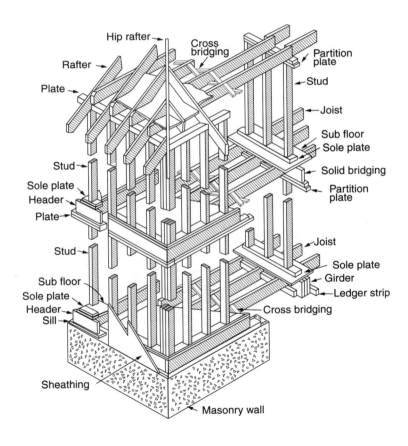

Figure 14–5
Platform-frame construction
U.S. Forest Service.

Exterior and Interior Studding

Studding is estimated by determining the number required, their size, and length. In residential framing, the studs are usually spaced 16 inches on center, as required by most building codes for 2-by-4-in. stock, and 24 inches on center for 2-by-6-in. stock. There are specific exceptions in many jurisdictions. Also, in many rural sections of the country, secondary buildings and some dwelling-type structures located outside the jurisdiction of building codes will have 2-by-4-in. studding on 18-, 20-, and sometimes 24-inch centers.

The number of studs, joists, and rafters, on various centers, is shown in Table 14–4 where the length of wall, floor, or roof is 8 ft. to 36 ft. Table 14–4A shows the BF in 2-by-4-in. stud walls and partitions, 8 ft. to 28 ft. long and 8 ft., 10 ft., and 12 ft. high; 12 in., 16 in., and 20 in. on center.

Table 14–5 shows the *multipliers* to use to calculate the *number* of studs, joists, and rafters in a given length of floor, wall, or roof.

Example
The floor in a room 28 ft. wide has 2-by-10-in. joists, 16 in. on center.

$$28 \times .75 = 21 + 1 = 22 \text{ joists}$$

A gable roof 36 ft. long has 2-by-8-in. rafters, 24 in. on center.

$$36 \times .50 = 18 + 1 = 19 \text{ rafters each side of the roof.}$$

Floor Joists

Joists in dwellings are usually of nominal 2-inch stock, and their depth ranges from 6 in. to 12 in. In other types of buildings the thickness may be 2 in., 3 in., or 4 in. depending on the floor load and the span. The depth of the larger joists may be as much as 16 in. Table 14–7 shows the maximum safe span for joists under various *live loads* with either plastered or unplastered ceilings below.

The *number* of joists may be obtained from Table 14–4 or Table 14–5. Where double or triple joists are required under load-bearing partitions or at stairwells, they should be added in the material list. The formulas for estimating the BF of joists are shown in Table 14–5A on page 264.

TABLE 14–4

Number of Studs, Joists, and Rafters Required in Walls, Floors, and Roofs

Length of Wall, Floor or Roof	Spacing on Center					
	12″	16″	20″	24″	30″	36″
8′	9	7	6	5	4	4
9	10	8	6	6	5	4
10	11	9	7	6	5	4
11	12	9	8	7	5	5
12	13	10	8	7	6	5
13	14	11	9	8	6	5
14	15	12	9	8	7	6
15	16	12	10	9	7	6
16	17	13	11	9	7	6
17	18	14	11	10	8	7
18	19	15	12	10	8	7
19	20	15	12	11	9	7
20	21	16	13	11	9	8
21	22	17	14	12	9	8
22	23	18	14	12	10	8
23	24	18	15	13	10	9
24	25	19	15	13	11	9
25	26	20	16	14	11	9
26	27	21	17	14	11	10
27	28	21	17	15	12	10
28	29	22	18	15	12	10
29	30	23	18	16	13	11
30	31	24	19	16	13	11
32	33	25	20	17	14	12
34	35	27	22	18	15	12
36	37	28	23	19	15	13

To the above there should be added members for doubling studs at windows, doors, corners; doubling joists under partitions, at stairwells and around chimneys; doubling rafters at roof openings around chimneys, etc.

TABLE 14-4A

Number of Board Feet Needed in 2" × 4" Stud Walls and Partitions 8' to 28' Long, End Studs Included. Add Plate and Shoe from End Columns, Add Window, and Door Framing.

(For studding 24" on center, use one-half the BF shown for studding 12" on center. Minor differences are due to end studs.)

Lin Ft. of Partition or Wall	12" Spacing Height of Wall			16" Spacing Height of Wall			20" Spacing Height of Wall			Plate and Shoe Single	Double
	8'	10'	12'	8'	10'	12'	8'	10'	12'	2" × 4"	2" × 4"
8	48	60	72	38	47	56	32	40	48	5	11
9	54	67	80	43	54	64	34	43	52	6	12
10	59	74	88	48	60	72	37	47	56	7	14
11	64	80	96	48	63	76	43	53	64	8	15
12	70	87	104	54	67	80	46	56	68	8	16
13	75	94	112	59	74	88	48	60	72	9	18
14	80	100	120	64	80	96	51	63	76	10	19
15	86	107	128	64	80	98	54	67	80	10	20
16	91	114	136	70	87	104	59	73	88	11	22
17	96	120	144	75	94	112	61	76	92	12	23
18	102	127	152	80	100	120	64	80	96	12	24
19	107	134	160	80	103	124	67	83	99	13	26
20	112	140	168	86	107	128	70	87	104	14	27
21	118	147	176	91	114	136	74	93	112	14	28
22	123	154	184	96	120	144	75	97	116	15	30
23	128	160	192	96	123	148	80	100	120	16	31
24	134	167	200	101	127	152	82	102	122	16	32
25	139	174	208	107	134	160	85	107	128	17	34
26	144	180	216	112	140	168	91	113	136	18	35
27	150	187	224	112	143	172	93	116	140	18	36
28	155	194	232	118	147	176	96	120	144	19	38

FORMULAS FOR BOARD FEET OF 2" × 4" STUDDING IN WALLS AND PARTITIONS INCLUDING END STUDS*

Where: L = length of wall
H = height of wall

Studding 12" on Center: $(L' + 1) \times .67 \times H' = BF$
Example: A wall is 20' long and 8' high
$(20' + 1) \times .67 \times 8' = 112.56\ BF$

Studding 16" on Center: $(L' \times .75 + 1) \times .67 \times H' = BF$
Example: A wall is 28' long and 10' high
$(28' \times .75 + 1) \times .67 \times 10' = 147.4\ BF$

Studding is 20" on Center: $(L' \times .6 + 1) \times .67 \times H' = BF$
Example: A wall is 24' long and 8' high
$(24' \times .6 + 1) \times .67 \times 8' = 82.5\ BF$

Studding is 24" on Center: $(L' \times .5 + 1) \times .67 \times H' = BF$
Example: A wall is 12' long and 8' high
$(12' \times .5 + 1) \times .67 \times 8' = 37.5\ BF$

FORMULAS FOR BOARD FEET OF 2" × 6" STUDDING IN WALLS AND PARTITIONS INCLUDING END STUDS

Where: L = length of wall
H = height of wall

Studding is 16" on Center: $(L' \times .75 + 1) \times H' = BF$
Studding is 18" on Center: $(L' \times .67 + 1) \times H' = BF$
Studding is 24" on Center: $(L' \times .5 + 1) \times H' = BF$
Studding is 30" on Center: $(L' \times .4 + 1) \times H' = BF$

---

*The results, using these formulas, may be slightly different in some instances from the figures shown in Table 14–4A which have been rounded out to whole numbers. Plates and shoes, and framing around openings should be added.

TABLE 14–5

Multipliers for Obtaining the Number of Studs, Joists and Rafters in the Length of Floors, Walls or Roofs

Spacing in Inches	Factor to Multiply the Width of Room, Length of Wall or Roof	
12	1.00	
16	.75	
18	.67	Add one for
20	.60	end in each
24	.50	case

TABLE 14–5A

**Formulas for Estimating the BF of Joists
of Various Sizes Per Floor**

(A = Area of the Floor or the Ground-floor Area of the
Dwelling) Quantities include an allowance for an end
joist and cutting and fitting waste.

Size of Joist	Inches Center to Center	Formula
2″ × 4″	16 20 24	.55A .44A .37A
2″ × 6″	16 20 24	.85A .68A .56A
2″ × 8″	16 20 24	1.10A .88A .73A
2″ × 10″	16 20 24	1.35A 1.08A .90A
2″ × 12″	16 20 24	1.65A 1.32A 1.10A

Illustrative Example

A 2-story dwelling, 32′ × 70′, has a *ground-floor* area of 2,240 sq ft. The first floor joists are 2″ × 12″, the second floor are 2″ × 10″, and the second floor ceiling joists are 2″ × 6″. All are on 16″ centers.

First floor joists	2″ × 12″	2,240 × 1.65 = 3,696 BF
Second ″	2″ × 10″	2,240 × 1.35 = 3,024 ″
Ceiling joists,	2″ × 6″	2,240 × .85 = 1,904 ″
		Total 8,624 ″

TABLE 14–5B

Board Feet of Joists 2" × 4" × 12" for Areas
200 to 3,000 Square Feet
(16", 20", and 24" on center)

Quantities include allowance for end joist, cutting,
and fitting waste.

Ground Floor Area	2" × 4" On Center			2" × 6" On Center			2" × 8" On Center			2" × 10" On Center			2" × 12" On Center		
	16"	20"	24"	16"	20"	24"	16"	20"	24"	16"	20"	24"	16"	20"	24"
200	110	88	73	170	136	114	220	176	147	270	216	180	330	264	220
300	165	132	110	255	204	170	330	264	220	405	324	270	495	396	330
400	220	176	147	340	272	227	440	352	293	540	432	360	660	528	440
500	275	220	183	425	340	283	550	440	367	695	540	450	825	660	550
600	330	264	220	510	408	340	660	528	440	810	648	540	990	792	660
700	385	308	257	595	476	397	770	616	514	945	756	630	1155	924	770
800	440	352	293	680	544	454	880	704	587	1080	864	720	1320	1056	880
900	495	396	330	765	612	510	990	792	660	1215	972	810	1485	1188	990
1000	550	440	367	850	680	567	1100	880	734	1350	1080	900	1650	1320	1100
1100	605	484	404	935	748	624	1210	968	807	1485	1188	990	1815	1415	1210
1200	660	528	440	1020	816	680	1320	1056	880	1620	1296	1080	1980	1584	1320
1300	715	572	477	1105	884	737	1430	1144	954	1755	1404	1170	2145	1716	1430
1400	770	616	514	1190	952	794	1540	1232	1027	1890	1512	1260	2310	1848	1540
1500	825	660	550	1275	1020	850	1650	1320	1100	2025	1620	1350	2475	1980	1650
1600	880	704	587	1360	1088	907	1760	1408	1177	2160	1728	1440	2640	2112	1760
1700	935	748	624	1445	1156	964	1870	1496	1247	2295	1836	1530	2805	2244	1870
1800	990	792	660	1530	1224	1021	1980	1584	1320	2430	1944	1620	2970	2376	1980
1900	1045	836	697	1615	1292	1077	2090	1672	1394	2565	2052	1710	3135	2508	2090
2000	1100	880	734	1700	1360	1134	2200	1760	1467	2700	2160	1800	3300	2640	2200
2100	1155	924	770	1785	1428	1190	2310	1848	1540	2835	2268	1890	3465	2772	2310
2200	1210	968	807	1870	1496	1247	2420	1936	1614	2970	2376	1980	3630	2904	2420
2300	1265	1012	844	1955	1564	1304	2530	2024	1688	3105	2484	2070	3795	3036	2530
2400	1320	1056	880	2040	1632	1360	2640	2112	1760	3240	2592	2160	3960	3168	2640
2500	1375	1100	917	2125	1700	1417	2750	2200	1834	3375	2700	2250	4125	3300	2750
2600	1430	1144	953	2210	1768	1474	2860	2288	1908	3510	2808	2340	4290	3432	2860
2700	1485	1188	990	2295	1836	1530	2970	2376	1980	3645	2916	2430	4455	3564	2970
2800	1540	1232	1027	2380	1904	1587	3080	2464	2054	3780	3024	2520	4620	3696	3080
2900	1595	1276	1064	2465	1972	1644	3190	2552	2128	3915	3132	2610	4785	3828	3190
3000	1650	1320	1100	2550	2040	1700	3300	2640	2200	4050	3240	2700	4950	3960	3300

The BF of joists of various sizes, 16 in., 20 in., and 24 in. on center, can be read directly from Table 14–5B for *floor areas* of 200 to 3,000 square feet. The number of board feet shown includes an allowance for one end joist, and also cutting and fitting waste.

Girder, Door, and Window Headers

Where an I-beam is used as a *girder* in frame construction, a 2-by-4 or a 2-by-6 is bolted to the top to which the joists are nailed. There tends to be less shrinkage and less sagging in this construction than in girders made up of doubled or tripled 2-inch stock. A recommended type of girder is made up of two 2-by-12 in. members with a steel plate between. The three members are bolted firmly together. Frequently, in place of the steel plate, a ½-in. piece of plywood is glued and nailed between the members.

Window and door headers may have two 2-by-4s *on edge* up to spans of 2 feet. Longer spans require stronger framing. As in girders, where 2-by-8-in. to 2-by-12-in. members are used, a piece of ¼-in. to ½-in. plywood is glued and nailed between two members. Table 14–6 shows header sizes and maximum spans recommended.

Figure 14–6 on page 270 illustrates girder and framed openings construction.

Bridging

Floor joists need to be reinforced so that the floor load is distributed over many joists and any deflection of the joists is uniform. To do this, bridging is placed between the joists. There are three types of bridging used: *wood cross bridging, wood solid bridging,* and *metal cross bridging.* (See Figure 14–7 on page 273.)

Wood cross bridging can be purchased at lumber yards in bundles of 50 pieces, or it can be cut on the job. The material used is usually 1 in. by 3 in., 1 in. by 4 in., or 2 in. by 2 in. The length of the pieces (struts) will depend on the depth and spacing of the joists as will the angle of the cut, since the piece is nailed even with the upper edge of one joist and even with the lower edge of the next joist. Two pieces of wood bridging are placed between each joist, crossing each other. In small rooms one line of bridging is adequate, but in larger rooms the lines of bridging should be spaced 6 in. to 8 in. apart. Bridging is put in place before the floor or subfloor is laid.

Solid wood bridging consists of a line of solid pieces of the same stock as the joists placed between the joists. The pieces are staggered in line in order to nail them securely between the joists.

TABLE 14–6

**Window and Door Opening Maximum
Spans for Various Headers**

Header Sizes	Maximum Span
2 - 2″ x 4″	2′
2 - 2 x 6	4′
2 - 2 x 8	5′
2 - 2 x 10	8′
2 - 2 x 12	10′

Note: The use of plywood or steel interlay is recommended (See illustrations in Figure 14–6.)

Metal cross bridging are factory-made 18 ga. steel braces that are placed in the manner as wood cross bridging. They are available in cartons of 200.

TABLE 14-7

Span Calculations Provide for Carrying the Live Loads Shown and the Additional Weight of the Joists and Double Flooring

SIZE	SPACING	20# L.L. Plaster Clg.	30# LIVE LOAD Plaster Clg.	30# LIVE LOAD No Plaster	40# LIVE LOAD Plaster Clg.	40# LIVE LOAD No Plaster	50# LIVE LOAD Plaster Clg.	50# LIVE LOAD No Plaster	60# LIVE LOAD Plaster Clg.	60# LIVE LOAD No Plaster
2 × 4	12"	8'–8"								
	16"	7'–11"								
	24"	6'–11"								
2 × 6	12"	13'–3"	11'–6"	14'–10"	10'–8"	13'–2"	10'–0"	12'–0"	9'–6"	11'–1"
	16"	12'–1"	10'–6"	12'–11"	9'–8"	11'–6"	9'–1"	10'–5"	8'–7"	9'–8"
	24"	10'–8"	9'–3"	10'–8"	8'–6"	9'–6"	8'–0"	8'–7"	7'–7"	7'–10"
2 × 8	12"	17'–6"	15'–3"	19'–7"	14'–1"	17'–5"	13'–3"	15'–10"	12'–7"	14'–8"
	16"	16'–0"	13'–11"	17'–1"	12'–11"	15'–3"	12'–1"	13'–10"	11'–5"	12'–9"
	24"	14'–2"	12'–3"	14'–2"	11'–4"	12'–6"	10'–7"	11'–4"	10'–1"	10'–6"

Size	Spacing									
2 × 10	12"	21'–11"	19'–2"	24'–6"	17'–9"	21'–10"	16'–8"	19'–11"	15'–10"	18'–5"
	16"	20'–2"	17'–6"	21'–6"	16'–3"	19'–2"	15'–3"	17'–5"	14'–6"	16'–1"
	24"	17'–10"	15'–6"	17'–10"	14'–3"	15'–10"	13'–5"	14'–4"	12'–8"	13'–3"
2 × 12	12"	26'–3"	23'–0"	29'–4"	21'–4"	26'–3"	20'–1"	24'–0"	19'–1"	22'–2"
	16"	24'–3"	21'–1"	25'–10"	19'–7"	23'–0"	18'–5"	21'–0"	17'–5"	19'–5"
	24"	21'–6"	18'–8"	21'–5"	17'–3"	19'–1"	16'–2"	17'–4"	15'–4"	16'–0"
3 × 8	12"	20'–0"	17'–7"	24'–3"	16'–4"	21'–8"	15'–4"	19'–10"	14'–7"	18'–4"
	16"	18'–6"	16'–1"	21'–4"	14'–11"	19'–1"	14'–1"	17'–4"	13'–4"	16'–0"
	24"	16'–5"	14'–3"	17'–9"	13'–2"	15'–9"	12'–4"	14'–4"	11'–9"	13'–3"
3 × 10	12"	25'–0"	22'–0"	30'–2"	20'–6"	27'–1"	19'–3"	24'–10"	18'–4"	23'–0"
	16"	23'–2"	20'–3"	26'–8"	18'–10"	23'–10"	17'–8"	21'–9"	16'–10"	20'–2"
	24"	20'–7"	17'–11"	22'–3"	16'–7"	19'–10"	15'–7"	18'–1"	14'–10"	16'–8"

Used by permission of Weyerhaeuser Company, Lumber and Plywood Division, St. Paul, Minnesota.

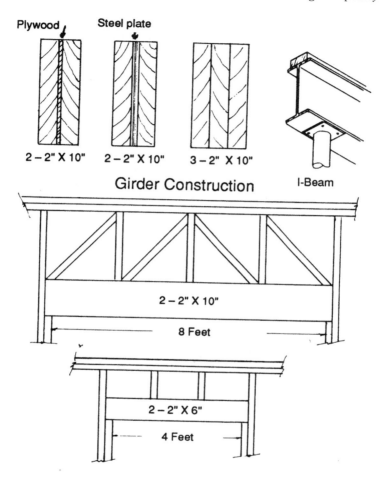

Plywood   Steel plate

2 – 2" X 10"   2 – 2" X 10"   3 – 2" X 10"   I-Beam

Girder Construction

2 – 2" X 10"

8 Feet

2 – 2" X 6"

4 Feet

Figure 14–6
Girder Construction and Framed
Openings

Table 14–9 on page 274 shows the amount of material required for wood bridging that is cut on the job.

Rafters

Rafters of 2 in. by 6 in. are common in residential and small commercial buildings, although on bungalows, porches, and many secondary buildings 2-by-4-in. rafters are used. On many commercial buildings with flat roofs, the size of the rafters may be 3 or 4 inches thick and 8 to 12 inches deep. Rafter dimensions depend largely on the span, the weight of the roof material, and the probable wind and snow load as shown in Table 14–8.

Table 14–10 on pages 276–277 shows the BF in gable and hip roofs with 2-by-4-in. rafters 16 in. and 24 in. on center with roof pitches of ⅛ to ½. Table 14–11 on pages 278–279 shows the BF in gable and hip roofs with 2-by-6-in. rafters, 16 in. and 24 in. on center with roof pitches of ⅛ to ½.

FORMULAS TO ESTIMATE THE BF OF RAFTER FRAMING

The formulas shown in Table 14–12 on page 280 may be used to quickly estimate the board feet (BF) of rafters in any gable or hip roof based on the square foot area under the roof. Multiply the factor in the Table for the pitch of the roof and the size and spacing of the rafters. Add 5 percent to the BF developed for each 12 inches of *horizontal* overhang. The *number* of rafters may be obtained from Table 14–4.

The *ridge pole*, the framing member at the ridge of the roof to which the rafters are attached, is required by

TABLE 14–8

SPAN OF RAFTERS

Span Calculations Provide for Carrying the Live Loads Shown and the Additional Weight of the Rafters, Sheathing, and Wood Shingles

SIZE	SPACING	15# L.L.		20# L.L.		30# L.L.		40# L.L.	
		No Plaster	Plaster	No Plaster	Plaster	No Plaster	Plaster	No Plaster	Plaster
2 × 4	12"	12'–7"	8'–11"	11'–4"	8'–4"	9'–8"	7'–6"	8'–7"	6'–11"
	16"	11'–0"	8'–2"	9'–11"	7'–7"	8'–5"	6'–10"	7'–5"	6'–3"
	24"	9'–1"	7'–2"	8'–2"	6'–8"	6'–11"	6'–0"	6'–1"	5'–6"
2 × 6	12"	19'–2"	13'–8"	17'–4"	12'–9"	14'–10"	11'–6"	13'–2"	10'–8"
	16"	16'–10"	12'–6"	15'–2"	11'–8"	12'–11"	10'–6"	11'–6"	9'–8"
	24"	13'–11"	11'–0"	13'–0"	10'–3"	10'–8"	9'–3"	9'–6"	8'–6"
2 × 8	12"	25'–1"	17'–11"	22'–9"	16'–10"	19'–7"	15'–3"	17'–5"	14'–1"
	16"	22'–1"	16'–6"	20'–0"	15'–5"	17'–1"	13'–11"	15'–3"	12'–11"
	24"	18'–5"	14'–7"	16'–7"	13'–8"	14'–2"	12'–3"	12'–6"	11'–4"

Used by permission of Weyerhaeuser Company, Lumber and Plywood Division, St. Paul, Minnesota.

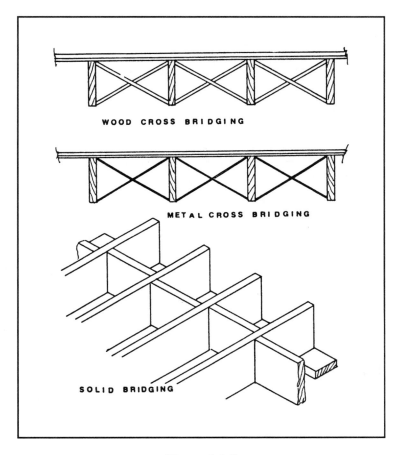

WOOD CROSS BRIDGING

METAL CROSS BRIDGING

SOLID BRIDGING

Figure 14–7
Bridging

most building codes to be of a depth not less than the cut end of the abutting rafters. The ridge pole is usually about 2 inches deeper than the rafters. For example, where 2-by-6-in. rafters are used, the ridge pole would be 2 in. by 8 in.

TABLE 14–9

**Material Required for Wood Bridging Cut
on the Job**

Size Joist In Inches	Spacing In Inches	Lineal Feet Per Set (2)	Lineal Feet Per-Foot-of-Row
2 × 6	16	2.57	1.92
2 × 8	16	2.70	2.02
2 × 10	16	2.87	2.15
2 × 12	16	3.06	2.30
2 × 8	20	3.31	2.00
2 × 10	20	3.45	2.07
2 × 12	20	3.61	2.17
2 × 8	24	3.94	2.00
2 × 10	24	4.05	1.97
2 × 12	24	4.19	2.10

Note: Add to the total lineal feet developed from the Table at least 10% cutting waste.

Example: A room 20 feet wide has two rows of bridging. The 2″ × 12″ joists are 16″ on center

2 × 20 lin ft × 2.30 ft per-foot of-row =	92.0	lin ft
Add 10% waste	9.2	″ ″
Total lin ft =	101.2	″ ″
Round out to	101	lin ft

(Per set method would be 30 sets × 3.06 = 91.8 to which must be added 10% waste.)

WOOD ROOF AND FLOOR TRUSSES

Truss Construction

Trusses may be pre-cut and assembled on the ground at the job site by carpenters, or they may be ordered from a prefabricating plant. The size of the members, that is, the bottom cord, rafter or top cord, and braces, depends on the span and load requirements. On residential roofs, where spans are usually less than 32 feet, the entire truss can be fabricated from 2-by-4-inch stock. Local building codes and FHA requirements govern the use of these light framing materials. Much depends on dead loads and wind and snow loads. An acceptable and practical design employs 2-by-6-in. top cord with other members of 2-by-4-in. material.

There are two basic methods of connecting the members of a truss. Earlier designs used half-inch plywood "gussets" which were applied to both sides and nailed and glued. More recently, truss components are attached with one on the many types of metal connector plates. These range from prepunched galvanized steel or toothed connectors to crimped, barbed, corrugated, or flared types. In truss assembly plants these are applied with presses, rollers, or pneumatic hammers.

Figure 14–8
Metal-plate type truss

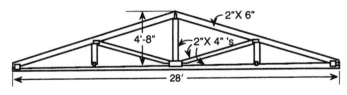

Illustrative Example

To build a truss with 2″ × 6″ rafters, 2″ × 4″ cord and braces, using metal connectors. The span is 28 feet and the pitch is 1/6, 4″ rise per-foot-of-run. Assume first truss fabricated and templates in place.

Using the truss in Figure 14–8 the length of the members can be scaled directly. (A line drawing or sketch to scale is recommended.)

Top cord	2–2″ × 6″ × 16′–0″	=	32.0	BF
Bottom Cord	2–2″ × 4″ × 16′–0″	=	21.3	″
Center post	1–2″ × 4″ × 5′–0″	=	3.3	″
Vertical braces	2–2″ × 4″ × 2′–6″	=	3.3	″
Diagonal braces	2–2″ × 4″ × 7′–0″	=	9.3	″
			69.2	BF

TABLE 14–10

2" × 4" Rafters Including Ridge Pole . . . Board Feet Required for Roof Pitch Shown. 16" and 24" Centers

Ground Floor Area	16" CENTER TO CENTER — ROOF PITCH						24" CENTER TO CENTER — ROOF PITCH					
	1/8	1/6	5/24	1/4	1/3	1/2	1/8	1/6	5/24	1/4	1/3	1/2
200	114	116	120	124	132	156	78	80	82	86	92	108
300	171	174	180	186	198	234	117	120	123	129	138	162
400	228	232	240	248	264	312	156	160	164	172	184	216
500	285	290	300	310	330	390	195	200	205	215	230	270
600	342	348	360	372	396	468	234	240	246	258	276	324
700	399	406	420	434	462	546	273	280	287	301	322	378
800	456	464	480	496	528	624	312	320	328	344	368	432
900	513	522	540	558	594	702	351	360	369	387	414	486
1000	570	580	600	620	660	780	390	400	410	430	460	540
1100	627	638	660	682	726	858	429	440	451	473	506	594
1200	684	696	720	744	792	936	468	480	492	516	552	648
1300	741	754	780	806	858	1014	507	520	533	559	598	702

1400	756	644	602	574	560	546		1092	924	868	840	812	798
1500	810	690	645	615	600	585		1170	990	930	900	870	855
1600	864	736	688	656	640	624		1248	1056	992	960	928	912
1700	918	782	731	697	680	663		1326	1122	1054	1020	986	969
1800	972	828	774	738	720	702		1404	1188	1116	1080	1044	1026
1900	1026	874	817	779	760	741		1482	1254	1178	1140	1102	1083
2000	1080	920	860	820	800	780		1560	1320	1240	1200	1160	1140
2100	1134	966	903	861	840	819		1638	1386	1302	1260	1218	1197
2200	1188	1012	946	902	880	858		1716	1452	1363	1320	1276	1254
2300	1242	1058	989	943	920	897		1794	1518	1426	1380	1334	1311
2400	1296	1104	1032	984	960	936		1872	1584	1488	1440	1392	1368
2500	1350	1150	1075	1025	1000	975		1950	1650	1550	1500	1450	1425
2600	1404	1196	1118	1066	1040	1014		2028	1716	1612	1560	1508	1482
2700	1458	1242	1161	1107	1080	1053		2106	1782	1674	1620	1566	1539
2800	1512	1288	1204	1148	1120	1092		2184	1848	1736	1680	1624	1596
2900	1566	1334	1247	1189	1160	1131		2262	1914	1798	1740	1682	1653
3000	1620	1380	1290	1230	1200	1170		2340	1980	1860	1800	1740	1710

TABLE 14–11

2″ × 6″ Rafters Including Ridge Pole Board Feet Required for Pitch Shown

Ground Floor Area	16″ CENTER TO CENTER ROOF PITCH						24″ CENTER TO CENTER ROOF PITCH					
	1/8	1/6	5/24	1/4	1/3	1/2	1/8	1/6	5/24	1/4	1/3	1/2
200	164	170	174	180	192	228	114	116	120	124	132	156
300	246	255	261	270	288	342	171	174	180	186	198	234
400	328	340	348	360	384	456	228	232	240	248	264	312
500	410	425	435	450	480	570	285	290	300	310	330	390
600	492	510	522	540	576	684	342	348	360	372	396	468
700	574	595	609	630	672	798	399	406	420	434	462	546
800	656	680	696	720	768	912	456	464	480	496	528	624
900	739	765	783	810	864	1026	513	522	540	558	594	702
1000	820	850	870	900	960	1140	570	580	600	620	660	780
1100	902	935	957	990	1056	1254	627	638	660	682	726	858
1200	984	1020	1044	1080	1152	1368	684	696	720	744	792	936
1300	1066	1105	1131	1170	1248	1482	741	754	780	806	858	1014

1400	1148	1190	1218	1260	1344	1596		798	812	840	868	924	1092
1500	1230	1274	1305	1350	1440	1710		855	870	900	930	990	1170
1600	1312	1360	1392	1440	1536	1824		912	928	960	992	1056	1248
1700	1394	1445	1479	1530	1632	1938		969	986	1020	1054	1122	1326
1800	1476	1530	1566	1620	1728	2052		1026	1044	1080	1116	1188	1404
1900	1558	1615	1653	1710	1824	2166		1083	1102	1140	1178	1254	1482
2000	1640	1700	1740	1800	1920	2280		1140	1160	1200	1240	1320	1560
2100	1722	1785	1827	1890	2016	2394		1197	1218	1260	1302	1386	1638
2200	1804	1870	1914	1980	2112	2508		1254	1276	1320	1364	1452	1716
2300	1886	1955	2001	2070	2208	2622		1311	1334	1380	1426	1518	1794
2400	1968	2040	2088	2160	2304	2736		1368	1392	1440	1488	1584	1872
2500	2050	2125	2175	2250	2400	2850		1425	1450	1500	1550	1650	1950
2600	2132	2210	2262	2340	2496	2964		1482	1508	1560	1612	1716	2028
2700	2214	2295	2349	2430	2592	3078		1539	1566	1620	1674	1782	2106
2800	2296	2380	2436	2520	2688	3192		1596	1624	1680	1736	1848	2184
2900	2378	2465	2523	2610	2784	3306		1653	1682	1740	1798	1914	2262
3000	2460	2550	2610	2700	2880	3420		1710	1740	1800	1860	1980	2340

TABLE 14–12

Formula for BF Rafters in Gable and Hip Roofs from Ground Floor Area

A = The Ground Floor (horizontal) Area

Size of Rafter and Centers	Pitch of Roof					
	⅛	⅙	⁵⁄₂₄	¼	⅓	½
2″ × 4″ −16″ on center	.57×A	.58×A	.60×A	.62×A	.46×A	.78×A
″　　−24″　″　　″	.39×A	.40×A	.41×A	.43×A	.46×A	.54×A
2″ × 6″ −16″　″　　″	.82×A	.85×A	.87×A	.90×A	.96×A	1.14×A
″　　−24″　″　　″	.57×A	.58×A	.60×A	.62×A	.66×A	.78×A

Illustrative Example 1

A dwelling 40′ × 50′ has a *ground-floor* area of 2,000 sq ft. The roof pitch is ⅓ and the roof has 2″ × 4″ rafters, 24″ on center. The total BF for the rafters is:

2,000 × .46 =　920 BF

If the rafters were 16″ on center, the answer would be:

2,000 × .66 = 1,320 BF

Where the roof extends over areas outside the perimeter of the dwelling, such as over a carport, breezeway, open porch, or other nondwelling area, use the horizontal area covered by the roof in place of the *ground-floor* area of the dwelling.

Prefabricated Trusses

It is estimated that 80 percent to 90 percent of trusses are made, delivered, and, many times, erected by fabricating plants located in all major cities and many smaller communities across the country. Prefabricated trusses eliminate job-site cutting and fabricating, saving time. The cost is considerably less than making them on the job.

One of the large roof and floor truss prefabricators is Gang-Nail Systems, Inc. They provide a detailed, engineered computer program based on the specifics of the job. Their service is "programmed to compare up to four truss configurations since many trusses can be used for a given span, yet one is often the most cost efficient."

TABLE 14–12A

Board Feet of Material for Trusses with 2" × 6" Top Cord, Other Members 2" × 4". No Overhang Contemplated

			SPAN IN FEET		
	20	24	26	28	30

Top Cord (2" × 6")

Pitch					
1/6	2–12'	2–14'	2–16'	2–16'	2–18'
5/24	2–12'	2–14'	2–16'	2–18'	2–18'
1/4	2–14'	2–16'	2–16'	2–18'	2–18'

Bottom Cord (2" × 4")

Pitch					
1/6	2–12'	2–14'	2–16'	2–16'	2–18'
5/24	2–12'	2–14'	2–16'	2–16'	2–18'
1/4	2–12'	2–14'	2–16'	2–16'	2–18'

Diagonals (2" × 4")

Pitch					
1/6	2–8'	2–10'	2–10'	2–12'	2–12'
5/24	2–10'	2–12'	2–12'	2–12'	2–14'
1/4	2–10'	2–12'	2–14'	2–14'	2–16'

(1/6 pitch = 4 inch per-ft-run, 5/24 = 5 inch per-ft-run, ¼ = 6 inch per-ft-run). Note: Add 2 or 3 ft of 1" × 4" to joint 2" × 4" bottom cord. Add metal connecting members, nails etc.

BF Framing Per Square Foot in Floors, Walls, and Roofs

Sometimes it is necessary to quickly estimate the board feet of framing in a given area of floor, wall, or roof when the size of the framing members and the center-to-center spacing is known. Table 14–13 shows the factor per square foot that the areas are to be multiplied by for size and center-to-center framing.

TABLE 14–12B

Roof Trusses—24″ On Center—Including Braces and Struts Board Feet Required for Rafter Size and Roof Pitch Shown (W-Type Trusses)

Ground Floor Area	2″ × 4″ RAFTER AND CORD						2″ × 6″ RAFTER . . 2″ × 4″ CORD					
	ROOF PITCH						ROOF PITCH					
	1/8	1/6	5/24	1/4	1/3	1/2	1/8	1/6	5/24	1/4	1/3	1/2
200	210	212	214	218	224	240	246	248	252	256	264	288
300	315	318	321	327	336	360	369	372	378	384	396	432
400	420	424	428	436	448	480	492	496	504	512	528	576
500	525	530	535	545	560	600	615	620	630	640	660	720
600	630	636	642	654	672	720	738	744	756	768	792	864
700	735	742	749	763	784	840	861	868	882	896	924	1008
800	840	848	856	872	896	960	984	992	1008	1024	1056	1152
900	945	954	963	981	1008	1080	1107	1116	1134	1152	1188	1296
1000	1050	1060	1070	1090	1120	1200	1230	1240	1260	1280	1320	1440
1100	1155	1166	1177	1199	1232	1320	1353	1364	1386	1408	1452	1584
1200	1260	1272	1284	1308	1344	1440	1476	1488	1512	1536	1584	1728
1300	1365	1378	1391	1417	1456	1560	1599	1612	1638	1664	1716	1872

1400	1470	1484	1498	1526	1568	1680		1722	1736	1764	1792	1848	2016
1500	1575	1590	1605	1635	1680	1800		1845	1860	1890	1920	1980	2160
1600	1680	1696	1712	1744	1792	1920		1968	1984	2016	2048	2112	2304
1700	1785	1802	1819	1853	1904	2040		2091	2108	2142	2176	2244	2448
1800	1890	1908	1926	1962	2016	2160		2214	2232	2268	2304	2376	2592
1900	1995	2014	2033	2071	2128	2280		2337	2356	2394	2432	2508	2736
2000	2100	2120	2140	2180	2240	2400		2460	2480	2520	2560	2640	2880
2100	2205	2226	2247	2289	2352	2520		2583	2604	2646	2688	2772	3024
2200	2310	2332	2354	2398	2464	2640		2706	2728	2772	2816	2904	3168
2300	2415	2438	2461	2507	2576	2760		2829	2852	2898	2944	3036	3312
2400	2520	2544	2568	2616	2688	2880		2952	2976	3024	3072	3168	3456
2500	2625	2650	2675	2725	2800	3000		3075	3100	3150	3200	3300	3600
2600	2730	2756	2782	2834	2912	3120		3198	3224	3276	3328	3432	3744
2700	2835	2862	2889	2943	3024	3240		3321	3348	3402	3456	3564	3888
2800	2940	2968	2996	3052	3136	3360		3444	3472	3528	3584	3696	4032
2900	3045	3074	3103	3161	3248	3480		3567	3596	3654	3712	3828	4176
3000	3150	3180	3210	3270	3360	3600		3690	3720	3780	3840	3960	4320

Examples

1. Exterior walls of a building 30' × 60' have 2" × 4" studs, 16"
   on center. The walls are 9' high.
   Area = 2(30' + 60') × 9' = 1620 sq. ft.
   BF   = 1620 sq. ft. × .50 = 810 BF

2. Floor joists in the same building are 2" × 10", 16" on center.
   Area = 30' × 60' = 1800 sq. ft.
   BF   = 1800 sq. ft. × 1.25 = 2250 BF

3. There is 200 lin ft of partitioning, 2" × 4" studs, 20" on cen-
   ter.
   Area = 200' × 9' = 1800 sq. ft.
   BF   = 1800 sq. ft. × .40 = 720 BF

TABLE 14–13

Board Feet of Framing Per Square Foot
of Walls, Floors, and Roofs

Size of Framing in Inches	Center to Center in Inches		
	12	16	20
1× 2	.167	.125	.10
2× 2	.333	.250	.20
2× 3	.500	.375	.30
2× 4	.667	.500	.40
2× 6	1.00	.750	.60
2× 8	1.33	1.00	.80
2×10	1.67	1.25	1.00
2×12	2.00	1.50	1.20
3× 6	1.50	1.13	.90
3× 8	2.00	1.50	1.20
3×10	2.50	1.88	1.50
3×12	3.00	2.25	1.80

Note: When spacing is 24 inches on center, use half the figure shown for
12 inches on center. Add for shoes and plates and also for framing around
openings.

NAILS AND NAILING

Figure 14–9 shows fourteen common nails with their gauge, length, and penny (d) designation. Most sizes are also available in finishing nails that have a small head that can be set below the surface of the finish carpentry.

Today there is a nail for almost every type of construction material; some have special heads, points, or sizes. Some have screw threads, spiral, annular, or knurled threads. There are nails for framing, roofing, drywall, underlayment, flooring, hardboard, trim, cedar shakes, asbestos siding, wood siding, and gypsum lath to name a few of the special purpose nails. Some are coated with rosin or cement for better holding power,

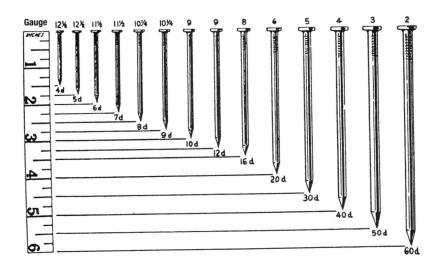

Figure 14–9
Common Nail Sizes

although the threaded nail, which can be driven, holds best. Table 14–14 lists the penny system of nails, both common and finishing, and shows the length, gauge, and approximate number of nails to the pound.

TABLE 14–14

Penny Nail System

Size	Length in Inches	Gauge Number	Approximate Number to Pound		
			Common	Finishing	Casing
2d	1	15	850		
3d	1¼	14	550	640	
4d	1½	12½	350	456	
5d	1¾	12½	230	328	
6d	2	11½	180	273	228
7d	2¼	11	140	170	178
8d	2½	10¼	100	151	133
9d	2¾	9½	80	125	100
10d	3	9	65	107	96
12d	3¼	9	50		60
16d	3½	8	40		50
20d	4	6	31		
30d	4½	5	22		
40d	5	4	18		
50d	5½	3	14		
60d	6	2	12		

Special Purpose Nails

There are a number of special purpose nails designed with shaped heads, threads, and points, depending upon their use. These three features are shown in Figure 14–10 on pages 288–289. Some of the more common special purpose nails are shown and described in Figures 14–11 and 14–12 (see pages 290–291), reprinted by courtesy of Independent Nail, Inc., Bridgewater, Massachusetts 02324.

Nails Required in Rough Carpentry

Table 14–15 on page 292 gives the approximate number of pounds and size of nail for various types of rough carpentry using the penny system.

SUBFLOORING AND ROOF DECKING

Board Subflooring and Roof Decking

Boards for subflooring and roof decking have square, shiplap, or tongue and groove edges and are nominally 4 in., 6 in., or 8 in. wide. The boards are laid at right angles to the floor joists and rafters. Occasionally they are laid diagonally to the joists in order for the finish floor to be laid parallel with the joists or at right angles to them. Board roof decking is usually at right angles to the rafters. Table 14–16 on page 293 shows the percentage to add to areas for milling and cutting waste for 1-inch boards.

Plywood Subflooring and Roof Decking

The use of plywood for subflooring and roof decking has all but replaced the use of boards. The plywood panels, 8 ft. by 8 ft. (other sizes are available), have identification index markings by the American Plywood Association (APA) to indicate the spacing of rafters and floor joists for the particular thickness of the subflooring or roof decking being used. See Tables 14–17, 14–18, and 14–19 on pages 294, 295, and 302 respectively.

Plywood panels ½-inch thick applied over joists 16 in. on center meet FHA requirements, although ⅝-inch thick panels are used more often when a finish floor or underlayment is intended.

Underlayments plywood panels are touch sanded to provide a smooth surface for applying carpeting, tile, and other nonconstructive floor finishes.

Figure 14-10
Nail Heads, Threads, and Points
Courtesy Independent Nail, Inc., Bridgewater, Massachusetts 02324.

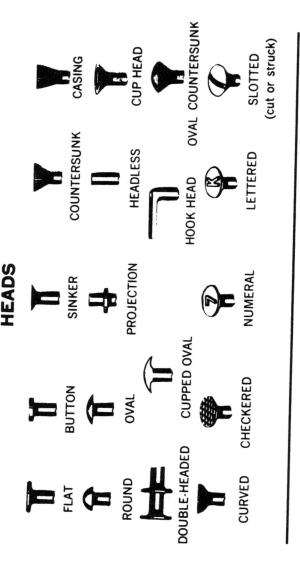

288

THREADS

SCREW-TITE® SPIRAL

STRONGHOLD® ANNULAR

STRONGHOLD SCREW THREAD

Combinations or variations of these threads are available.

POINTS

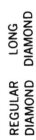

REGULAR
DIAMOND

LONG
DIAMOND

BLUNT

CONICAL

NEEDLE

SHEARED
BEVEL

SHEARED
CHISEL

SHEARED
SQUARE

INTERIOR HARDBOARD NAILS

- **STRONGHOLD®** Annular Thread for tight, smooth decorative, pre-finished plywood and interior hardboard paneling applications.
- Small, button head for flush driving helps make head inconspicuous.
- Color-matching eliminates need for nail-setting and puttying.
- Hardened Steel — heat treated, quenched, and tempered steel to make this slender shank nail easily driveable.

EXTERIOR HARDBOARD SIDING NAILS

- **SCREWTITE®** Spiral Thread for weathertight siding applications and ease of driving into hardboard siding.
- Stiff Stock Steel — high carbon, high tensile steel aids driving.
- Pilot point to help start the nail straight and ease a path for the thread.
- Flat, countersunk head for flush driving and best apearance of installed hardboard siding.
- Hot-dipped Galvanized for maximum corrosion resistance for a steel nail.

BLUNT-PQINT WOOD SIDING NAILS

- **STRONGHOLD®** Annular Thread for weathertight siding applications.
- Stiff Stock Steel — high carbon, high tensile steel aids driving.
- Slender shank minimizes splitting.
- Large, slightly countersunk head (checkered) for ease of flush driving.
- Blunt diamond point crushes wood fibers, effectively minimizing splitting in wood siding patters, or close to end or edge of wood siding patterns.
- Hot-dipped Galvanized for maximum corrosion resistance for a steel nail.

COPPER SLATING NAILS

- **STRONGHOLD®** Annular Thread for maximum holding power in wood or plywood.
- **SCREWTITE®** Spiral Thread for maximum shear, racking, lateral load resistance and high holding power.

FLOORING NAILS

- **SCREWTITE®** Spiral Thread effectively minimizes "squeaks" in floor systems by pulling flooring tight. Turning action provided by thread minimizes splitting of hardwood flooring.
- Flat, countersunk head permits flush driving.
- Stiff Stock Steel — high carbon, high tensile steel for driving into less dense flooring.
- Hardened Steel — heat treated, quenched, tempered steel for ease of driving in dense hardwood flooring.

ASBESTOS SIDING FACE NAILS

- **STRONGHOLD®** Annular Thre for weathertight siding appli tion.

Figure 14–11
Special Purpose Nails
Courtesy Independent Nail, Inc., Bridgewater, Massachusetts 02324.

UNDERLAYMENT NAILS

- **STRONGHOLD®** Annular Thread effectively minimizes "squeaks" in floor systems by minimizing nail-popping, which can also mar floor coverings.
- Flat, slightly countersunk head permits flush driving for smooth underlayment surface.
- Regular steel for driving into less dense underlayment and subflooring.
- Hardened Steel — heat treated, quenched and tempered steel for ease of driving into denser underlayment and subflooring.

TRUSSED RAFTER POLE BARN NAILS

- **STRONGHOLD®** Annular Thread for maximum withdrawal resistance in engineered wood construction.
- **SCREWTITE®** Spiral Thread for maximum shear resistance in engineered wood construction.
- The generally smaller wire diameters than those of comparable common nails minimize splitting of drier, thinner lumber, framing, and decking.
- Stiff Stock Steel — high carbon, high tensile steel for driving into medium density wood species.
- Hardened Steel — heat treated, quenched, and tempered steel for driving into high density, dry wood species and for maximum rigidity.

ROOFING NAILS

- **STRONGHOLD®** Annular Thread for best holding power when fastening asphalt shingles and roll roofing over wood or plywood decking or sheathing.
- Large, flat head for maximum bearing area.
- Hot-dipped Galvanized for maximum corrosion resistance for a steel nail.

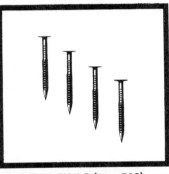

ASBESTOS SIDING FACE NAILS

- **STRONGHOLD®** Annular Thread for weathertight siding application.
- Hot-dipped Galvanized for maximum corrosion resistance for a steel nail.

CEDAR SHAKE NAILS

- **STRONGHOLD®** Annular Thread for weathertight cedar shingle and shake applications.
- Slender shank minimizes shake and shingle splitting.
- Small, checkered, button head blends with "textured" shingle and shake surfaces.
- Hot-dipped Galvanized for maximum corrosion resistance for a steel nail.

DRYWALL NAILS (type 500)

- **STRONGHOLD®** Annular Thread effectively minimizes nail-popping that can mar finished drywall appearance.
- Long diamond point for the "thumb" start.
- Flat, slightly countersunk, thin-rimmed head aids flush driving with minimum disruption of drywall surfaces.

Figure 14–12

More Special Purpose Nails

Courtesy Independent Nail, Inc., Bridgewater, Massachusetts 02324.

TABLE 14-15

Common Nails Required for Rough Carpentry
(Based on Framing Members 16" o.c.)

Type Material	Kind of Nail	Pounds
Studding 16" o.c.	10d, 16d	10 per 1,000 BF
Joists and Rafters		
2" × 6"	16d	9 per 1,000"
2" × 8"	"	8 per 1,000"
2" × 10"	"	7 per 1,000"
2" × 12"	"	6 per 1,000"
Average house framing	8d, 10d, 16d, 20d	10–15 per 1,000"
Wood bridging 1" × 3"	8d	1 per 12 sets
Shiplap and boards for walls, floors and roof		
1" × 6"	8d	25–30 per 1,000 BF
1" × 8"	"	20–25 per 1,000"
1" × 10"	"	15–20 per 1,000"
Plywood (Plyscord) sheathing and subfloor		
5/16", 3/8", 1/2"	6d	10 per 1,000 Sq ft
5/8", ¾"	8d	20 per 1,000 " "
Insulating board (Composition)	2d galv.	10 per 1,000 " "
Furring on masonry	8d masonry	½ per 100 lin ft
Furring on studs	8d common	½ per 100 " "

Subfloor-underlayment systems utilize a special ply-wood with tongue and groove edges. The panels are 1⅛ in. to 1¼ in. thick. This is a single application of subfloor and underlayment combined that eliminates the labor cost of a separate underlayment. Blocking under panel edges is also eliminated because the panel edges are tongue and grooved. The system provides for spans up to 48 in. between supports consisting of built-up or laminated girders. (See Figure 14–15, and Table 14–20 on page 303.) Field gluing of subfloor-underlayment increases the stiffness of the floor. The glue is so strong that floor and joists behave like integral T-beam units.

Roof decking over which wood or asphalt shingles are to be laid should be 5/16 in. thick for rafters 16 in. o.c., and 3/8 in. thick for rafters 24 in. o.c. For slate, asbestos, cement, or tile shingles, ½ in. thick panels are recommended for rafters 16 in. o.c., and ⅝ in. thick for rafters 24 in. o.c.

TABLE 14–16

Approximate Milling and Cutting Waste*
in 1-inch Boards, to Be Added to Area

Item	Nominal Size, Inches	Laid Horizontal	Laid Diagonal
Shiplap	6	22%	27%
	8	21	26
	10	18	24
	12	15	21
Tongue & Groove	4	33	38
	6	22	27
	8	20	26
	10	18	23
	12	15	20
Square edge	4	19	24
	6	14	20
	8	15	20
	10	13	18
	12	12	17
Matched Solid Paneling Patterns	6	24	29
	8	24	29
	10	19	24
	12	17	22

*This includes 5 percent cutting and fitting waste which should be adjusted to the particular operation being considered.

EXTERIOR WALL SHEATHING

Plywood, fiberboard, waferboard, composite, and insulating panels are commonly used for exterior wall sheathing. While 4-by-8-inch panels are generally used, 2-by-8-inch and 4-by-10-inch panels are available. Manufacturers' specifications should be followed carefully.

TABLE 14–17*

APA Panel Subflooring(a)
(APA RATED SHEATHING)

Panel Span Rating (or Group Number)	Panel Thickness (In.)	Maximum Span (In.)	Nail Size & Type(g)	Nail Spacing (In.)	
				Supported Panel Edges	Intermediate Supports(f)
24/16	7/16	16	6d common	6	12
32/16	15/32, 1/2, 5/8	16(b)	8d common(c)	6	12
40/20	9/16, 19/32, 5/8, 3/4, 7/8	20(d)	8d common	6	12
48/24	23/32, 3/4, 7/8	24	8d common	6	12
1-1/8″ Groups 1 & 2(e)	1-1/8	48	10d common	6	6

(a) For subfloor recommendations under ceramic tile, refer to Table 11. For subfloor recommendations under gypsum concrete, contact manufacturer of floor topping.

(b) Span may be 24 inches if 3/4-inch wood strip flooring is installed at right angles to joists.

(c) 6d common nail permitted if panel is 1/2 inch or thinner.

(d) Span may be 24 inches if 3/4-inch wood strip flooring is installed at right angles to joists or if a minimum 1-1/2 inches of lightweight concrete is applied over panels.

(e) Check dealer for availability.

(f) Applicable building codes may require 10″ oc nail spacing at intermediate supports for floors.

(g) Other code-approved fasteners may be used.

*Reprinted with permission of The American Plywood Association.

TABLE 14–18*

APA Plywood Systems for Ceramic Tile Flooring

(Based on ANSI Standard A108 and recommendations of the Tile Council of America)

Joist Spacing (in.)	Minimum Panel Thickness (in.)		Tile Installation
	Subfloor[a]	Underlayment[b]	
Residential			
16	15/32	(d)	"Dry-Set" mortar; or latex – Portland Cement mortar
16	19/32	—	Cement mortar (3/4" – 1-1/4")
16	19/32	11/32	Organic adhesive
16	19/32	15/32[e]	Epoxy mortar
16	19/32 T&G[c][e]	—	Epoxy mortar
Commercial			
16	15/32	(d)	"Dry-Set" mortar; or latex – Portland Cement mortar
16	19/32	—	Cement mortar (3/4" – 1-1/4")
16	19/32	19/32[c][e]	Epoxy mortar

(a) APA RATED SHEATHING with subfloor Span Rating of 16" oc (15/32" panel) or 20" oc (19/32" panel). except as noted.

(b) APA Underlayment or sanded Exterior grade, except as noted

(c) APA RATED STURD-I-FLOOR with 20" oc Span Rating.

(d) Bond glass mesh mortar units to subfloor with latex – Portland Cement mortar, prior to spreading mortar for setting ceramic tile.

(e) Leave 1/4" space at panel ends and edges. trim panels as necessary to maintain end spacing and panel support on framing. Fill joints with epoxy mortar when it is spread for setting tile. With single-layer residential floors, use solid lumber blocking or framing under all panel end and edge joints (including T&G joints).

(f) See Table 6, 9 or 10, as applicable, for panel fastening recommendations.

*Reprinted with permission of The American Plywood Association.

ESTIMATING ROUGH SURFACE MATERIALS

There is a difference of opinion among estimators as to whether window, door, and other openings should be deducted in full, in part, or not at all. We believe that the most accurate method is to deduct openings in full, add an appropriate cutting waste to the net square feet or board feet estimated, and also make an allowance for the extra labor working around the openings.

Roof Decking Materials

Tables 14–21 through Table 14–26 on pages 304–315 show the quantities of different roof decking materials for *gable* and *hip* roofs (including intersecting roofs) with pitches ⅛ to ½, and ground floor areas 200 sq. ft. to 3,000 sq. ft. Quantities may be read directly from the tables. Milling and cutting waste is included. Roof surface areas are shown in the first column to enable the reader to estimate other types of roof decking not shown in these tables.

Flooring and Subflooring Materials

Table 14–27 on pages 316–317 shows the quantities of different flooring and subflooring materials for areas of 200 sq. ft. to 3,000 sq. ft. Milling and cutting waste is included. For types of flooring and subflooring that are not shown in the table, use the Floor Area column and add the appropriate milling and cutting waste.

Exterior Wall Sheathing Materials

To calculate the quantity of exterior wall sheathing, multiply the length or the perimeter of the wall by the height. Deduct all door and window openings and add the applicable milling and cutting waste shown in Table 14–16 on page 293. The approximate milling and cutting waste for plywood, fiberboard, waferboard, and other panel-type sheathing will average about 10 percent. Remember to include the areas of gable ends and dormer sides.

Portable and Mechanical Nailers

Figures 14–13 and 14–14 on pages 298–301 show workers using pneumatic nailing tools in the framing of a building. Mechanical nailers also save considerable time in laying subflooring, roof decking, and applying sheathing.

ESTIMATING ROUGH CARPENTRY LABOR

The principal factors that affect labor production are discussed under "Estimating Labor," Chapter 10. Each of these should be reviewed and carefully considered when estimating carpentry labor.

Table 14–28 on pages 318–319 shows the approximate hours of labor to install or apply rough carpentry materials. The hours shown are the combined hours of a carpenter and helper. The ratio of carpenter hours to helper hours varies with the kind of work being done and the contractor's labor organization setup. Some operations require a laborer's or helper's assistance for efficiency as well as economy. In other situations, such as installing wood bridging or furring, a carpenter works more efficiently alone.

The reader is advised to consider all circumstances and use good judgment based on experience in apportioning the hours shown in Table 14–28 between carpenter and helper.

Figure 14–13
Pneumatic nailing tool being used in
framing a building.
Courtesy Senco Products, Inc., Cincinnati, Ohio 45244.

Figure 14–13
(continued)

Figure 14–14
Pneumatic nailing tools being used
to install bridging.
Courtesy Senco Products, Inc., Cincinnati, Ohio 45244.

Figure 14–14
(continued)

TABLE 14–19*

Recommended Minimum Fastening Schedule for APA Panel Roof Sheathing

Panel Thickness[b] (in.)	Nailing[c][d] Size	Spacing (in.) Panel Edges	Spacing (in.) Intermediate
5/16	6d	6	12
3/8	6d	6	12
7/16, 15/32, 1/2	6d	6	12
19/32, 5/8, 23/32, 3/4, 7/8, 1	8d	6	12[a]
1-1/8	8d or 10d	6	12[a]

(a) For spans 48 inches or greater, space nails 6 inches at all supports.

(b) For stapling asphalt shingles to 5/16-inch and thicker panels, use staples with a 3/4-inch minimum crown width and a 3/4-inch leg length. Space according to shingle manufacturer's recommendations.

(c) Use common smooth or deformed shank nails with panels to 1 inch thick. For 1-1/8-inch panels use 8d ring- or screw-shank or 10d common smooth-shank nails.

(d) Other code approved fasteners may be used.

*Reprinted with permission of The American Plywood Association.

APA RATED STURD-I-FLOOR 16, 20 and 24 oc

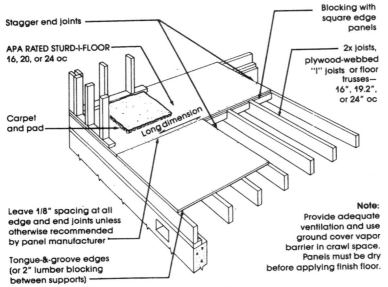

Stagger end joints

APA RATED STURD-I-FLOOR 16, 20, or 24 oc

Carpet and pad

Long dimension

Blocking with square edge panels

2x joists, plywood-webbed "I" joists or floor trusses— 16", 19.2", or 24" oc

Leave 1/8" spacing at all edge and end joints unless otherwise recommended by panel manufacturer

Tongue-&-groove edges (or 2" lumber blocking between supports)

Note: Provide adequate ventilation and use ground cover vapor barrier in crawl space. Panels must be dry before applying finish floor.

Figure 14–15

TABLE 14–20

APA RATED STURD-I-FLOOR[a]

Span Rating (Maximum Joist Spacing) (in.)	Panel Thickness[b] (in.)	Fastening: Glue-Nailed[c]			Fastening: Nailed-Only		
		Nail Size and Type	Spacing (in.)		Nail Size and Type	Spacing (in.)	
			Supported Panel Edges	Intermediate Supports		Supported Panel Edges	Intermediate Supports[g]
16	19/32, 5/8, 21/32	6d ring- or screw-shank[d]	12	12	6d ring- or screw-shank	6	12
20	19/32, 5/8, 23/32, 3/4	6d ring- or screw-shank[d]	12	12	6d ring- or screw-shank	6	12
24	11/16, 23/32, 3/4	6d ring- or screw-shank[d]	12	12	6d ring- or screw-shank	6	12
	7/8, 1	8d ring- or screw-shank[d]	6	12	8d ring- or screw-shank	6	12
48 (2-4-1)	1-1/8	8d ring- or screw-shank[e]	6	(f)	8d ring- or screw-shank[e]	6	(f)

(a) Special conditions may impose heavy traffic and concentrated loads that require construction in excess of the minimums shown. See page 23 for heavy duty floor recommendations.

(b) As indicated above, panels in a given thickness may be manufactured in more than one Span Rating. Panels with a Span Rating greater than the actual joist spacing may be substituted for panels of the same thickness with a Span Rating matching the actual joist spacing. For example, 19/32-inch-thick Sturd-I-Floor 20 oc may be substituted for 19/32-inch-thick Sturd-I-Floor 16 oc over joists 16 inches on center.

(c) Use only adhesives conforming to APA Specification AFG-01, applied in accordance with the manufacturer's recommendations. If non-veneered panels with sealed surfaces and edges are to be used, use only solvent-based glues; check with panel manufacturer.

(d) 8d common nails may be substituted if ring- or screw-shank nails are not available.

(e) 10d common nails may be substituted with 1-1/8-inch panels if supports are well seasoned.

(f) Space nails 6 inches for 48-inch spans and 12 inches for 32-inch spans.

(g) Applicable building codes may require 10" oc nail spacing at intermediate supports for floors.

Courtesy of The American Plywood Association

TABLE 14–21

Decking for Gable and Hip Roofs with Pitch of 1/8
(Includes Milling and Cutting Waste)

Ground Floor Area Sq. Ft.	Roof Surface Area Sq. Ft.	Shiplap 1" by 6" & 8" Board Ft.	Shiplap 1" by 10" Board Ft.	Tongue & Groove Boards 1" by 4" Board Ft.	Tongue & Groove Boards 1" by 6" & 8" Board Ft.	Tongue & Groove Boards 1" by 10" Board Ft.	Square Edge Boards 1" by 4" Board Ft.	Square Edge Boards 1" by 6" & 8" Board Ft.	Plywood Insulation & Particleboard Square Feet
200	206	252	244	274	252	244	246	236	226
300	309	378	366	411	378	366	369	354	339
400	412	504	488	548	504	488	492	472	452
500	515	630	610	685	630	610	615	590	565
600	618	756	632	822	756	732	738	708	678
700	721	882	854	959	882	854	861	826	791
800	824	1008	976	1096	1008	976	984	944	904
900	927	1134	1098	1233	1134	1098	1107	1062	1017
1000	1030	1260	1220	1370	1260	1220	1230	1180	1130
1100	1133	1386	1342	1507	1386	1342	1353	1298	1243
1200	1236	1512	1464	1644	1512	1464	1476	1416	1356

1300	1339	1638	1586	1781	1638	1586	1600	1534	1469
1400	1442	1764	1708	1918	1764	1708	1722	1652	1582
1500	1545	1890	1830	2055	1890	1830	1845	1770	1695
1600	1648	2016	1952	2192	2016	1952	1968	1888	1808
1700	1751	2142	2074	2329	2142	2074	2091	2006	1921
1800	1854	2268	2196	2466	2268	2196	2214	2124	2034
1900	1957	2394	2318	2603	2394	2318	2337	2242	2147
2000	2060	2520	2440	2740	2520	2440	2460	2360	2260
2100	2163	2646	2562	2877	2646	2562	2583	2478	2373
2200	2266	2772	2684	3014	2772	2684	2706	2596	2486
2300	2369	2898	2806	3151	2898	2806	2829	2714	2600
2400	2472	3024	2928	3288	3024	2928	2953	2832	2712
2500	2575	3150	3050	3425	3150	3050	3075	2950	2825
2600	2678	3276	3172	3562	3276	3172	3198	3068	2938
2700	2781	3402	3294	3699	3402	3294	3321	3186	3051
2800	2884	3528	3416	3836	3528	3416	3440	3304	3164
2900	2987	3654	3538	3973	3654	3538	3567	3422	3277
3000	3090	3780	3660	4110	3780	3660	3690	3540	3390

TABLE 14–22

Decking for Gable and Hip Roofs with Pitch of 1/16
(Includes Milling and Cutting Waste)

Ground Floor Area Sq. Ft.	Roof Surface Area Sq. Ft.	Shiplap 1" by 6" & 8" Board Ft.	Shiplap 1" by 10" Board Ft.	Tongue & Groove Boards 1" by 4" Board Ft.	Tongue & Groove Boards 1" by 6" & 8" Board Ft.	Tongue & Groove Boards 1" by 10" Board Ft.	Square Edge Boards 1" by 4" Board Ft.	Square Edge Boards 1" by 6" & 8" Board Ft.	Plywood Insulation & Particleboard Square Feet
200	212	258	250	282	258	250	252	244	234
300	318	387	375	423	387	375	378	366	351
400	424	516	500	564	516	500	504	488	468
500	530	645	625	705	645	625	630	610	585
600	636	774	750	846	774	750	756	732	702
700	742	900	875	987	900	875	882	854	819
800	848	1030	1000	1128	1030	1000	1008	976	936
900	954	1160	1125	1269	1160	1125	1134	1098	1053
1000	1060	1290	1250	1410	1290	1250	1260	1220	1170
1100	1166	1420	1375	1551	1420	1375	1386	1342	1287
1200	1272	1551	1500	1692	1550	1500	1512	1464	1404
1300	1378	1680	1625	1833	1680	1625	1638	1586	1521
1400	1484	1810	1750	1974	1810	1750	1764	1708	1638
1500	1590	1935	1875	2115	1935	1875	1890	1830	1755

Ground Floor Area Sq. Ft.	Roof Surface Area Sq. Ft.	Shiplap 1" by		Tongue & Groove Boards 1" by			Square Edge Boards 1" by		Plywood Insulation & Particleboard Square Feet
		6" & 8" Board Ft.	10" Board Ft.	4" Board Ft.	6" & 8" Board Ft.	10" Board Ft.	4" Board Ft.	6" & 8" Board Ft.	
1600	1696	2064	2000	2256	2064	2000	2016	1952	1872
1700	1802	2195	2125	2397	2195	2125	2142	2074	1989
1800	1908	2322	2250	2538	2322	2250	2268	2196	2106
1900	2014	2450	2375	2679	2450	2375	2394	2318	2223
2000	2120	2580	2500	2820	2580	2500	2520	2440	2340
2100	2226	2710	2625	2961	2710	2625	2646	2562	2457
2200	2332	2840	2750	3102	2840	2750	2772	2684	2574
2300	2438	2970	2875	3243	2970	2875	2898	2806	2691
2400	2544	3100	3000	3384	3100	3000	3024	2928	2808
2500	2650	3225	3125	3525	3225	3125	3150	3050	2925
2600	2756	3354	3250	3666	3354	3250	3276	3172	3042
2700	2862	3483	3375	3807	3483	3375	3402	3294	3159
2800	2968	3610	3500	3948	3610	3500	3528	3416	3276
2900	3074	3740	3625	4089	3740	3625	3654	3528	3393
3000	3180	3870	3750	4230	3870	3750	3780	3660	3510

TABLE 14–23
Decking for Gable and Hip Roofs with Pitch of 5/24
(Includes Milling and Cutting Waste)

Ground Floor Area	Roof Surface Area	Shiplap 1" by		Tongue & Groove Boards 1" by			Square Edge Boards 1" by		Plywood Insulation & Particleboard
		6" & 8"	10"	4"	6" & 8"	10"	4"	6" & 8"	
Sq. Ft.	Sq. Ft.	Board Ft.	Board Ft.	Board Ft.	Board Ft.	Board Ft.	Board Ft.	Board Ft.	Square Feet
200	218	266	258	290	266	258	260	250	240
300	327	400	387	435	400	387	390	375	360
400	426	532	516	580	532	516	520	500	480
500	546	665	645	725	665	645	650	625	600
600	654	800	774	870	800	774	780	750	720
700	763	930	900	1015	930	900	910	875	840
800	872	1064	1030	1160	1064	1030	1040	1000	960
900	981	1200	1160	1305	1200	1160	1170	1125	1080
1000	1090	1330	1290	1450	1330	1290	1300	1250	1200
1100	1199	1463	1420	1595	1463	1420	1430	1375	1320
1200	1308	1600	1550	1740	1600	1550	1560	1500	1440
1300	1417	1730	1680	1885	1730	1680	1690	1625	1560
1400	1526	1860	1810	2030	1860	1810	1820	1750	1680
1500	1635	2000	1935	2175	2000	1935	1950	1875	1800

Ground Floor Area Sq. Ft.	Roof Surface Area Sq. Ft.	Shiplap 1" by		Tongue & Groove Boards 1" by			Square Edge Boards 1" by		Plywood Insulation & Particleboard
		6" & 8" Board Ft.	10" Board Ft.	4" Board Ft.	6" & 8" Board Ft.	10" Board Ft.	4" Board Ft.	6" & 8" Board Ft.	Square Feet
1600	1744	2130	2064	2320	2130	2064	2080	2000	1920
1700	1853	2260	2195	2465	2260	2195	2210	2125	2040
1800	1962	2400	2322	2610	2400	2322	2340	2250	2160
1900	2071	2530	2450	2755	2530	2450	2470	2375	2280
2000	2180	2660	2580	2900	2660	2580	2600	2500	2400
2100	2289	2800	2710	3045	2800	2710	2730	2625	2520
2200	2398	2930	2840	3190	2930	2840	2860	2750	2640
2300	2507	2060	2970	3335	3060	2970	2990	2875	2760
2400	2616	3200	3100	3480	3200	3100	3120	3000	2880
2500	2725	3325	3225	3625	3325	3225	3250	3125	3000
2600	2834	3460	3354	3770	3460	3354	3380	3250	3120
2700	2943	3590	3483.	3915	3590	3483	3510	3375	3240
2800	3052	3724	3610	4060	3724	3610	3640	3500	3360
2900	3161	3860	3740	4205	3860	3740	3770	3625	3480
3000	3270	3990	3870	4350	3990	3870	3900	3750	3600

TABLE 14-24
Decking for Gable and Hip Roofs with Pitch of 1/4
(Includes Milling and Cutting Waste)

Ground Floor Area Sq. Ft.	Roof Surface Area Sq. Ft.	Shiplap 1" by		Tongue & Groove Boards 1" by			Square Edge Boards 1" by		Plywood Insulation & Particleboard
		6" & 8" Board Ft.	10" Board Ft.	4" Board Ft.	6" & 8" Board Ft.	10" Board Ft.	4" Board Ft.	6" & 8" Board Ft.	Square Feet
200	224	274	264	298	274	264	266	258	246
300	336	411	396	447	411	396	400	387	369
400	448	548	528	600	548	528	532	516	492
500	560	685	660	745	685	660	665	645	615
600	672	822	792	894	822	792	800	774	738
700	784	959	924	1043	959	924	930	900	861
800	896	1096	1056	1192	1096	1056	1064	1030	984
900	1008	1233	1188	1340	1233	1188	1200	1160	1107
1000	1120	1370	1320	1490	1370	1320	1330	1290	1230
1100	1232	1507	1452	1639	1507	1452	1463	1420	1353
1200	1344	1644	1584	1788	1644	1584	1600	1550	1476
1300	1456	1781	1716	1937	1781	1716	1730	1680	1600
1400	1568	1918	1848	2086	1918	1848	1860	1810	1722
1500	1680	2055	1980	2235	2055	1980	2000	1935	1845
1600	1792	2192	2112	2384	2192	2112	2130	2064	1968

Ground Floor Area Sq. Ft.	Roof Surface Area Sq. Ft.	Shiplap 1" by		Tongue & Groove Boards 1" by			Square Edge Boards 1" by		Plywood Insulation & Particleboard
		6" & 8" Board Ft.	10" Board Ft.	4" Board Ft.	6" & 8" Board Ft.	10" Board Ft.	4" Board Ft.	6" & 8" Board Ft.	Square Feet
1700	1904	2329	2244	2533	2329	2244	2260	2195	2091
1800	2016	2466	2376	2682	2466	2376	2400	2322	2214
1900	2128	2603	2508	2830	2603	2508	2530	2450	2337
2000	2240	2740	2640	2980	2740	2640	2660	2580	2460
2100	2352	2877	2772	3130	2877	2772	2800	2710	2583
2200	2464	3014	2904	3278	3014	2904	2930	2840	2706
2300	2576	3151	3036	3427	3151	3036	3060	2970	2829
2400	2688	3288	3168	3576	3288	3168	3200	3100	2952
2500	2800	3425	3300	3725	3425	3300	3325	3225	3075
2600	2912	3562	3432	3874	3562	3432	3460	3354	3198
2700	3024	3700	3564	4023	3700	3564	3590	3483	3321
2800	3136	3836	3696	4172	3836	3696	3724	3610	3440
2900	3248	3973	3828	4321	3973	3828	3860	3740	3567
3000	3360	4110	3960	4470	4110	3960	3990	3870	3690

TABLE 14-25

Decking for Gable and Hip Roofs with Pitch of 1/3
(Includes Milling and Cutting Waste)

Ground Floor Area Sq. Ft.	Roof Surface Area Sq. Ft.	Shiplap 1" by 6" & 8" Board Ft.	Shiplap 1" by 10" Board Ft.	Tongue & Groove Boards 1" by 4" Board Ft.	Tongue & Groove Boards 1" by 6" & 10" Board Ft.	Tongue & Groove Boards 1" by 10" Board Ft.	Square Edge Boards 1" by 4" Board Ft.	Square Edge Boards 1" by 6" & 8" Board Ft.	Plywood, Insulation & Particleboard Square Feet
200	240	292	284	320	292	284	286	276	264
300	360	438	426	480	438	426	429	414	396
400	480	584	569	640	584	568	572	552	528
500	600	730	710	800	730	710	715	690	660
600	720	876	852	960	876	852	858	828	792
700	840	1022	994	1120	1022	994	1000	966	924
800	960	1170	1136	1280	1170	1136	1144	1104	1056
900	1080	1314	1278	1440	1314	1278	1287	1242	1186
1000	1200	1460	1420	1600	1460	1420	1430	1380	1320
1100	1320	1610	1562	1760	1610	1562	1573	1518	1452
1200	1440	1752	1704	1920	1752	1704	1716	1656	1584
1300	1560	1900	1846	2080	1900	1846	1859	1794	1716
1400	1680	2044	1990	2240	2044	1990	2000	1932	1848
1500	1800	2190	2130	2400	2190	2130	2145	2070	1980
1600	1920	2336	2272	2560	2336	2272	2288	2208	2112

| Ground Floor Area Sq. Ft. | Roof Surface Area Sq. Ft. | Shiplap 1" by | | Tongue & Groove Boards 1" by | | | Square Edge Boards 1" by | | Plywood Insulation & Particleboard |
		6" & 8" Board Ft.	10" Board Ft.	4" Board Ft.	6" & 8" Board Ft.	10" Board Ft.	4" Board Ft.	6" & 8" Board Ft.	Square Feet
1700	2040	2480	2414	2720	2480	2414	2430	2346	2244
1800	2160	2630	2556	2880	2630	2556	2574	2484	2376
1900	2280	2774	2700	3040	2774	2700	2717	2622	2508
2000	2400	2920	2840	3200	2920	2840	2860	2760	2640
2100	2520	3066	2982	3360	3066	2982	3000	2898	2772
2200	2740	3212	3124	3520	3212	3124	3146	3036	2904
2300	2760	3358	3266	3680	3358	3266	3290	3174	3036
2400	2880	3504	3410	3840	3504	3410	3430	3312	3168
2500	3000	3650	3550	4000	3650	3550	3575	3450	3300
2600	3120	3800	3692	4160	3800	3692	3720	3588	3432
2700	3240	3942	3834	4320	3942	3834	3860	3726	3564
2800	3360	4090	3976	4480	4090	3976	4000	3864	3696
2900	3480	4234	4118	4640	4234	4118	4150	4000	3828
3000	3600	4380	4260	4800	4380	4260	4290	4140	3960

TABLE 14-26

Decking for Gable and Hip Roofs with Pitch of 1/2
(Includes Milling and Cutting Waste)

Ground Floor Area Sq. Ft.	Roof Surface Area Sq. Ft.	Shiplap 1" by		Tongue & Groove Boards 1" by			Square Edge Boards 1" by		Plywood, Insulation & Particleboard Square Feet
		6" & 8" Board Ft.	10" Board Ft.	4" Board Ft.	6" & 8" Board Ft.	10" Board Ft.	4" Board Ft.	6" & 8" Board Ft.	
200	284	346	336	378	346	336	338	326	312
300	426	520	504	567	520	504	507	489	468
400	568	690	672	756	690	672	676	652	624
500	710	865	840	945	865	840	845	815	780
600	852	1040	1008	1134	1040	1008	1014	978	936
700	994	1210	1176	1323	1210	1176	1183	1141	1092
800	1136	1385	1344	1512	1385	1344	1352	1304	1248
900	1278	1560	1512	1700	1560	1512	1521	1467	1404
1000	1420	1730	1680	1890	1730	1680	1690	1630	1560
1100	1562	1900	1848	2080	1900	1848	1859	1793	1716
1200	1704	2076	2016	2268	2076	2016	2028	1956	1872
1300	1846	2250	2184	2457	2250	2184	2200	2119	2028
1400	1990	2422	2352	2646	2422	2352	2370	2282	2184
1500	2130	2595	2520	2835	2595	2520	2535	2445	2340
1600	2272	2770	2688	3024	2770	2688	2705	2608	2500

Ground Floor Area Sq. Ft.	Roof Surface Area Sq. Ft.	Shiplap 1" by 6" & 8" Board Ft.	Shiplap 1" by 10" Board Ft.	Tongue & Groove Boards 1" by 4" Board Ft.	Tongue & Groove Boards 1" by 6" & 8" Board Ft.	Tongue & Groove Boards 1" by 10" Board Ft.	Square Edge Boards 1" by 4" Board Ft.	Square Edge Boards 1" by 6" & 8" Board Ft.	Plywood Insulation & Particleboard Square Feet
1700	2414	2940	2856	3213	2940	2856	2875	2771	2652
1800	2556	3115	3024	3400	3115	3024	3040	2934	2810
1900	2700	3290	3192	3590	3290	3192	3210	3097	2964
2000	2840	3460	3360	3780	3460	3360	3380	3260	3120
2100	2982	3633	3528	3970	3633	3528	3550	3423	3276
2200	3124	3810	3696	4158	3810	3696	3720	3586	3432
2300	3266	3980	3864	4350	3980	3864	3890	3749	3590
2400	3410	4150	4032	4536	4150	4032	4056	3912	3744
2500	3550	4325	4200	4725	4325	4200	4225	4075	3900
2600	3692	4500	4368	4914	4500	4368	4395	4238	4056
2700	3834	4670	4536	5100	4670	4536	4565	4400	4212
2800	3976	4845	4704	5292	4845	4704	4730	4564	4368
2900	4118	5020	4872	5480	5020	4872	4900	4727	4524
3000	4260	5190	5040	5670	5190	5040	5070	4890	4680

TABLE 14–27
Rough and Finish Floor Materials, Including Milling and Cutting Waste. Quantities may be read directly for all room and floor areas 200 square feet to 3,000 square feet. Interpolate for areas in between.

Floor Area Sq. Ft.	Rough Flooring, Incl. Milling and Cutting Waste				Fin. Floor Incl. Milling & Cutting Waste	
	Shiplap & Tongue & Groove Bds. 1" × 6", 1" × 8"	Square Edge Boards		Plywood, Particle-Board, etc.	1" × 3"	1" × 4"
	Board Ft.	1" × 4" Board Ft.	1" × 6" Board Ft.	Square Ft.	Bd. Ft.	Bd. Ft.
200	242	238	228	220	276	258
300	363	357	342	330	414	387
400	484	476	456	440	552	516
500	605	595	570	550	690	645
600	726	714	684	660	828	774
700	847	833	800	770	966	900
800	968	952	912	880	1104	1030
900	1089	1071	1026	990	1242	1160
1000	1210	1190	1140	1100	1380	1290
1100	1331	1309	1254	1210	1518	1420
1200	1452	1428	1368	1320	1656	1550
1300	1573	1547	1482	1430	1794	1680
1400	1694	1666	1600	1540	1932	1800
1500	1815	1785	1710	1650	2070	1935

| Floor Area Sq. Ft. | Rough Flooring, Incl. Milling and Cutting Waste | | | | Fin. Floor Incl. Milling & Cutting Waste | |
	Shiplap & Tongue & Groove Bds. 1" × 6", 1" × 8" Board Ft.	Square Edge Boards 1" × 4" Board Ft.	Square Edge Boards 1" × 6" Board Ft.	Plywood, Particle-Board, etc. Square Ft.	1" × 3" Bd. Ft.	1" × 4" Bd. Ft.
1600	1936	1904	1824	1760	2208	2064
1700	2057	2023	1938	1870	2346	2195
1800	2178	2142	2052	1980	2484	2322
1900	2300	2261	2166	2090	2622	2450
2000	2420	2380	2280	2200	2760	2580
2100	2541	2500	2400	2310	2898	2710
2200	2662	2618	2500	2420	3036	2840
2300	2783	2737	2622	2530	3174	2970
2400	2904	2856	2736	2640	3312	3100
2500	3025	2975	2850	2750	3450	3225
2600	3146	3094	2964	2860	3580	3354
2700	3267	3213	3078	2970	3726	3483
2800	3388	3332	3200	3080	3864	3610
2900	3509	3450	3300	3190	4000	3740
3000	3630	3570	3420	3300	4140	3870

TABLE 14–28

Approximate Hours of Labor to Install
Rough Carpentry

Kind of Work	Hours Per 1,000 BF
Sills and plates, bolted and grouted	20–24
Exterior stud walls, 2″ × 4″, 2″ × 6″, 16″ o.c.	20–22
Partition stud walls, 2″ × 4″, 2″ × 6″, 16″ o.c.	18–24
Floor joists, 2″ × 6″, 2″ × 8″, 16″ o.c.	18–22
Floor joists, 2″ × 10″, 2″ × 12″, 16″ o.c.	16–20
Girders, built-up from 2″ stock	20–24
Rafters, 2″ × 6″, 2″ × 8″, plain gable roofs	22–30
Rafters, 2″ × 6″, 2″ × 8″, plain hip roofs	30–36
Rafters, 2″ × 6″, 2″ × 8″, flat roofs	26–30
Average hours to completely frame a dwelling	20–30
Exterior wall sheathing, 1″ × 6″, 1″ × 8″, square edge	16–18
Subflooring, 1″ boards, unmatched	15–20
Subflooring, 1″ shiplap or tongue and groove	16–20
Subflooring laid diagonally	18–22
Roof decking, flat roof, 1″ boards unmatched	15–20
Roof decking, flat roof, 1″ boards matched	16–18
Roof decking, gable roof, 1″ boards matched	20–22
Roof decking, hip or cut up roof, 1″ boards matched	22–26

TABLE 14–28 (continued)

Kind of Work	Hours Per 1,000 Sq. Ft.
Plywood subfloor, 5/8" to 3/4" tongue and groove	12–14
Plywood subfloor-underlayment (2-4-1), 1 1/8"	22–24
Plywood sheathing, 1/2" to 3/4"	14–16
Plywood roof decking, flat roof, 5/8"	12–14
Plywood roof decking, gable roof, 5/8"	14–16
Plywood roof decking, cut up roof, 5/8"	16–18
Insulating (composition) sheathing, 1/2"	12–20
Insulating (composition) roof decking, 1 1/2"	20–30
Bridging, wood, 1" × 3", cut and nailed	1 hour per 10 sets
Bridging, wood, 1× 3", redicut	1 hour per 15 sets
Bridging, wood, solid, 2" × 10", 2" × 12"	1 hour per 10 pieces
Bridging, metal on wood joists	1 hour per 20 sets
Furring on studding, including shimming	2–3 hours per 100 lin. ft.
Furring on masonry, including shimming	4–5 hours per 100 lin. ft.

SELECTED REFERENCES

Carpentry Estimating, W. P. Jackson, Craftsman Book Company, Carlsbad, CA, 1987.

Roof Framing, Marshall Gross, Craftsman Book Company, Carlsbad, CA, 1984.

Wood Engineering and Construction Handbook, Keith Faherty and Thomas Williamson, McGraw-Hill Publishing Company, New York, 1999.

CHAPTER 15

FINISH CARPENTRY

While there is a general agreement among contractors on what constitutes "rough" and what constitutes "finish" carpentry, there are some types of work that could be placed in either one or the other classification. For purposes of convenience and consistency, all carpentry that requires finished materials and perhaps greater skill in installation than rough carpentry will be classified here as finish carpentry. Such items include interior and exterior trim, doors and windows including frames, screens, wood siding, stair building, cabinet work, and finish floors. Exterior finish carpentry will include such items as cornice enclosures.

FINISH WOOD FLOORING

The top layer of flooring, the wearing surface, is called the finish flooring. Finish flooring may be pine, oak, maple, or beech, depending on appearance and quality desired and whether the project is residential or commercial. There are different grades of each kind of wood flooring, so it is important when estimating costs to be certain of the kind of wood, the size specified, and the quality or grade.

It should be noted that wood floors can be laid over concrete slabs satisfactorily. Proper installation meets FHA requirements and has been tested by the National Oak Manufacturers' Association. A double layer of 1-by-2-inch wood sleepers are nailed together with a moisture barrier of 4 mil polyethylene film between them. The bottom sleeper layer is secured to the slab by mastic and concrete nails. The strip hardwood flooring is nailed at right angles to the sleepers, with one nail at each bearing point. The slab, when poured, should have a 4 mil or 6 mil polyethylene over the base or fill.

In building with wood joist construction, finish wood flooring is laid over subflooring.

Nails

Table 15–1 shows a nail schedule for finish wood flooring. The type of nail conforms to the recommendations of the National Oak Flooring Manufacturers' Association.

Kinds of Flooring

Strip Flooring is one of the most popular styles of wood flooring. It comes in various widths but the nominal 1-by-

TABLE 15–1
Nail Schedule For Finishing Flooring

Nominal Size in Inches	Finish Size in Inches	Kind and Size Nail	Lbs of Nails Per 1,000 BF
1 × 2	25/32 × 1¼	8d cut	65–70*
1 × 3	25/32 × 2¼	8d cut	60–65
1 × 4	25/32 × 3¼	8d cut	30–40
1 × 2	3/8 × 1½	4d cut	18–20
1 × 2½	3/8 × 2	4d cut	16–18

*Note: Machine driven barb fasteners are acceptable, follow manufacturer's recommendations.

3-inch size (actual size: ¾ in. by 2¼ in.) is the most widely used. It is tongue and grooved, side matched and end matched, and is available in red or white oak. Strip flooring is also available in maple, beech, and birch. These woods are very dense and hard and are graded under the Maple Flooring Manufacturers' Association and the National Oak Flooring Manufacturers' Association. Southern pine strip flooring is available in *flat grain* or *edge grain*, side and end matched, or end matched only. Grading is under the rules of the Southern Pine Inspection Bureau. Douglas fir strip flooring is also manufactured in *edge grain* and *flat grain*. Figure 15–1 shows three types of strip flooring.

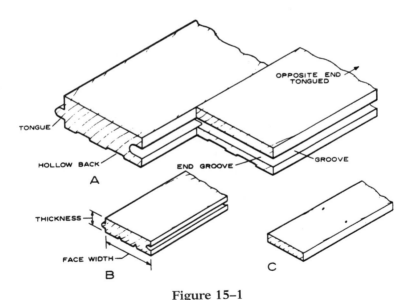

Figure 15–1
Types of strip flooring
A: Side and end matched—²⁵⁄₃₂ inch
B: Thin flooring strips—matched
C: Thin flooring strips—square edged

Prefinished strip flooring is finished, buffed, and polished at the factory. Its advantage is that it speeds up construction; a disadvantage is that great care is needed in laying the flooring to avoid hammer and nail marks.

Wood block flooring is made in several patterns ranging in size from 4 in. by 4 in. to 9 in. by 9 in. and larger. Thicknesses vary by type from $25/32$ in. for laminated blocking on plywood block tile to $\frac{1}{2}$ in. for stabilized veneer. Many wood block floors are factory finished and require only waxing after installation. Wood block floors are usually laid in adhesive over wood subflooring or concrete slab. This type of flooring is usually delivered in the exact quantities needed, making waste a minimum.

Parquet flooring, also a wood block flooring, is available in 6-in., 8-in., or 12-in. square patterns. The woods are hardwoods such as oak, maple, cherry, walnut, hickory, or exotic woods like teak.

Determining Quantities of Flooring

Table 15–2 shows the percentage of milling and cutting waste to be added to the measured floor area to obtain the quantity required to cover the area. Refer also to Table 14–27 which shows the quantity of 1-by-3-inch and 1-by-4-inch (nominal) finish flooring for areas 200 to 3,000 square feet.

Estimating Labor to Lay Wood Flooring

Table 15–3 shows the approximate average hours of labor to lay different kinds and sizes of wood flooring. The hours are for quality work done by carpenters—no hammer or nail marks. When subcontractors with experienced floor layers do the work, the hours in the Table can be reasonably reduced. All of the factors that affect labor production, as discussed in Chapter 10, should be given proper consideration. Also, the hours in Table 15–3 are total man-hours. Where carpenter helpers or laborers are used, the hours that are shown should be prorated.

TABLE 15–2

Estimating Flooring Quantities

Nominal Size in Inches	Actual Size in Inches	Milling + 5% Cutting Waste	Multiply Area by	Floor Needed Per 1,000 BF
1 × 2	25/32 × 1½	55%	1.55	1,550
2 × 2½	" × 2	42	1.42	1,420
1 × 3	" × 2¼	38	1.38	1,380
1 × 4	" × 3¼	29	1.29	1,290
1 × 2	3/8 × 1½	38	1.38	1,380
1 × 2½	" × 2	30	1.30	1,300
	½ × 1½	38	1.38	1,380
	" × 2	30	1.30	1,300

TABLE 15–3

Approximate Hours Labor to Lay Wood Floors

Size and Type of Flooring	Man-Hours Per 100 Sq. Ft.	Man-Hours Per 100 BF
Strip Flooring		
25/32" × 3 1/4" softwood	2.5–3.0	3.3–4.0
25/32" × 2 1/4" softwood	3.0–3.5	4.0–4.8
25/32" × 1 1/2" oak, maple, birch		
Commercial	2.5–3.0	3.9–4.6
Residential	4.0–4.5	6.2–7.0
25/32" × 2 1/4" oak, maple, birch	4.5–5.0	6.2–7.0
Parquet		
9" × 9" × 5/16", prefinished, in mastic	3.0–4.0	——

Sanding Wood Floors

After finish floors are laid, they are sanded by machine before shellac, varnish, or other finishes are applied. The work may be done by floor layers or by subcontractors. The cost will vary with the kind of flooring, the size of the rooms, and total area involved.

WINDOWS, STORM SASH, AND SCREENS

There is a variety of styles of windows available for residential and commercial buildings. Figure 15–2 shows several typical residential windows. Each type can be obtained in wood or aluminum. Casement, gliding, and stationary windows are also available in steel. Windows are generally sold as complete units, frequently with hardware and screens. Prices vary by type, size, material, quality, and also by locality.

Table 15–4 shows the approximate hours of labor to install window parts, complete units, screens, and storm sash. Where difficult or unusual working conditions exist, the labor hours should be adjusted upward.

DOORS AND SCREENS

Component parts of doors are the jambs (called bucks in exterior doors), the door, door stops, casings, hardware, and for some exterior doors, a saddle. The parts may be obtained separately, though doors as complete units with all parts including hardware are available.

Doors should be described, when estimating, as to kind of wood, style and size, thickness, and type of casings. Hardware for doors varies in cost depending on type, material, and manufacturer. Table 15–5 shows the

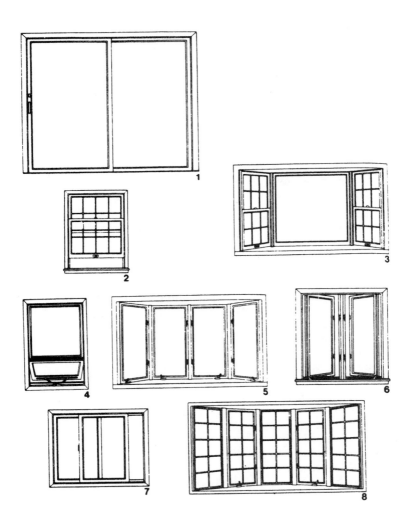

Typical window styles readily available in residential construction. 1. Gliding door. 2. Double-hung window. 3. Double-hung angle bay window. 4. Awning window with fixed sash above. 5. Casement angle bay window. 6. Casement window. 7. Gliding window. 8. Bow window. Windows and gliding door shown are available in low-maintenance Perma-Shield line from Andersen Corporation, Bayport, MN 55003.

Figure 15–12

approximate hours of labor to install parts of a door or entire units. Also shown are the hours required to fit and hang garage doors, screen doors, and storm doors and to weather-strip exterior doors.

TABLE 15-4

Average Hours Required to Install Windows, Storm Sash, and Screens

Type of Work Done	Hours Required
Assemble wooden frames on the job	1
Install single wooden frame in frame building . .	0.5
Install single wooden frame in masonry building .	1
Additional time for double frames with mullion .	.5
Fit and hang wood sash complete (pair)	1.5
Fit and hang casement sash complete (each) . . .	1.5
Interior window trim installed—softwood	1
Interior window trim installed—hardwood. . . .	1.5
Complete window, wood frame, and sash—factory assembled	2
Complete window, wood frame, and sash—knockdown.	4
Complete basement window.	0.5
Wood window screens.	1
Wood storm windows	1
Wood shutters—stationary (per pair)	2
Wood shutters—hinged (per pair)	3
Complete D.H. aluminum window	2
Complete aluminum window with sliding sash . .	2
Picture window, wood frame, trim, and sash . . .	4
(Thermopane same labor plus cost of material)	
Aluminum combination screens and storm sash	2
Installing factory steel sash (each)	3
Glazing factory steel sash	5
Weather stripping D.H. windows.	2
Placing back band per window	0.75

TABLE 15–5

Hours Required to Install Interior and Exterior Doors and Their Parts

Type of Work Done	Hours Required
Set interior or exterior jambs (frames)	.75
Install mortise lock set	2.0
Cylinder lock on front door	.75
Fit and hang interior door—softwood	1.5
Fit and hang interior door—hardwood	2
Put on stops and casings (two sides)—softwood	1
Put on stops and casings (two sides)—hardwood	1.25
Install frame, fit and hang door trim complete—interior	4
Install frame, fit and hang door trim complete—exterior	5
Install one pre-assembled interior door complete	1.5
Installing several units of pre-assembled doors	1.
Make and hang 3' × 7' batten door	3
Garage swinging doors 8' × 8' fit and hang per pair	4
Garage overhead door 8' × 8' × 13/8"—hang complete	6
Garage overhead door 15' × 7' × 1¼ —hang complete	8
Storm door—fit and hang complete, or combination storm and screen	2
Screen door—fit and hang complete	2
Weather-stripping exterior door	2
Trimmed opening, set jams and trim 7' × 8'	2
Placing back-band per one side door opening	.75
Placing back-band per one side trimmed opening	1

MISCELLANEOUS INTERIOR TRIM

There are several items of interior trim, the materials for which are sold by the lineal foot. These items consist mainly of different sizes and styles of baseboards, base moldings, floor moldings, picture moldings, chair and plate rails, and ceiling moldings.

Baseboards may be one of the standard single-member molded stock styles, ranging from 2 in. to 4 in. high, or they may be three-member baseboards. This type is made up of a 4-in. or 6-in. board, nominally 1 in. thick. A quarter round floor molding, and a wall molding on top of the baseboard complete the three members. A two-member baseboard omits the wall molding.

Figure 15–3
Popular Door Styles
Courtesy of National Wood Windows and Door Association

TABLE 15–6

Hours Required to Install 100 Lineal Feet of Miscellaneous Items of Interior Trim

Type of Work Done		Kind of Wood	Hours Required Per 100 Lineal Feet	Lineal Feet Per Hour
Baseboard	1 member	Softwood	6	17
Baseboard	2 member	Softwood	7	14
Baseboard	3 member	Softwood	8	13
Baseboard	3 member	Hardwood	10	10
Chair rail	1 member	Softwood	5	20
Chair rail	1 member	Hardwood	6	17
Plate rail	2 member	Hardwood	12	8
Plate rail	3 member	Hardwood	16	6
Picture molding		Softwood	5	20
Ceiling molding		Softwood	6	17
Wood paneling strips		7/8" Hardwood	8	13

Plate rails, sometimes installed in residential properties, consist of two, and occasionally three members.

Ceiling moldings, usually a cove type, and *picture moldings* are one-member units.

Table 15–6 shows the approximate hours of labor to install 100 lineal feet of the different types of interior trim.

INTERIOR STAIRS AND STAIRCASES

The construction of stairs and stairways with platform landings and winding turns, particularly main staircases in fine residential buildings, requires skillful and experienced carpentry. Because stair building is such a specialized art, most stairs or their parts, are made in millwork plants and delivered to the job where they are installed by the contractor's carpenters or by subcontractors. Most of the stair manufacturers provide catalogs that show various designs including balusters, handrails, and hardware.

TABLE 15-7

Hours Required to Cut and Erect Staircases

Type of Work Done	Hours Required Per Tread
Cutting out open stringer	.3
Housing out box stringer	.8
Erecting box or open stringers	.8
Placing treads or risers on open stringer	.1
Placing treads or risers on housed stringer	.4
Placing rail and balusters	1.0
*Complete erection of open staircase	1.5
*Complete erection of boxed stairs	2.5
Erection of plank cellar stairs without risers	1.0

*Add handrail and balusters as required.

The main parts of a staircase are the *treads,* the vertical *risers,* and the *stringers,* which are the side pieces to which treads and risers are attached. When one or both sides of the staircase is open, a balustrade, or handrail, and balusters are installed.

An open, or plain, stringer is one where the stringers are notched out and the treads and risers are nailed to them. Rough, open-string basement stairs are an example.

Box stairs have their stringers routed out, or housed out. The treads and risers are slid into place from the back side and then wedged and glued with hardwood wedges.

Because of the numerous architectural designs of staircases and the different kinds of wood used in their construction, it is recommended that local stair builders be consulted for accurate estimates. Table 15-7 shows the approximate hours of labor required to cut, assemble, and install the various parts of staircases.

EXTERIOR WOOD TRIM

Most items of exterior trim are both ornamental and functional. They include *water tables, corner boards, wood gutters,* and *cornice trim.* Quantities of trim are estimated by the lineal foot.

Water table, the baseboard at which the siding sometimes begins, is at the base of the building where the foundation ends and the frame wall begins. It is a projection permitting rain water that runs down the outside surface of the building to drip away from the foundation. It usually consists of two parts: a 1-by-6-in. or 1-by-8-in. board and a beveled cap or top called a *drip cap.*

Corner boards, as the name implies, attach vertically to the outside corners of a building, forming the finish up against which the exterior horizontal siding butts. The stock is usually 1½ in. thick and 3 in. or 4 in. wide. Frequently a 4-in. board laps a 3-in. board at the corner, forming a square edge. Another type of corner board is formed by two equally wide boards at the corner with a quarter round molding where they meet, forming a rounded corner.

Wood gutters of fir or cedar are popular in some parts of the country in place of metal or plastic. They are generally less expensive and, for some types of construction, are better suited architecturally. They are available at most building supply houses. Wood gutters are attached with galvanized nails or screws.

Wood cornices are formed where the roof meets the side walls of a building. There are many different architectural designs, but most fall within three classifications:

- Closed or flush cornices
- Open cornices
- Box cornices

A cornice that has no overhang, the rafters being cut off flush with the exterior wall sheathing, is called a *closed* or *flush cornice.* The members consist of a frieze board and a crown molding. See Figure 15–4. A cornice with open or exposed rafters projecting is called an *open cornice.* See Figure 15–5.

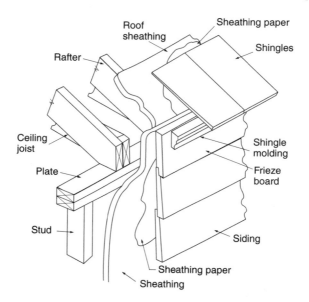

Figure 15–4
Construction details of a "closed cornice"

There are two kinds of *box cornice*. Both types are boxed in with a frieze board on the face of the building, a soffit board horizontally extending from the building to the edge of the extending rafters, and a fascia board closing the box. When a box cornice has a wide overhang, it is necessary to nail cleats from the end of the rafter to the wall of the building for the soffit boards. The cleats are called *lookouts*. Illustrations of box cornices with and without lookouts are shown in Figures 15–6 and 15–7.

Verge or barge board is the member used to finish off the gable end of a building. It may be planted on top of the siding or nailed first to the sheathing and have the siding butt to it.

Table 15–8 shows the approximate hours of labor to apply and install exterior trim items.

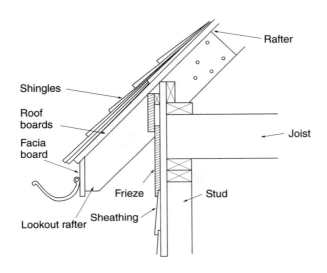

Figure 15–5
Construction details of an "open cornice"

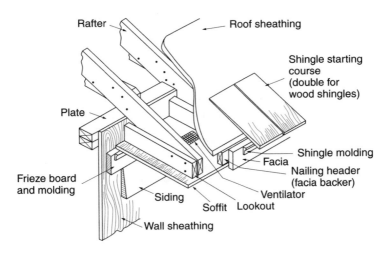

Figure 15–6
Construction details of a "wide box cornice"
(with horizontal lookouts)

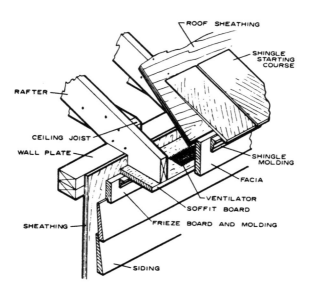

ROOF SHEATHING

SHINGLE STARTING COURSE

RAFTER

CEILING JOIST

WALL PLATE

SHINGLE MOLDING

FACIA

VENTILATOR

SOFFIT BOARD

SHEATHING

FRIEZE BOARD AND MOLDING

SIDING

Figure 15–7
Construction details of a "narrow box cornice"
(without lookouts)

TABLE 15–8

**Approximate Hours Labor to Install 100
Lineal Feet of Various Kinds of Exterior
Trim**

Item of Trim	Hours Labor Per 100 Lin. Ft.	Lineal Feet Per Hour
Water table and drip cap	4– 5	20–25
Corner boards (2 members)	5– 6	17–20
Verge board	4– 5	20–25
2-member closed cornice	8–10	10–13
3-member boxed cornice	12–14	7– 8
Wood gutters	10–12	8–10

SELECTED REFERENCE

Walker's Building Estimator's Reference Book, Frank R. Walker Company, 5030 N. Harlem Avenue, Chicago, IL 60656, 1999.

CHAPTER 16

INTERIOR WALLBOARDS AND TILE

This chapter deals with the various types of interior composition and wood wall finish commonly encountered, excluding wet plaster finishes (see Chapter 17, Lathing and Plastering). These wall finishes include gypsum board; the different fiber boards made of wood fiber, cane fiber, and wood pulp; the several kinds of plywood finishes; solid wood paneling; and tile.

Interior wall finishes may be applied directly to wood or metal studding; over 1-by-2-in. or 1-by-3-in. furring strips, usually spaced 12 in., 16 in., or 24 in. on center; or over backing boards. If attached to concrete or masonry surfaces, special anchor or self-clinching nails are recommended.

GYPSUM BOARD PRODUCTS

The several types of gypsum board products are listed in Table 16–1 with sizes and edge designs. Edge details are illustrated in Figure 16–1 on page 342.

Wallboards

Wallboards are primarily *regular, predecorated, foil-backed, fire resistant (Type X),* and *moisture resistant (MR).*

TABLE 16-1

Summary of Gypsum Board Products

Type	Thickness (in.)	Thickness (mm)	Width	Length	Edge Detail
WALLBOARD (ASTM C36, FS SSL30d) Regular	1/4"	6.4	48"	8', 10', 12'	Tapered
	5/16"	7.9	48"	8', 10', 12', 14'	
	3/8"	9.5	48"	7', 8', 9', 10', 12'	
	1/2"	12.7	48"	7', 8', 9', 10', 12', 14'	
	5/8"	15.9	48"	8', 10', 12'	
Foil-backed	3/8"	9.5	48"	8', 10', 12'	
	1/2"	12.7	48"	8', 9', 10', 12'	
Type X	1/2"	12.7	48"	8', 9', 10', 12'	
	5/8"	15.9	48"	8', 9', 10', 12'	
LATH (ASTM C37, FS SSL30d) Plain	3/8"	9.5	16", 16 1/5", 24"	4', 8', 12'	Round
	1/2"	12.7	16", 16 1/5", 24"	4'	
Perforated	3/8"	9.5	16"	4'	
	1/2"	12.7	16"	4'	
Foil-backed	3/8"	9.5	16"	4', 8'	
	1/2"	12.7	16"	4'	
Type X	3/8"	9.5	16"	4', 8'	

Type	Thickness (in.)	Thickness (mm)	Width	Length	Edge Detail
BACKING BOARD (ASTM C442, FS SSL30d) Regular	¼"	6.4	48"	8'	Square
	⅜"	9.5	48"	8'	Square
	½"	12.7	24"	8'	Tongue-and-grooved
Foil-backed	⅜"	9.5	48"	8'	Square
Type X	⅝"	15.9	24"	7'	Tongue-and-grooved
	1"	25.4	24"	7'	Tongue-and-grooved
SHEATHING (ASTM C79, FS SSL30d) Regular	⅜"	9.5	24"	8'	Tongue-and-grooved
			48"	8', 9'	Square
	4/10"	10.0	24"	8', 9'	Tongue-and-grooved
			48"	8', 9'	Square
	½"	12.7	24"	8', 9'	Tongue-and-grooved
			48"	8', 9'	Square
	⅝"	15.9	24"	8', 9'	Tongue-and-grooved
			48"	8'	Square
Type X	⅝"	15.9	24"	8', 9'	Tongue-and-grooved
			48"	8'	Square
FORMBOARD (ASTM C318, FS SSL30d)	½"	12.7	32"	8', 10'	Square

Figure 16–1
Edge detail of gypsum board

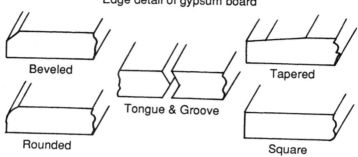

Regular wallboard has a surface that can be painted or that can have wall covering or tile applied. The sheets are ¼ in., 5⁄16 in., 3⁄8 in., ½ in., or 5⁄8 in. thick, and are available in 4-ft. widths, and in 6-ft. to 14-ft. lengths. The edges are commonly tapered to permit taping and finishing of the joints. Sheets are nailed, screwed, or adhesive bonded to wood or metal studs, to backing board, or to furring.

Predecorated wallboard is a regular wallboard with factory-decorated finish on one side, usually paper, vinyl, or wood grained. Edge design is tapered or square, and joint treatment is not required.

Foil-backed (insulating) wallboard is regular wallboard that has aluminum foil laminated to the back. This acts as a vapor barrier and reflective-type of insulation that eliminates the need for a separate vapor barrier in the exterior wall construction. Sheets are available 3⁄8 in. or ½ in. thick, 4 ft. wide, and in lengths of 8 ft., 10 ft., and 12 ft. Edge designs are usually tapered.

Fire-resistant wallboard (Type X), also called fire-rated wallboard, is similar to regular wallboard in appearance and edge treatment, but its core is specially formulated with additives and glass fibers for greater fire resistance. The ½-in. board has a 45-minute fire rating; the 5⁄8-in.

board has a 1-hour fire rating. The sheets are usually 48 in. wide and are available in lengths of 8 ft. to 12 ft.

Moisture resistant (MR) wallboard has a chemically treated light green paper and a gypsum core formulated with asphaltic additives. Edges are usually tapered, and thicknesses of ½ in. and ⅝ in. are common. The sheets are 4 ft. wide and 8 ft. to 12 ft. long. MR wallboards are used in bathrooms, powder rooms, kitchens, and other high-moisture areas, and where ceramic, metal, or plastic wall tiles are applied with adhesives.

Backing boards have a gray liner paper on both sides that is not suited for finishing or decorating. These boards serve as a base for multilayer applications of gypsum board, or as a backing for finish materials such as ceramic, metal, and plastic wall or ceiling tile applied with adhesives. Backing boards are graded *regular, foil-backed,* and *fire-resistive (Type X).* Thicknesses are usually ⅜ in. or ½ in.; widths are 24 in. and 48 in.; length is 8 ft. Type X may be ⅝ in. to 1 in. thick, 24 in. wide, and 7 ft. long.

Coreboard is a 1-in. factory-laminated board, usually two layers of regular ½-in. backing board. It is used as the core in the construction of prefabricated gypsum partitions and shaft walls. It is available in 24-in. widths.

Gypsum Board Fasteners

Special nails and screws are used to attach gypsum board products to wood, metal, and masonry. Adhesives are also used, but they are generally supplemented with mechanical fasteners. Staples and clips are sometimes used in backing boards and base plies. Detailed specifications and recommendations are set forth in the Gypsum Association's publication (GA-201-85), *Using Gypsum Board For Walls and Ceilings.* In all cases, the manufacturer's recommendations should be followed.

Joint Treatment

The joints formed by the tapered edges where the gypsum boards meet are concealed by applying fabric over joint compound. Approximately 370 ft. of fabric and 50 lb. of joint compound are needed to finish 1,000 sq. ft. of gypsum wallboard. Table 16–2 shows the approximate average quantities needed.

TABLE 16–2*

Tape and Compound Schedule

Square Feet Wallboard	Joint Compound	Rolls of Fabric
100–200	1 gallon	2 60-ft. rolls
100–400	2 gallon	3 60-ft. rolls
500–600	3 gallon	1 250-ft. roll
700–800	4 gallon	1 250-ft. roll and 1 60-ft. roll
900–1000	5 gallon	1 500-ft. roll

Note: An alternative of 50 lb. of powder joint compound per 1,000 sq. ft. may be used.

How to Estimate Building Losses and Construction Costs, Paul I. Thomas, reprinted by permission from Craftsman Book Company.

Stilts and Taping Machines

Adjustable stilts for workers replace scaffolding. (See Figure 16–2.) The workers quickly become accustomed to these, and they considerably reduce the time to apply gypsum wallboard.

Trim Accessories

To provide true and straight lines for smooth finishing at outside corners there are metal trim accessories avail-

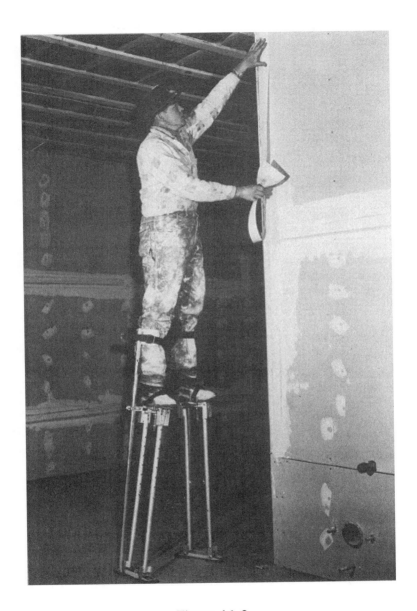

Figure 16–2
Applying tape to drywall, working on adjustable stilts.
Courtesy Goldblatt Trowel Trade Tools, Kansas City, Kansas

able for corner reinforcement, casing beads, metal trim, and decorative moldings.

FIBER BOARD

The various types and finishes on fiber-base wallboards are too numerous to list. Some of the well-known trade names are Celotex™, Fir-tex™, Masonite™, Upson-Board™, and Hardboard™. Sheets are usually 4 ft. wide and 8 ft. long, with 6-ft., 10-ft., and 12-ft. lengths available on order. Thicknesses are ¼ in. to ¾ in.

Application varies with the type of board; it may be with nails colored to match the finish, with adhesive, or with both. Matching inside and outside corner moldings, base molding, and joint strips are available in plastic, wood, or metal.

WALL AND CEILING INSULATION TILE

Fiber composition tiles are used for insulating as well as for decorative purposes on walls and ceilings. Sizes range from 12 in. by 12 in. to 16 in. by 32 in., by ½ in. thick. The tiles are applied with adhesives or with concealed nails or staples. Numerous designs and finishes are available. The tiles may be applied to furring on studding, rafters, and ceiling joists.

PLYWOOD PANELS

Unfinished and prefinished panels for interior wall finish come in a wide variety of veneers. The sheets most commonly used are 4 ft. wide and 8 ft. long, although narrower and longer sheets are available. Thicknesses of ¼ in., ⅜ in., ½ in. and ⅝ in. are most common. Where studding or furring is 16 in. on center, a minimum thickness of ¼ in. is recommended. On 20-in. or 24-in. centers, ⅜ in. minimum thickness is recommended.

Panels may be applied directly to studding, to studding with ¾-in. or ½-in. plywood or composition board backing, or over furring on studs or masonry walls. Adhesives or nails are used, or both. Joints may be flush or V-grooved, although they are often concealed with panel strips.

SOLID WOOD PANELING

Solid wood paneling of knotty pine, cherry, cyprus, and so forth is usually nailed to furring applied over studding or masonry. There may or may not be an underlayment of composition wallboard for sealing and insulating purposes.

Most solid wood paneling consists of ¾-in. or ⅞-in. boards, 6 in., 8-in., or 10 in. wide. The panels may all be the same width or in random widths. The boards have tongue and grooved or shiplapped edges for joining and also for concealing nailing. Where it is necessary to face nail, 6- or 8-penny finish nails are used.

ESTIMATING MATERIALS

When estimating materials to apply interior composition wallboards or wood paneling, all necessary furring and blocking should be included.

The area to be covered is computed after deducting door, window, and other openings to obtain the actual number of square feet of wall and ceiling area. An allowance should be made for fitting and cutting waste; 10 percent is usually adequate for average conditions. Much depends on the shape, number of openings, size, and layout of the room. In actual practice a sketch is made, from which the number and size of each piece of wallboard or solid wood paneling is determined.

ESTIMATING LABOR

Table 16–3 shows the approximate average hours of labor required to apply composition wallboard, plywood and solid wood paneling, and ceiling and wall tile under normal working conditions. As in estimating all labor productivity, the experience of the workers and other factors outlined in Chapter 10 must be taken into consideration.

CERAMIC WALL AND FLOOR TILE

Types of Ceramic Tile

Ceramic tile for walls and floors is manufactured in a number of types, sizes, and finishes. Tile may be glazed or unglazed. There are four major types of ceramic tile: *glazed wall tile, ceramic mosaic tile, quarry tile,* and *paver tile.*

Glazed wall tile, 1/4 in. or 5/16 in. thick, is most popular in sizes 4 1/4 in. by 4 1/4 in., 6 in. by 4 1/4 in., and 6 in. by 6 in., back-mounted or unmounted. Larger sizes are available.

Ceramic mosaic tile, glazed and unglazed, is formed by pressing refined clays and is customarily used on the floors of residential kitchens, bathrooms, and showers. It also is used on walls of commercial and institutional buildings and in swimming pools. Conductive mosaic tile contains additives to prevent or minimize electrostatic buildup, and is used in hospital rooms and laboratories. Ceramic mosaic tile is usually 1/4 in. thick and comes in sizes of 1 in. by 1 in., 2 in. by 1 in., and 2 in. by 2 in. It is face-mounted in sheets 2 feet square, or mounted on a mesh backing.

TABLE 16–3

Approximate Hours Labor to Apply Furring, Wallboards, Paneling, and Wall and Ceiling Tile

Type of Work	Hours Labor Per 100 Sq. Ft.
Furring on studding, 1″ × 2″, 12″ on center	1–1.5
Furring on masonry, 1″ × 2″, 12″ on center	1½–2
3/8″ or 1/2″ Gypsum wallboard on studs	1½–2.5
3/8″ or 1/2″ Gypsum wallboard on ceiling joists	2–3
2-Ply, 3/8″ Gypsum wallboard on studs	3–5
Taping and finishing Gypsum wallboard joints	1–2
Plywood, prefinished, 1/4″ on backing board. stripped joints and corners	2–3
Plywood, prefinished, 1/2″ flush joints or tongue and grooved, using nails and adhesives	6–8
Solid wood paneling, 1″ × 6″ and 1″ × 8″	3–4
Wood strips over solid wood paneling joints	1–1.5
Fiber boards, (Masonite™, Upsom-Board™, Celotex™, etc.)	2–3
Insulating and accoustical wall tile, 12″ × 12″, stapled	3–4
Insulating and accoustical ceiling tile, adhesive	4–5

Quarry tile, glazed or unglazed, is formed by extruding clay and is used for heavy-duty floors in all types of construction, both interior and exterior. It is usually ½ in. thick and 6 in. square, although other sizes are available.

Paver tile, glazed or unglazed, is similar to ceramic mosaic tile, but is usually ¾ in. or thicker and 4 in. by 4 in. or larger.

Grades of Ceramic Tile

There are four grades of ceramic tile that are recognized by the American National Standard Specifications for Ceramic Tile. The two major grades are *Standard* and *Second.* Standard grade tile may vary in shade but must be free of spots and defects. Seconds may have minor blemishes and defects that are not structural. Tiles are shipped in sealed cartons with grade and contents indicated by grade seals. The seals are grade colored: blue for standard and yellow for seconds. Quarry tile is shipped in cartons with the grade stamp imprinted on the carton instead of grade seals. Decorative Thin Wall Tile and Special Purpose Tile are the other two grades. Decorative Thin Wall Tile must meet all the requirements of standard grade with the exception of breaking strength. Special Purpose Tile is tile that cannot be classified in any of the other categories.

Materials for Setting Ceramic Tile

The following information on the most widely used materials for setting ceramic tile is reprinted with permission of the Tile Council of America from its *Handbook for Ceramic Tile Installation.*

Portland Cement Mortar:
A mixture of portland cement and sand, roughly in proportions of 1:6 on floors and of portland cement, sand and lime in proportions of 1:5:½ to 1:7:1 for walls.

Portland cement mortar is suitable for most surfaces and ordinary types of installation. The thick bed, ¾" to 1" on walls and nominally 1-¼" on floors, facilitates accurate slopes on planes in the finished tile work.

Complete installation and material specifications are contained in ANSI A108.1 ANSI. 1430 Broadway, New York, NY 10018.

Dry-Set Mortar:

A mixture of portland cement with sand and additives imparting water retentivity which is used as a bond coat for setting tile.

Dry-set mortar is suitable for use over a variety of surfaces. It is used in one layer, as thin as $3/32''$, after tile are beat in, has excellent water and impact resistance, is water-cleanable, non-flammable, good for exterior work, and does not require soaking of tile.

Complete installation specifications and material specifications are contained in ANSI A108.5 and ANSI A118.1. For conductivity dry-set mortar see ANSI A108.7 and ANSI A118.2.

Latex-Portland Cement Mortar:

A mixture of portland cement, sand and special latex additive which is used as a bond coat for setting tile.

The uses of latex-portland cement mortar are similar to those of dry-set mortar. It is less rigid than portland cement mortar.

Complete installation specifications and material specifications are contained in ANSI A108.5 and ANSI A118.4.

Epoxy Mortar:

A mortar system designed for chemical resistance employing epoxy resin and epoxy hardener portions.

Epoxy mortar is suitable for use where chemical resistance of floors, high bond strength and high impact resistance are important considerations. High temperature resistant formulas are also available. Complete installation and material specifications are contained in ANSI A108.6 and ANSI A118.3.

Modified Epoxy Emulsion Mortars:

A mortar system employing emulsified epoxy resins and hardeners with portland cement and silica sand.

Modified epoxy emulsion mortars are formulated for thin-set installations of ceramic tile on floors and walls, interior and exterior. Their features include high bond strength, ease of application, little or no shrinkage, and economical epoxy application. *They are not designed for chemical resistance.*

Complete installation and material specifications are contained in ANSI A108.9 and ANSI A118.8.

Furan Mortar:

A mortar system designed for chemical resistance consisting of furan resin and furan hardener portions.

Furan mortar is suitable for use where chemical resistance of floors is an important consideration.

Complete installation and material specifications are contained in ANSI A108.8 and ANSI A118.5.

Epoxy Adhesive:

An adhesive system employing epoxy resin and epoxy hardener portions.

Epoxy adhesive is formulated for thin-setting of tile on floors, walls, and counters. It is designed primarily for high bond strength and ease of application and not for optimum chemical resistance. However, its chemical and solvent resistance tends to be better than that of organic adhesives.

Complete installation and material specifications are contained in ANSI A108.4 and ANSI A118.3.

Organic Adhesive:

A prepared organic material for interior use only, ready to use with no further addition of liquid or powder, which cures or sets by evaporation.

Organic adhesives are suitable for installing tile on floors, walls and counters where surfaces are appropriate and properly prepared—in accordance with adhesive manufacturers' directions.

Adherence to ANSI A136.1 is the minimum criterion for selecting an organic adhesive. Complete installation specifications are contained in ANSI A108.4. Organic adhesives are not recommended in areas exposed to temperatures exceeding 140°F. Some backing materials may require lower temperatures.

Materials for Grouting Ceramic Tile

The following statement regarding tile grouting materials is reprinted with permission of the Tile Council of America from its *Handbook for Ceramic Tile Installation.*

Grouting materials for ceramic tile are available in many forms to meet the requirements of the different kinds of tile and types of exposures. Portland cement is the base for most grouts and is modified to provide specific qualities such as whiteness, mildew resistance, uniformity, hardness, flexibility and water retentivity. Noncement–based grouts such as epoxies, furans, and silicone rubber offer properties not possible with cement grouts. However, special skills on the part of the tile setter are required. *These materials can be appreciably greater in cost than cement-based grouts.* Complete installation and material specifications are contained in ANSI A108.10 and ANSI A118.6.

Estimating Materials

The unit of measurement for estimating ceramic wall and floor tile is the square foot. Caps, corners, and base tile are usually figured by the lineal foot. Bathroom, shower, and kitchen accessories such as soap dishes, towel bars, grab bars, and toothbrush holders are estimated by the unit.

In measuring walls, all window and door openings are deducted, but the trim for such openings should be included in the materials estimate. Tile prices vary considerably; therefore, the specifications must be explicit as to type, grade, size, shape, color, pattern, and whether glazed or unglazed, abrasive, conductive, and so forth.

Estimating Labor Installing Ceramic Tile

As with other building trades, estimating the labor to install ceramic tile on walls and floors requires a thorough knowledge of the particular job conditions the tiler will encounter. Refer to "Estimating Labor" in Chapter 10 for a discussion of the many factors that affect labor productivity.

Because the physical area and its surroundings in which a tile setter often works restricts freedom of movement, thought must be given to whether he can work efficiently alone or whether a helper is required part or full time.

The man-hours of labor to install ceramic tile is dependent also on whether the tile will be set individually, or on face-mounted or back-mounted sheets.

Table 16–4 lists some of the more common ceramic tile applications and the approximate man-hours of labor required, under average working conditions. It should be used with judgment; it is only a guide for estimating labor.

TABLE 16–4

Approximate Average Hours of Labor to Apply Ceramic Tile to Walls and Floors

Type of Installation	Adhesive	Man-Hours Per 100 Sq. Ft.
Floor Tile, Ceramic Mosaic*		
1″ × 1″	(face-mounted) portland cement mortar	12–15
1″ × 1″	(face-mounted) organic adhesive	8–10
2″ × 2″	(face-mounted) portland cement mortar	10–12
2″ × 2″	(face-mounted) organic adhesive	9–11
4″ × 4″	(face-mounted) portland cement mortar	12–14
4″ × 4″	(face-mounted) organic adhesive	10–12
Wall Tile*		
4¼″ × 4¼″	(unmounted) portland cement mortar	14–16
4¼″ × 4¼″	(unmounted) organic adhesive	9–12
4¼″ × 4¼″	(face-mounted) portland cement mortar	12–14
4¼″ × 4¼″	(face mounted) organic adhesive	8–10
6″ × 6″	(unmounted) portland cement mortar	10–14
1″ × 1″	(face-mounted) portland cement mortar	10–12
1″ × 1″	(face-mounted) organic adhesive	8–9
Shower and tub backs	portland cement mortar	14–16
	organic adhesive	9–12

Ceramic Tile Cove, Base and Cap	Per 10 lin. ft.	1–2
Ceramic Tile Accessories		
Soap dishes, towel bars, paper holders, etc.	Per ea. 1/2 to 3/4	

*Reduce hours 15% to 20% for large open areas.

SELECTED REFERENCES

Drywall, W. Robert Harris, American Technical Publishers, Inc., 1155 W. 175th Street, Homewood, IL 60430, 1997.

Drywall Contracting, James T. Frane, Craftsman Book Company, Carlsbad, CA, 1987.

Recommended Specifications for the APPLICATION and FINISHING of GYPSUM BOARD, Gypsum Association, 1993.

CHAPTER 17

LATHING AND PLASTERING

Originally *plaster lath* denoted narrow strips of wood placed close together to provide a base for the application of plaster. Wood lath has been replaced by a number of substitutes. Perforated sheets of steel, and gypsum board either solid or perforated, are known in the trade as *metal lath* and *gypsum lath,* respectively.

LATHING

Measurement of Areas

As in measuring areas for applying other interior materials, there are different practices and opinions on the question of deducting window and door openings. One estimator may deduct all such openings in full, while another may deduct only half the areas. Still others do not deduct any openings. The latter two methods may be justified and have merit, but we recommend deducting openings in full and adding 10 percent to the *net* wall area for waste. Whichever method is used, judgment should be exercised to make certain that an adequate amount of material is allowed and sufficient labor is included. For the square foot areas of sidewalls and ceilings, see Tables 17–1 and 17–2.

TABLE 17–1

Square Foot Area of the Sidewalls of Rooms
Measuring 2' × 3' to 20' × 30' with 10' ceilings

FEET	2	3	4	5	6	7	8	9	10
3	100	120	140	160	180	200	220	240	260
4	120	140	160	180	200	220	240	260	280
5	140	160	180	200	220	240	260	280	300
6	160	180	200	220	240	260	280	300	320
7	180	200	220	240	260	280	300	320	340
8	200	220	240	260	280	300	320	340	360
9	220	240	260	280	300	320	340	360	380
10	240	260	280	300	320	340	360	380	400
11	260	280	300	320	340	360	380	400	420
12	280	300	320	340	360	380	400	420	440
13	300	320	340	360	380	400	420	440	460
14	320	340	360	380	400	420	440	460	480
15	340	360	380	400	420	440	460	480	500
16	360	380	400	420	440	460	480	500	520
17	380	400	420	440	460	480	500	520	540
18	400	420	440	460	480	500	520	540	560
19	420	440	460	480	500	520	540	560	580
20	440	460	480	500	520	540	560	580	600
21	460	480	500	520	540	560	580	600	620
22	480	500	520	540	560	580	600	620	640
23	500	520	540	560	580	600	620	640	660
24	520	540	560	580	600	620	640	660	680
25	540	560	580	600	620	640	660	680	700
26	560	580	600	620	640	660	680	700	720
27	580	600	620	640	660	680	700	720	740
28	600	620	640	660	680	700	720	740	760
29	620	640	660	680	700	720	740	760	780
30	640	660	680	700	720	740	760	780	800

For 7'6" ceiling height multiply by .75
" 8'0 " " " " .80
" 8'6" " " " " .85
" 9'0" " " " " .90
" 9'6" " " " " .95
" 11'0" " " " " 1.10
" 12'0" " " " " 1.20

11	12	13	14	15	16	17	18	19	20
280	300	320	340	360	380	400	420	440	460
300	320	340	360	380	400	420	440	460	480
320	340	360	380	400	420	440	460	480	500
340	360	380	400	420	440	460	480	500	520
360	380	400	420	440	460	480	500	520	540
380	400	420	440	460	480	500	520	540	560
400	420	440	460	480	500	520	540	560	580
420	440	460	480	500	520	540	560	580	600
440	460	480	500	520	540	560	580	600	620
460	480	500	520	540	560	580	600	620	640
480	500	520	540	560	580	600	620	640	660
500	520	540	560	580	600	620	640	660	680
520	540	560	580	600	620	640	660	680	700
540	560	580	600	620	640	660	680	700	720
560	580	600	620	640	660	680	700	720	740
580	600	620	640	660	680	700	720	740	760
600	620	640	660	680	700	720	740	760	780
620	640	660	680	700	720	740	760	780	800
640	660	680	700	720	740	760	780	800	820
660	680	700	720	740	760	780	800	820	840
680	700	720	740	760	780	800	820	840	860
700	720	740	760	780	800	820	840	860	880
720	740	760	780	800	820	840	860	880	900
740	760	780	800	820	840	860	880	900	920
760	780	800	820	840	860	880	900	920	940
780	800	820	840	860	880	900	920	940	960
800	820	840	860	880	900	920	940	960	980
820	840	860	880	900	920	940	960	980	1000

TABLE 17–2

Square Foot Area of Ceilings and Floors of Rooms Measuring 2' × 3' to 20' × 30' with 10' ceilings

FEET	2	3	4	5	6	7	8	9	10
3	6	9	12	15	18	21	24	27	30
4	8	12	16	20	24	28	32	36	40
5	10	15	20	25	30	35	40	45	50
6	12	18	24	30	36	42	48	54	60
7	14	21	28	35	42	49	56	63	70
8	16	24	32	40	48	56	64	72	80
9	18	27	36	45	54	63	72	81	90
10	20	30	40	50	60	70	80	90	100
11	22	33	44	55	66	77	88	99	110
12	24	36	48	60	72	84	96	108	120
13	26	39	52	65	78	91	104	117	130
14	28	42	56	70	84	98	112	126	140
15	30	45	60	75	90	105	120	135	150
16	32	48	64	80	96	112	128	144	160
17	34	51	68	85	102	119	136	153	170
18	36	54	72	90	108	126	144	162	180
19	38	57	76	95	114	133	152	171	190
20	40	60	80	100	120	140	160	180	200
21	42	63	84	105	126	147	168	189	210
22	44	66	88	110	132	154	176	198	220
23	46	69	92	115	138	161	184	207	230
24	48	72	96	120	144	168	192	216	240
25	50	75	100	125	150	175	200	225	250
26	52	78	104	130	156	182	208	234	260
27	54	81	108	135	162	189	216	243	270
28	56	84	112	140	168	196	224	252	280
29	58	87	116	145	174	203	232	261	290
30	60	90	120	150	180	210	240	270	300

11	12	13	14	15	16	17	18	19	20
33	36	39	42	45	48	51	54	57	60
44	48	52	56	60	64	68	72	76	80
55	60	65	70	75	80	85	90	95	100
66	72	78	84	90	96	102	108	114	120
77	84	91	98	105	112	119	126	133	140
88	96	104	112	120	128	136	144	152	160
99	108	117	126	135	144	153	162	171	180
110	120	130	140	150	160	170	180	190	200
121	132	143	154	165	176	187	198	209	220
132	144	156	168	180	192	204	216	228	240
143	156	169	182	195	208	221	234	247	260
154	168	182	196	210	224	238	252	266	280
165	180	195	210	225	240	255	270	285	300
176	192	208	224	240	256	272	288	304	320
187	204	221	238	255	272	289	306	323	340
198	216	234	252	270	288	306	324	342	360
209	228	247	266	285	304	323	342	361	380
220	240	260	280	300	320	340	360	380	400
231	252	273	294	315	336	357	378	399	420
242	264	286	308	330	352	374	396	418	440
253	276	299	322	345	368	391	414	437	460
264	288	312	336	360	384	408	432	456	480
275	300	325	350	375	400	425	450	475	500
286	312	338	364	390	416	442	468	494	520
297	324	351	378	405	432	459	486	513	540
308	336	364	392	420	448	476	504	532	560
319	348	377	406	435	464	493	522	551	580
330	360	390	420	450	480	510	540	570	600

GYPSUM LATH

A widely used plaster base is gypsum lath (see Table 16–1). It is relatively easy to apply, and cutting and fitting around openings can be done quickly with a sharp knife. It is fire-resistive, offers desirable insulating qualities, and provides a better adhesive bond than wood or steel lath.

Gypsum lath is available in thicknesses of ⅜ in. and ½ in. for 16-inch and 24-inch stud spacing, with perforated sheets of 16 in. wide and 4 ft. long. The perforations are ¾-inch holes every 16 square inches. It also comes in solid sheets ⅜ in. thick, 16 in. and 24 in. wide, and 4 ft., 8 ft., and 12 ft. long. The perforations permit the plaster to form a key, or mechanical bond, on the back side. This type is often preferred for ceilings or is specified where a better fire-resistive rating is required. Some building codes require perforations. Corner beads of expanded metal lath or perforated metal should be installed at all exterior corners for reinforcement and to act as a leveling gauge for the plasterer.

Estimating Quantities

The unit for estimating quantities of gypsum lath is the square yard. Special flathead blued nails, 1⅛-inches long, No. 13 gauge, are used in nailing ⅜-inch thick gypsum lath. The nails are spaced about 5 in. apart with approximately 7 to 9 pounds required for 100 square yards. For gypsum lath that is ½ in. thick, nails 1¼ in. long are used and spaced 4 in. apart.

The gross area of the surface to be lathed is measured, and all openings are deducted. An allowance of about 10 percent is added for cutting and fitting waste. Corner beads, cornerites, and other accessories should be added.

Estimating Labor

A lather should be able to apply gypsum lath on frame walls at an average rate of 100 square yards in approximately 8 to 9 hours. When application is with pneumatic staplers, the rate should increase to about 6 hours for 100 square yards. When applied to ceilings, the rate will drop to about 9 to 10 hours for 100 square yards.

METAL LATH

Metal lath is made of steel. It comes either painted or galvanized to protect it against rusting. Metal lath provides the best mechanical bond of all laths. Portland cement does not adhere well to gypsum lath; where that type of mortar is required, metal lath is used.

Metal lath is made in a variety of patterns that may be generally classified as "expanded," "ribbed," "sheet," or "wire" laths. It comes in sheets measuring 27 in. by 96 in., with 10 sheets to a bundle. One bundle will cover 20 square yards. Metal lath is fastened to wood framing and furring by 1½-inch galvanized roofing nails. Where sheets are joined between studs and joists, the lath is wired together with 18 ga. wire.

Estimating Quantities

The unit for estimating metal lath, like gypsum lath, is the square yard. Approximately 6 to 8 pounds of 1½-inch flat head blued or galvanized nails are required for 100 square yards of lath. *In measuring areas for metal lath, all openings are deducted in full.* An allowance of 10 percent should be added for waste in cutting, fitting, and lapping.

Estimating Labor

Under average working conditions, a lather can apply approximately 100 square yards of metal lath to wood studding in 7 to 8 hours. When applied to wood ceiling joists, the rate will drop to 8 to 9 hours. Wiring metal lath to steel studs and joists will run approximately 10 hours per 100 square yards.

GROUNDS AND SCREEDS

Grounds are wooden or metal strips that are nailed around rough interior door openings and some window openings to establish the thickness of the plaster when it is applied. They are leveling guides for the mason. Grounds are also nailed along the base where the plaster is to stop. Grounds may be nominal 1-by-2-inch furring. Usually they are left in place when they are flush with the plaster, and the baseboard is nailed over them.

Screeds are plaster strips placed by the plasterer on the walls and ceilings to function as guidelines upon which the straight edge may ride when rodding the plaster on walls and ceilings.

The cost of screeds is included in the cost of the plastering labor. The cost of grounds at door and window openings may be estimated at approximately ½ hour per average size opening. The material cost is not significant. Baseboard grounds may be estimated at the rate of 100 lineal feet in 2 hours, and the cost of the material should be added.

Corner Bead

Galvanized steel corner bead is used at all corners for the protection of the finished plaster edge. It also serves as a ground and leveling guide. Corner bead can be used with all types of lath. The cost of corner bead is estimated on the basis of a lather's installation at the rate of approximately 100 lineal feet in 2 hours.

INTERIOR PLASTERING

Plaster is mortar that is applied in successive coats to wall and ceiling surfaces. It is composed of a fine aggregate held together by either gypsum or portland cement as a cementatious material. Water is added to form a workable plastic mass that can be spread over the surface and smoothed out with a trowel or float. Gypsum is used for interior plaster, and portland cement is used for exterior plaster because it is weather resistant. Portland cement is also used for certain kinds of interior plastering work such as basement walls and backing for ceramic tile.

A mechanic who applies plaster is known in the building trades as a *plasterer*. Because their work is somewhat related to masonry, some plasterers are also masons and *vice versa*.

Plaster is applied in one, two, or three coats. Three-coat plaster consists of a *scratch coat,* a *brown coat,* and a *finish coat*. Three-coat is considered to be the highest grade of plastering, but two-coat work is becoming increasingly common, especially in house building. One reason is the lower labor cost; another reason is that the solid gypsum board lath enables a plasterer to obtain a true surface with two coats rather than three, which is usually necessary on metal lath.

Both gypsum and portland cement plaster may be applied to masonry walls. Generally two-coat work is considered adequate, although local building codes may require three coats in certain instances.

Kinds of Interior Plastering

Three-Coat Plastering

Three-coat plaster is applied over metal lath. It consists of two base coats and one finish coat. The first coat is called a *scratch coat* because it is scored, or scratched, while still wet to provide a bond for the second coat. Much of the first-coat plaster is forced through the spaces in the lath, giving the plaster a mechanical bond behind the lath. A second, or *brown coat* is applied after the scratch coat has set firm but not dried out. This coat is troweled level and smooth in preparation for the finish coat. The finish coat usually is a mortar of gypsum and hydrated lime. It is called a putty, or white coat, and is about 1/16-inch thick.

Two-Coat Plastering

Two-coat plaster is applied over solid plaster bases such as gypsum board or masonry. The first coat is put on in either one operation or by the double back method. In the *double back* method a thin coat is first applied. Before it has set, the plasterer goes back over the area filling in and leveling the surface to the full thickness of the grounds. A finish coat is then applied in the same manner as in three-coat work after the base coat has set.

White Putty Finish

The most common plaster finish coat is the white or lime-putty coat which is applied over the base in a thickness of about 1/16 in. It produces a smooth, hard white surface, which may be papered or painted.

Double hydrated finishing lime requires no slaking on the job because it is 92 percent hydrated during manufacture. Neat gypsum plaster is added to the lime putty as a setting agent to produce hardness and also to resist shrinking which is a characteristic of hydrated lime. The proportions specified by the ASA are:

> One 100-pound bag of gypsum gauging plaster to not more than
> Four 50-pound bags of hydrated lime.

Keene's Cement Finish

During the calcination of natural gypsum, if all of the water is chemically removed, instead of 75 percent as in the case of plaster of Paris, the resultant product is called Keene's cement. It is a form of gauging plaster that, when mixed with lime putty, produces a very hard and durable surface.

The ASA (American Standards Association) specifications for mixing proportions of *medium hard finish* Keene's cement are:

> One 100-pound bag of Keene's cement to not more than
> One 50-pound bag of hydrated lime.

The ASA specifications for *hard finish* Keene's cement are:

> One 100-pound bag of Keene's cement to not more than
> One-half (50-pound) bag of hydrated lime.

Veneer (Thin-Coat) Plaster

Where economy is a major factor, a thin coat of lime putty plaster is applied *directly* to gypsum board after the joints have been taped. The thickness of the skim-coat is similar to the $\frac{1}{16}$-inch finish coat applied over regular two- and three-coat plastering.

Sand Finish

A *sand finish* employs mortar similar to the scratch and brown coats, except that a very fine sand (sometimes white sand) is used. A cork float is used to obtain a sand-paper-like surface. Occasionally, colors are added to tint the sand finish, making it unnecessary to paint or decorate.

Plaster Materials

Gypsum is a gray or white calcium, a rocklike mineral that is found in most parts of the world. It contains, by weight, approximately 21 percent of chemically combined water. When the gypsum rock is pulverized and subjected to heat, water is driven off. This is a process called *calcination*. When 75 percent of the water is removed, the resulting product is *plaster of Paris*. The calcined gypsum when remixed with water sets and returns to gypsum rock.

Calcined gypsum is the cementatious base material for most interior plaster mortars. Mixed with sand, perlite, or vermiculite, it is used as a base plaster for scratch, brown, or sand-finish coats. When mixed with hydrated lime, it is used as a putty or white finish coat.

Job-Mixed Plaster

Gypsum plaster for base coats may be mixed on the job by machine or by hand in a mortar box. When job-mixed, the following proportions of calcined gypsum to aggregate are recommended by the Gypsum Association.

- *For two-coat or "double back" plastering:* 100 pounds of calcined gypsum plaster to 250 pounds of damp, loose sand, or 2½ cubic feet of perlite or vermiculite.

- *For three-coat application of plasters:* The first (scratch) coat is 100 pounds of calcined gypsum plaster to 200 pounds of damp, loose sand, or 2 cubic feet of perlite or vermiculite. The second (brown) coat is 100 pounds of gypsum to 300 pounds of damp, loose sand, or 3 cubic feet of perlite or vermiculite. In lieu of the above, the scratch coat may consist of 100 pounds of gypsum to 250 pounds of sand, or 2½ cubic feet of perlite or vermiculite providing the brown coat is of similar proportion.

- It is common practice to designate proportions by ratios such as *1:2*, meaning 100 pounds of gypsum to 200 pounds of sand (or 2 cubic feet of perlite or vermiculite depending on the aggregate indicated). Similarly, *1:2½* means 100 pounds of gypsum to 250 pounds of sand (or 2½ cubic feet of perlite or vermiculite).

- Where double ratios occur, such as *1:2, 1:3*, the first symbol refers to the scratch coat proportion and the second symbol to the brown coat. A single designation such as *1:2½* indicates that proportion is the same for both coats.

The use of sand, perlite, or vermiculite as an aggregate in job-mixed plaster is a matter of choice and also is dependent upon several conditions. For ordinary plastering, sand is usually used where an ample supply of clean, sharp sand is available at low cost.

Because of the better insulating and fire-resistive qualities of the lightweight aggregates, they may be preferred over sand. Where it is necessary to comply with specific regulations on fire ratings, perlite and vermiculite are used. When plaster thicknesses are specified in excess of those normally used, lightweight aggregates are preferred.

Neat gypsum is the term applied to gypsum with no aggregate added. It comes in 100-pound sacks. Perlite and vermiculite are sold in sacks containing 3 to 4 cubic feet. Their weight varies between 7½ and 10 pounds per cubic foot. Sand is sold by the ton or the cubic yard. A cubic yard contains 27 cubic feet and weighs between 2,600 and 2,800 pounds.

Ready-Mixed Plasters

Much of the gypsum for scratch and brown coats is mill-mixed. It is delivered to the job in 80-pound sacks for lath bases and 67-pound sacks for masonry bases. At one time sand was the principal aggregate used in mill-mixed plasters, but it has almost been superseded by the lightweight perlite and vermiculite aggregates. These lightweight aggregates weigh about 10 percent as much as sand. They have superior sound and heat insulating qualities. They handle with great ease and can be applied more rapidly by the plasterer. An important advantage of mill-mixed plasters is their uniformity of mixture, which is not always easy to control in job mixing.

Estimating Quantities

The unit of measurement in plastering is a square yard. The gross wall area of the room is computed, and *all window and door openings are deducted in full.*

When estimating the quantity of materials required for one-, two-, or three-coat plaster, it is essential to know the thickness of the plaster. Because metal lath has openings through which the plaster is forced to provide a key, or mechanical bond, more plaster is required than for gypsum or masonry bases. Therefore, it is more difficult to control the amount of plaster on metal lath. Too much pressure on the trowel forces more plaster than is necessary through the openings, and it drops behind the lath. Table 17–3 shows the approximate number of sacks of ready-mix perlite gypsum plaster required for 100 square yards on lath, and the number of sacks of masonry plaster for 100 square yards of masonry base. The plaster thickness shown includes $\frac{1}{16}$-inch white coat finish.

The approximate quantities of materials for job-mixed, sanded, or perlite gypsum plaster for 100 square yards are shown in Table 17–4.

TABLE 17–3

Approximate Number of Sacks of Perlite Gypsum Plaster Required to Cover 100 Square Yards

Kind of Base	Plaster Thickness (Inches)	Number of Sacks
Metal lath	3/4	35 (80-lb. sacks)
Gypsum lath	1/2	20 (80-lb. sacks)
Masonry walls	5/8	24 (67-lb. sacks)

TABLE 17–4

Approximate Quantities of Materials Required for 100 Square Yards of Job-Mixed Plaster

Kind of Base	Plaster Thickness in Inches	100-lb. Sacks of Neat Gypsum	Cu. Yds. Sand 1:2½ Mix	Aggregate Perlite Cu. Ft.	4 Cu. Ft. Sacks	Mix
Metal lath	3/4	20	2.0	50	12.5	1:2½
Gypsum lath	1/2	10	1.0	25	6.5	1:2½
Masonry walls	5/8	12	1.2	30	7.5	1:3

ASA permit 250 lb. damp, loose sand or 2½ cu. ft., of vermiculite or perlite provided this proportioning is used for both scratch and brown coats on three-coat work.

The number of 4-cu. ft. sacks of perlite are shown to the nearest ½ sack.

Table 17–5 shows the approximate quantities of gypsum and hydrated lime required to cover 100 square yards with white coat finish $\frac{1}{16}$-inch thick.

The quantities indicated in Tables 17–4 and 17–5 are within the coverage ranges published by the U.S. Gypsum Company.

TABLE 17–5

Approximate Quantities of Materials Required for 100 Square Yards White Coat Finish 1/16-Inch Thick

Kind of Finish	100-Lb. Sacks Neat Gypsum	100-Lb. Sacks Keene's Cement	50-Lb. Sacks Hydrated Lime
Lime putty	2		8
Keene's Cement		4	4

Based on ASA mix of 1 (100-lb.) sack of neat gypsum gauging plaster to 4 (50-lb.) sacks of hydrated lime, and 1 (100–lb) sack of Keene's cement to 1 (50–lb.) sack hydrated lime for medium hard finish.

Estimating Labor

When estimating the labor costs for plastering, careful attention should be given to the physical working conditions the plasterer will be confronted with on each job. The labor cost on small areas is usually higher than where large areas are involved and where the workers can move about freely. The labor cost for plastering ceilings, in most instances, is greater per square yard than for sidewalls. Closets and built-in cupboards, counters, and so forth slow workers down when applying plaster.

All of the factors that affect the productivity of labor are outlined and discussed in Chapter 10 and we recommend that they be reviewed from the standpoint of their influence when plastering. On most jobs, a plasterer's helper or laborer works with the plasterer 50 percent of the time. With present-day lightweight aggregates and machines for mixing plaster as it is needed, the helper's time is greatly reduced over what it was when wood lath and mostly sand aggregate was used. There are still certain working conditions in which a plasterer's helper is needed full time. These include situations where the work is done from scaffolds, in confined small areas, and where materials have to be brought up on elevators to keep the plasterer supplied. Careful consideration must be given to the probable number of hours helpers are needed in each instance.

Table 17–6 shows the approximate number of hours required for a plasterer and a plasterer's helper to apply 100 square yards of different kinds of plaster on several types of plaster base. It is assumed that the plaster base has been prepared for the plasterer.

TABLE 17–6

Approximate Hours to Apply 100 Square Yards of Gypsum Plaster on Various Kinds of Plaster Base

Kind of Base	Plaster Thickness in Inches	No. of Coats Excluding White Coat	Plasterer	Laborer
Metal lath	3/4	2	15	12
Gypsum lath	1/2	1	10	8
Gypsum lath	3/4	2	14	10
Masonry walls	1/2	1	10	8
Masonry walls	5/8	2	12	9
White Finish Coat				
Lime putty	1/16	1	10	8
Keene's cement	1/16	1	13	9

Machine Plastering

Plastering machines are available for applying rough or basecoat plasters. There are portable units operated by electricity and also ones that have gasoline motors. These machines actually spray mortar on the walls and ceiling, but a plasterer must still darby and trowel the surface after he has deposited the mortar.

Machine application uses more mortar but requires much less labor, which is the most expensive factor in plastering. It is a specialized operation and is used mainly where large quantities are contemplated.

EXTERIOR PLASTERING

Stucco, a frequently used term for exterior plaster, is composed of a mortar made with portland cement rather than gypsum because it is exposed to the weather. The aggregate in the mortar is sand, although a fine pebble gravel, marble dust, white sand, and colors are sometimes used in the finish coat. Hydrated lime is added as a plasticizer to make the mortar workable when being applied.

Exterior plaster can be applied directly to brick, concrete, concrete block, hollow terra cotta tile, and other kinds of masonry. It is also applied to wire lath nailed directly to studding, or over sheathing. Three-coat exterior plaster is considered the best grade of work; under many building codes it is a requirement. However, there is some two-coat work being done, both on masonry and lath bases in low-priced dwellings and on secondary buildings. As with interior plastering, the successive coats are scratch, brown, and finish coats. The usual thickness of exterior plaster is about ¾ in., but it may be found in thicknesses that vary from ⅝ in. to 1 in.

Measurement of Areas

Please refer to the section on *Lathing* for a discussion of the different customs for measuring area; the comments also apply to exterior plaster work. In general, it is recommended that gross areas be computed and *all window and door openings be deducted in full.* An allowance of about 10 percent should be added to the net area for waste.

Metal Lath Base

Metal lath applied to frame structures as an exterior plaster base comes in several forms. The three most common types are: (1) the expanded metal lath, (2) the welded wire lath, and (3) the woven wire lath. Woven wire lath, because of its similarity, is frequently called *chicken wire* and is used rather extensively. Because it is not good practice to nail lath solidly to studding or sheathing, a special 1½-inch self furring nail is used. This nail holds the wire lath ¼ in. to ⅜ in. away from the surface, which allows the mortar to flow in back of the lath and form a key at all points.

Whether the metal lath is applied over open studding or solid wood sheathing, an underlayment of 15-lb. saturated felt paper should first be applied. In the case of open sheathing, it acts as a backing behind the lath for the plaster. When applied over solid wood sheathing, it prevents the wood from soaking up the water in the plaster. When asphalt-coated composition sheathing is used, it is not necessary to apply saturated felt paper, as the coating prevents absorption of the water from the mortar.

Before applying saturated felt paper on open studding, 18-gauge steel wires are stretched horizontally, spaced about 6 or 8 inches apart. In some sections of the country it is the custom to draw the wires tightly in the horizontal position; in other sections it is the practice to staple the wires on every other stud. At the alternate studs the wires are then drawn upward until taught and stapled to the stud. The effect of either method is the same. The wires provide rigidity in back of the wire lath and paper against pressure from the plastering trowel when the mortar is being applied.

To estimate the quantity of wire lath and the labor to apply it, please refer to that subject under *Lathing* in this chapter. Costs may be obtained by using the formula shown for metal lath.

The labor cost of stringing 18-gauge wire every 8 inches should be estimated at an average of 6 hours per 100 square yards of surface area. The number of lineal feet of wire is obtained by multiplying the area by 1.5. It requires approximately 8 pounds of 18-gauge wire per 100 square yards.

Another type of metal lath used on exterior open-stud stucco work is a paper-backed product of 2-by-2-inch mesh 16-gauge steel wire. The trade name is *Steeltex.* When this type of metal lath base is applied over open studding, neither the wire back nor the 15-lb. saturated felt paper are required.

The labor to apply 100 square yards of paper-backed stucco lath averages approximately 10 hours.

Masonry Base

All masonry bases should be completely cleaned to remove dirt and grease before plastering; otherwise, the mortar may not adhere. In some cases, clipping, scratching, or roughing the surface is necessary to obtain a good mechanical bond.

Exterior Plaster Mortar

For estimating the cost of portland cement mortar, the proportions of materials for a cubic yard are identical to those used for masonry cement mortar (Chapter 13) employing a 1:3 mix with hydrated lime equal to 10 percent of the weight of cement added.

Estimating Quantities of Mortar

The unit used for quantities of portland cement plastering is 1 square yard. The area to be covered is computed in square feet and converted to square yards by dividing by nine. Table 17–7 shows the number of cubic yards of cement mortar required for 100 square yards of plastering of the thickness indicated.

TABLE 17–7

Cubic Yards Portland Cement Plaster Required for 100 Square Yards for the Thicknesses Shown (1:3 Mix)

Plaster Thickness in Inches	Cubic Yards	
	Masonry Base	Wire Lath Base
1/4	0.75	0.90
1/2	1.50	1.80
5/8	1.87	2.25
3/4	2.25	2.70
1	3.00	3.60

Estimating Labor

Table 17–8 shows the approximate number of hours required for a mason and a helper to apply 100 square yards of plaster of various thicknesses and numbers of coats on both masonry and wire lath base. A plasterer and a helper work together on exterior plastering in the same manner as in interior plastering.

TABLE 17–8

**Approximate Hours to Apply 100 Square
Yards Portland Cement Plaster on
Wire Lath and Masonry Bases
(Float Finish)**

Base	Plaster Thickness in Inches	Number of Coats	Plasterer	Laborer
Masonry	1/2	1	13	6
	5/8	2	21	10
	3/4	3	28	20
Wire Lath	5/8	2	30	20
	3/4	3	40	30

Add approximately 4 hours to plasterer's time for a troweled finish.

Scaffolding and Hoisting

When scaffolding, either patented or structural, has to be erected, the cost should be added as a separate item. Also, the cost of raising materials to upper floors by crane or elevator should be treated as a separate item.

CHAPTER 18

EXTERIOR WALL FINISHES

SOLID WOOD SIDING

There are several types of solid wood siding, as shown in Figure 18–1. The most commonly used are bevel siding, drop siding, and board siding.

Bevel Siding

Bevel siding is applied horizontally, one board overlapping the one below by a minimum of 1 inch for 4-inch and 6-inch widths, and 1¼ inches for wider boards. At building corners bevel siding may be joined by mitering, by overlapping the ends, or by using patented aluminum clips that conceal the end cuts of both boards. The use of corner boards is not so common as when applying drop siding, although mitering the corners takes more time and increases the cost.

The quantity of bevel siding required is determined by computing the square foot area to be covered; deducting door, window, and other openings; and then adding a percentage for waste to take care of milling, lapping, cutting, and fitting. (See Table 18–1.)

Bevel siding is available in 4-inch, 5-inch, and 6-inch nominal widths with ½-inch butts; and in 6-inch, 8-inch,

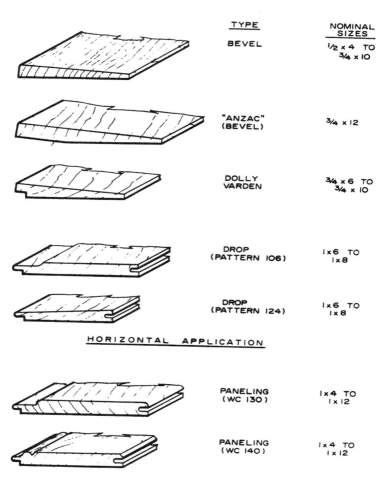

TYPE	NOMINAL SIZES
BEVEL	1/2 x 4 TO 3/4 x 10
"ANZAC" (BEVEL)	3/4 x 12
DOLLY VARDEN	3/4 x 6 TO 3/4 x 10
DROP (PATTERN 106)	1 x 6 TO 1 x 8
DROP (PATTERN 124)	1 x 6 TO 1 x 8

HORIZONTAL APPLICATION

TYPE	NOMINAL SIZES
PANELING (WC 130)	1 x 4 TO 1 x 12
PANELING (WC 140)	1 x 4 TO 1 x 12

HORIZONTAL OR VERTICAL APPLICATION

Figure 18–1
Wood Siding Types
U.S. Department of Agriculture, Forest Service, Agriculture Handbook No. 73

TABLE 18–1

Milling, Lap, and Cutting Waste
for Solid Wood Siding

Type Siding	Nominal Size Inches	Actual Size Inches	Edge Type	Milling and/or Lap Waste Percent	Sq. Ft. Per 100 Sq. Ft. Wall*
	1/2 × 4	1/2 × 3¼	Plain	44	149
	1/2 × 5	1/2 × 4¼	Plain	35	140
Bevel	1/2 × 6	1/2 × 5¼	Plain	29	134
Siding	1/2 × 8	1/2 × 7¼	Plain	22	127
1-Inch	5/8 × 8	5/8 × 7¼	Plain	22	127
Lap	3/4 × 8	3/4 × 7¼	Rabbeted	22	127
	3/4 × 10	3/4 × 9¼	Rabbeted	18	123
Drop	1 × 6	3/4 × 5¼	Tongue	13	118
Siding	1 × 8	3/4 × 7¼	& Groove	9	114
	1 × 6	5/8 × 5 3/16	Matched	16	121
	1 × 8	5/8 × 6 7/8	Matched	16	121
Board	1 × 10	5/8 × 8 7/8	Matched	13	118
Siding	1 × 6	5/8 × 5 1/2	S4S	9	114
	1 × 8	5/8 × 7 1/4	S4S	10	115
	1 × 10	5/8 × 9 1/4	S4S	8	113

*5% has been added for cutting and fitting waste.

and 10-inch nominal widths with ⅝-inch butts; and in 8-inch, 10-inch, and 12-inch nominal widths with ¾-inch butts.

Drop Siding

Drop siding is known by various names such as *novelty siding, rustic siding, log cabin siding,* and *dolly varden,* which has the appearance of bevel siding but has the drop siding edge treatment. The edges are matched or shiplapped. Corner boards are used with the different types of drop siding rather than metal clips or mitering.

Board Siding

Solid board siding is usually dressed on all four sides. It may be applied horizontally or vertically. The sizes generally used are nominal 1 in. thick and 4 in., 6 in., and 8 in. wide. When applied vertically the edges may be matched or the square edge joints covered with 1-by-2-inch battens. When applied horizontally the boards may be lapped over one another a minimum of 1-inch. This method is frequently referred to as *lapboard siding*.

Materials and Labor

Table 18–1 shows milling and cutting waste, edge treatment, and the square feet of solid wood siding per 100 square feet of wall surface. Table 18–2 shows the approximate hours of labor to apply the various types of solid wood siding.

PLYWOOD SIDING

The use of plywood panels for exterior siding provides an increasingly popular exterior finish. The panels are made in a variety of weather-resistant patterns that are easy to install, with considerable savings in labor costs over many other types of siding.

Plywood siding includes APA Sturd-i-Wall system which consists of APA 303 plywood panel siding applied directly to studs or over nonstructural fiberboard, gypsum, or rigid foam insulation sheathing. The system is accepted by HUD and model codes. It can meet the thermal requirements of HUD/FHA, FmHA, and state and regional building codes.

Because a single layer of plywood panel siding is strong and rack-resistant, it eliminates the cost of installing separate structural sheathing or diagonal wall bracing.

TABLE 18–2

Approximate Hours Labor to Install
Various Types of Wood Siding

Type Siding	Nominal Size Inches	Nails Per 100 Sq. Ft.	Approximate Hours Labor Per 1,000 BF
	1/2 × 4	1 1/2 lbs.	22–33
Bevel with	1/2 × 5	1 1/2 lbs.	20–30
mitered	1/2 × 6	1 lb.	18–28
corners	1/2 × 8	3/4 lb.	16–24
(Deduct 10%	5/8 × 8	3/4 lb.	16–24
for fitted	3/4 × 8	3/4 lb.	16–24
ends)	3/4 × 10	1/2 lb.	18–27
Drop siding	1 × 6	2 1/2 lbs.	20–24
with fitted	1 × 8	2 lbs.	18–20
ends			
Batten board	1 × 4	30 lbs.	22–30
siding	1 × 6	25 lbs.	20–26
	1 × 8	20 lbs.	18–24
Plywood Siding			Per 1,000 Sq. Ft.
4′ × 8′ panels	3/8″ thick		14–16
	5/8″ thick		16–18

Note: When estimating labor productivity, working conditions, the experience of the workers, and other factors outlined and describe in Chapter 10, pages 116–117, must be taken into consideration.

Panel sidings are usually installed vertically, but they may also be placed horizontally (face grain across supports) if horizontal joints are blocked. The panel joints may be matched, flush, or concealed with battens.

Detailed descriptions and specifications for plywood siding can be obtained from the American Plywood Association, P.O. Box 11700, Tacoma, Washington 98411-0700.

ALUMINUM SIDING

Aluminum siding is sold in panels or strips 8 inches wide, or double 4-inch widths for a narrow clapboard effect. The length of the panels is usually 12½ feet. They may be backed with an insulating material, or they may be unbacked. Backboard material is usually fiberboard or foamed plastic. Panels are factory-finished in acrylic enamel modified with silicones to resist fading. Color selection varies with each manufacturer. Color-matching nails, inside and outside corners, window trim, and moldings are available. Metal thicknesses vary, but generally .019-inch gauge is adequate when properly backed; the minimum is .024-inch gauge for unbacked siding. A slightly curved panel is designed for vertical application.

Backed aluminum siding may be applied directly to studs over sheathing paper. Unbacked aluminum siding should be installed only over solid sheathing.

Aluminum shingles or shakes, although primarily designed for roofing, are sometimes used as exterior siding. (See Roofing, Chapter 19.)

Labor to apply aluminum panel siding horizontally, once the surface has been prepared, will range between 25 and 30 hours per 1,000 square feet. When it is applied vertically, the labor will approximate 30 to 35 hours per 1,000 square feet.

VINYL SIDING

Solid vinyl siding is made of extruded polyvinyl chloride (PVC). It is produced in several colors that are solid through the material and is available with a full line of color-matching accessories. The surface finishes are usually natural wood grain texture. Widths are 8 in. to simulate the wider clapboards; double 4 in. to simulate the narrow clapboards; double 6 in. to provide more coverage per panel; and triple 4 in. vertical solid vinyl siding. Panels are 12 ft. 6 in. in length, except triple 4-inch, which is approximately 10 feet long. There are 8 pieces (100 linear feet) per square; 16 pieces to the carton of 2 squares. The approximate weight per square is 50 lbs. or ½ lb. per square foot. Vinyl siding and accessories are similar in many respects to aluminum siding in both shape and in application procedures.

The labor to apply vinyl siding horizontally ranges between 25 and 30 hours per 1,000 square feet, and between 30 and 35 hours per 1,000 square feet when applied vertically.

HARDBOARD SIDING

Hardboard siding is a high-density composition board of wood fibers. It is strong, durable, and weather resistant. It is available in a variety of styles for horizontal and vertical applications. The surfaces may be smooth, textured, stainable, paintable, or prefinished. *Lap* and *drop* siding are also available in thicknesses of ½ in. and 7/16 in.; widths range from 6 in. to 12 in.; and length is usually 16 ft.

Hardboard panel siding, 7/16 in. by 4 in. by 8 in., is a popular siding that comes in a variety of finishes and textures. Hardboard sidings may be applied over sheathed or unsheathed walls with studs spaced not more than 16 in. on center. Application is similar to solid wood bevel and drop siding. Metal inside and outside corner clips are used, although some prefer wood inside and outside corner boards.

Carpenters can apply hardboard lap and drop siding at approximately 20 to 25 hours per 1,000 square feet. Hardboard panel siding can be applied at approximately 16 to 18 hours per 1,000 square feet.

WOOD SHINGLE SIDING

(See Roofing, Chapter 19.)

CHAPTER 19

ROOFING

MEASUREMENT OF ROOF AREAS

Roofing materials are measured, estimated, and sold by the *square*. A square is an area of 100 square feet. In estimating, to specify a particular roofing material by type, such as *wood shingles* or *built-up roofing*, is not sufficient to determine its cost. There are different grades, weights, and qualities of each type. It makes considerable difference whether a built-up roof is 3-ply or 5-ply, and whether it has a smooth surface, slag, or crushed marble. Spanish tile may be terra cotta, or it may be made of colored cement to simulate terra cotta. Metal roofs, corrugated or sheet metal, are made in different thicknesses or gauges.

Roofing Terminology (See Figure 19–1.)

- *Birdsmouth* is the part of the rafter that has a vertical and level cut that fits and is attached to the wall plates.
- *Overhang* is the projection of the rafter, or the roof, beyond the outside wall of the building.
- *Pitch* is the slope that the roof surface makes with a horizontal plane. It is expressed as the ratio of the

rise to the span per foot. A ⅓-pitch roof has a rise equal to ⅓ of its span.

- *Ridge* is the horizontal member to which the upper ends of the rafters are attached.
- *Rise* is the vertical distance from the rafter supports (plates) to the ridge.
- *Run* is one half of the span, the horizontal distance over which a rafter passes.
- *Span* is the distance between the outside rafter supports (plates).
- *Tail cut* is the shape of the cut at the lower end of the rafter or overhang, sometimes trimmed to produce an ornamental effect.
- *Wall plate* is the member to which the rafters are attached at the top of the wall.

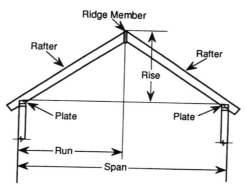

Figure 19–1

Gable Roof Areas

The area of a gable roof is the rafter length from the ridge to the lower edge or eave line, including the overhang, multiplied by the horizontal length of the roof.

Computing Rafter Lengths

- Method 1: The rafter length is equal to the square root of the sum of the square of the *run* and the square of the *rise*.

Example
In Figure 19–2, the run is 16 ft. and the rise is 12 ft.

$$\text{Rafter length} = \sqrt{16^2 + 12^2}$$
$$= \sqrt{256 + 144}$$
$$= \sqrt{400}$$
$$= 20 \text{ ft.}$$

- Method 2: When the rise and the run are known, the rise-per-foot-of-run is obtained by dividing the rise in inches by the run in feet. Table 19–1 shows the factor for the rise-per-foot-of-run by which the run is multiplied to obtain the rafter length.

Example
In Figure 19–2, the rise is 144 inches (12 × 12 in.), and the run is 16 ft.

$$\text{Rise-per-foot-of-run} = \frac{144}{16} = 9 \text{ in.}$$

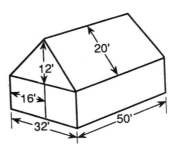

Figure 19–2

Table 19–1 shows a factor of 1.250 for a rise-per-foot-of-run of 9 inches.

$$\text{Rafter length} = 1.250 \times 16 = 20 \text{ ft.}$$

TABLE 19–1

Factor for Multiplying Run to Obtain Rafter Length When Rise-Per-Foot-of-Run Is Known

Rise-per-foot-of-run in inches	Pitch*	Factor for multiplying run in feet for rafter length**
3	1/8	1.031
3½		1.042
4	1/6	1.054
4½		1.068
5	5/24	1.083
5½		1.100
6	1/4	1.118
6½		1.137
7	7/24	1.158
7½		1.179
8	1/3	1.202
8½		1.225
9	3/8	1.250
9½		1.275
10	5/12	1.302
10½		1.329
11	11/24	1.357
11½		1.385
12	1/2	1.414

*Pitch is determined by dividing rise-per-foot-of-run in inches by 24. (Most roofs of dwellings are built with a pitch of 1/8, 1/6, 1/4, 1/3, or 1/2.)
**Formula for developing the factors in Table 19–1:

$$\frac{\sqrt{(\text{Rise in inches})^2 + 12^2}}{12} = \text{Factor}$$

Example: Factor for a pitch of 1/4: $\dfrac{\sqrt{6^2 + 12^2}}{12} = \dfrac{13.416}{12}$

$$= 1.118$$

- Method 3: Using the same example, and Table 19–2, for a rise of 12 ft. and a run of 16 ft. (32/2), the rafter length is 20 ft.

When the rafter length has been determined by Method 1, 2, or 3, it should be carried to the next half or full foot. The area of the gable roof is then computed by multiplying the rafter length, *plus any overhang*, by the horizontal length of the roof. Gable dormers of the same pitch as the main roof may be ignored for all practical purposes, except for the dormer overhang, because the dormer roof area is equivalent to the area taken up by the dormer in the main roof. In Figure 19–2, the rafter length has been determined to be 20 feet. The length of the roof (excluding any overhang) is 50 feet. The roof area is 20 ft. × 50 ft. × 2 = 2,000 square feet.

Gable Roof Areas from Ground Area

The area of a gable or an intersecting-gable roof can be obtained sufficiently closely by computing the plan or ground area covered by the roof. The inches rise-per-foot-of-run is determined, and from Table 19–1 the nearest corresponding factor is selected. The roof area is obtained by multiplying that factor by the ground or plan area. Using Figure 19–2 again, the ground area of the building is 32 ft. × 50 ft. = 1,600 sq.ft. The factor from Table 19–1 for the rise-per-foot-of-run of 9 inches is 1.250. The area of the roof is 1,600 × 1.250 = 2,000 square feet.

Intersecting Gable Roof Areas

When one gable roof joins another, with each having the same pitch, the area of the main roof is computed without deducting the area that is taken up by the intersecting roof. The area of the intersecting roof is computed by measuring its gable end to the point where it joins the main roof *at the eaves*. (See Figure 19–3.)

TABLE 19–2

Rafter Lengths

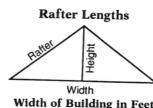

Width of Building in Feet

Height (Feet and Inches)	18	20	22	24	26	28	30	32	34	36
	Length of Rafters in Feet and Inches									
4'6"	10 2	10 11	11 9	12 10	13 9	14 9	15 8	16 8	17 7	18 7
5'0"	10 4	11 3	12 1	13 0	13 11	14 11	15 10	16 9	17 9	18 8
5'6"	10 6	11 5	12 4	13 3	14 2	15 1	16 0	16 11	17 11	18 10
6'0"	10 9	11 8	12 6	13 5	14 4	15 3	16 2	17 1	18 1	19 0
6'6"	11 0	11 11	12 9	13 8	14 6	15 5	16 4	17 3	18 3	19 2
7'0"	11 4	12 3	13 0	13 11	14 8	15 8	16 7	17 5	18 5	19 4
7'6"	11 8	12 6	13 4	14 2	15 0	15 11	16 9	17 8	18 7	19 6
8'0"	12 0	12 10	13 8	14 5	15 3	16 2	17 0	17 10	18 9	19 8
8'6"	12 4	13 2	13 11	14 8	15 6	16 5	17 3	18 2	18 11	19 11
9'0"	12 9	13 5	14 3	15 0	15 10	16 8	17 6	18 5	19 3	20 2
9'6"	13 2	13 10	14 6	15 4	16 1	16 11	17 9	18 8	19 6	20 5
10'0"	13 6	14 2	14 10	15 8	16 5	17 3	18 1	18 11	19 9	20 7
10'6"	13 10	14 6	15 3	15 11	16 9	17 6	18 4	19 3	20 0	20 10
11'0"	14 3	14 10	15 7	16 4	17 0	17 10	18 8	19 6	20 3	21 1
11'6"	14 7	15 3	15 11	16 8	17 4	18 2	18 11	19 9	20 6	21 4
12'0"	15 0	15 8	16 4	17 0	17 9	18 5	19 3	20 0	20 10	21 8
12'6"	15 5	16 0	16 8	17 4	18 0	18 9	19 7	20 4	21 1	21 11
13'0"	15 10	16 5	17 0	17 8	18 5	19 2	19 10	20 8	21 5	22 2
13'6"	16 3	16 10	17 5	18 0	18 9	19 6	20 3	20 11	21 9	22 6
14'0"	16 8	17 3	17 10	18 5	19 2	19 10	20 7	21 3	22 0	22 10
14'6"	17 1	17 8	18 3	18 10	19 6	20 2	20 10	21 7	22 4	23 2
15'0"	17 6	18 0	18 8	19 3	19 11	20 7	21 3	21 11	22 8	23 6
15'6"	17 11	18 5	19 0	19 8	20 3	20 11	21 7	22 4	23 0	23 10
16'0"	18 4	18 10	19 5	20 0	20 8	21 4	21 11	22 8	23 4	24 1
16'6"	18 10	19 4	19 10	20 5	21 0	21 8	22 4	23 0	23 8	24 5
17'0"	19 3	19 9	20 3	20 11	21 5	22 1	22 8	23 4	24 1	24 9
17'6"	19 9	20 2	20 8	21 3	21 10	22 5	23 0	23 8	24 5	25 1
18'0"	20 1	20 7	21 0	21 8	22 3	22 10	23 5	24 1	24 9	25 6
18'6"	20 7	21 0	21 6	22 0	22 8	23 3	23 10	24 5	25 2	25 10
19'0"	21 0	21 5	22 0	22 6	23 1	23 8	24 3	24 10	25 6	26 2
19'6"	21 5	21 11	22 5	22 11	23 5	24 0	24 8	25 3	25 10	26 7
20'0"	21 11	22 4	22 10	23 4	23 10	24 5	25 0	25 8	26 3	26 11
20'6"	22 5	22 10	23 3	23 9	24 4	24 10	25 5	26 0	26 8	27 4

Estimating Tables for Home Building, Craftsman Book Co., 1986. Reprinted here with permission.

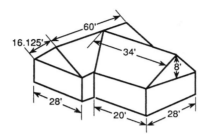

Figure 19–3

Example

Area of Roof Obtained from Ground Area (Figure 19–3)

Main roof ground area	28 ft. × 60 ft. =	1,680 sq. ft.
Intersecting roof ground area	20 ft. × 28 ft. =	560 sq. ft.
Total ground area		= 2,240 sq. ft.

Inches rise-per-foot-of-run $\dfrac{96\text{ in.}}{14\text{ ft.}}$ = 6.85

From Table 19–1, the factor by interpolation = 1.152

Total area of roofs (2,240 sq. ft. × 1.152) = 2,580 sq. ft.

Area of Roof by Actual Measurement (Figure 19–3)

One side of main roof: 16.125 ft. × 60 ft. = 967.50 sq. ft.

Intersected side of roof: $\dfrac{2(16\text{ ft.} + 30\text{ ft.}) \times 16.125}{2}$ = 741.75 sq. ft.

Intersecting roof: $\dfrac{2(20\text{ ft.} + 34\text{ ft.}) \times 16.125}{2}$ = 870.75 sq. ft.

Total area = 2,580 sq. ft.

Dormer Roof Areas

The area of a *flat* dormer roof, excluding overhang, is slightly less than the area the dormer itself occupies in the main roof. Some estimators figure the gross area of the entire roof and do not take into account the area of the dormer except to allow for any overhang.

When the pitch of either a gable or hip dormer is the same as that of the main roof, the area of the dormer roof is equal to the area the dormer occupies in the main roof, plus the area of any overhang. In other words, where dormers are encountered, the areas of their roofs may be ignored except for their overhang.

Hip Roof Areas

The pitch of the four sides of hip roofs are usually the same. The total area of a hip roof is identical to the area of a gable roof that has the same pitch and covers the same plan or ground area.

Example
In Figure 19–4, the area of the *hip* roof is:

$$\frac{2(32 \text{ ft.} \times 20 \text{ ft.})}{2} \quad = \quad 640 \text{ sq. ft.}$$

$$2\left(\frac{28 \text{ ft.} + 60 \text{ ft.}}{2} \times 20\right) = 1{,}760 \text{ sq. ft.}$$

$$\text{Total area} = 2{,}400 \text{ sq. ft.}$$

In Figure 19–4, the area of the *gable* roof is:

$$2(20 \text{ ft.} \times 60 \text{ ft.}) = 2{,}400 \text{ sq. ft.}$$

Hip Roof Areas from Ground Area

The area of any hip roof may also be computed from the plan or ground area covered by the roof including any horizontal overhang. The procedure is the same as that used to compute the area of gable roofs from the ground area covered by the roof.

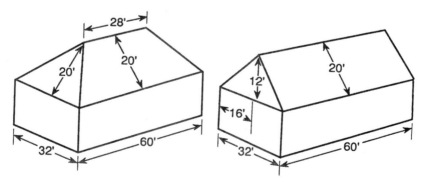

Figure 19–4

From the inches rise-per-foot-of-run, the nearest corresponding factor is taken from Table 19–1. Multiply the plan or ground area, including the horizontal overhang, by that factor.

Example
Using the hip roof in Figure 19–4, the inches rise-per-foot-of-run is 144/16 = 9 inches. Table 19–1 shows a factor of 1.250. The ground area is:

$$32 \text{ ft.} \times 60 \text{ ft.} = 1,920 \text{ sq. ft.}$$

The area of the hip roof is:

$$1,920 \times 1.250 = 2,400 \text{ sq. ft.}$$

Gambrel Roof Areas

Gambrel roofs, often found on barns and occasionally on new dwellings, are frequently referred to as "Dutch Colonial" roofs. The pitch of this roof is broken between the plate and the ridge to form a double pitch roof as shown in Figure 19–5. While the proportions of gambrel roofs vary, the outer and steeply sloped sides generally have a pitch of 1 which means the rise is twice the run, and which is often one fifth of the width of the building. The upper roof has a pitch of 1/4, which means that the rise is one half the run.

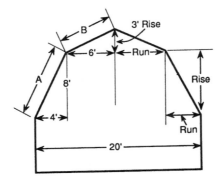

Figure 19–5

The area of this type of roof is best calculated by determining the length of the rafters A and B (Figure 19–5) and by measuring the rise and the run of both slopes.

Example

Assume that the roof in Figure 19–5 is 40 feet long:

Lower slope roof (A) length	$= \sqrt{4^2 + 8^2}$	= 8.94 lin. ft.
Upper slope roof (B) length	$= \sqrt{3^2 + 6^2}$	= 6.71 lin. ft.
Area of the lower sides	= 2(40 ft. × 8.94 ft.)	= 715 sq. ft.
Area of the upper sides	= 2(40 ft. × 6.7 ft.)	= <u>536</u> sq. ft.
	Total area	= 1,251 sq. ft.

Shed Roof Areas

The areas of shed-type roofs are the least difficult to figure. Areas are also relatively easy to determine from plan or ground measurements of a building.

A shed roof is a single pitch roof the slope length of which can be obtained from the rise and the run. In Figure 19–6, for example, the *run* (one-half the span) is 12 ft. and the rise is 3 ft.

The rise-per-foot-of-run is $\dfrac{3' \times 12''}{12'}$ = 3 inches

From Table 19–1, the factor by which the run is multiplied to obtain the length of the rafter is 1.031.

Rafter length = 12 ft. × 1.031 = 12.37 ft.
Area of roof = 12.37 ft. × 20 ft. = 247 sq. ft.
By actual measurement:
Rafter length = $\sqrt{3^2 + 12^2} = \sqrt{153}$ = 12.37 ft.
Area of roof = 12.37 ft. × 20 ft. = 247 sq. ft.

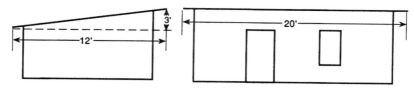

Figure 19–6

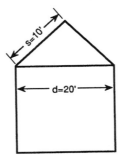

Figure 19–7

Conical Roof Areas

The roofs of such structures as silos and water tanks are often conical in shape. The area of a conical roof is one-half the slope length, *s* (Figure 19–7) × *d*, the diameter × 3.1416 π.

$$\text{Area} = \frac{s \times (3.1416 \times d)}{2} s \times (3.1416 \times d)$$

Area of the conical roof in Figure 19–7

$$= \frac{(10) \times (3.1416 \times 20 \text{ ft.})}{2}$$

$$= 314 \text{ sq. ft.}$$

Gothic-Shape Roof Areas

A gothic-shape roof (Figure 19–8) is formed by two circles passing through the wall plates and intersecting at the ridge. A frequently used rule for laying out a gothic roof is to scribe two circles, each with a radius of two thirds the width of the building, with the centers of the circles at A and A_1 as shown.

If the length of the gothic roof in Figure 19–8 is 50 feet, the area of the roof is the distance BC + CD times 50 feet.

The distance BC is the circumference of the circle times .2087. The circumference of the circle is the radius

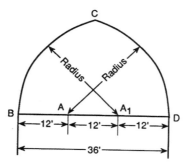

Figure 19–8

(2/3 of the width of the building) times 6.2832. Since BC and CD are equal, the formula for obtaining the roof area would be:

$$
\begin{aligned}
\text{Area} &= 2(\text{radius} \times 6.2832) \times .2087 \times 50 \text{ ft.} \\
&= 2(24 \text{ ft.} \times 6.2832) \times .2087 \times 50 \text{ ft.} \\
&= 3147.13 \text{ sq. ft.}
\end{aligned}
$$

In terms of the width and the length of the building, the area for a gothic roof of these proportions may be expressed for practical application, where W = width and L = length, as follows:

$$
\begin{aligned}
\text{Area} &= 1.75 \, W \times L \\
&= 1.75 \times 36 \text{ ft.} \times 50 \text{ ft.} \\
&= 3,150 \text{ sq. ft.}
\end{aligned}
$$

SHINGLE ROOFING IN GENERAL

The most commonly used shingles, other than wood shingles, are asphalt, fiberglass, slate, tile or its cement imitation, and aluminum. Shingles are suitable for inclined roof surfaces with sufficient slope to prevent water from getting under the lapped ends or butts. In the application of the various kinds of shingles, there are several important things to consider.

Type of Shingle

Each type of shingle is available in different grades, sizes, shapes, weights, and color or finish. Wood shingles, for example, are made in three sizes: 16-, 18-, and 24-inch, and each size can be obtained in three or more grades. Asphalt shingles come in several sizes, shapes, and colors with weights ranging from 145 to 330 pounds per square. Slate shingles are sold in different sizes, thicknesses, colors, and grades.

Type of Roof Decking

Shingles, with the exception of wood shingles, should be laid over solid roof decking. When nominal 1-inch boards are used, they should be matched to provide rigidity and firmness between rafters. Plywood decking should be a minimum of 5/16 in. thick, and the joints should be reinforced with blocking.

Usually open sheathing is recommended as a base for wood shingles. This may be 1-inch boards, 4 or 6 inches wide and spaced equal to the exposure.

Felt Underlay

A layer of asphalt-saturated roofing paper should be applied over the roof decking before laying all but wood shingles. The purpose is to seal the roof surface against moisture. Under heavy shingles like tile or slate, a 30-pound felt is recommended.

Exposure

Exposure is the number of inches of a shingle that is exposed to the weather. It is the distance from the butt edge of one shingle to the butt edge of the shingle above or below it.

When shingles are laid with the exposure specified by the manufacturer, one square will cover 100 square feet. If the exposure is increased, the area covered by a square of shingles is increased proportionately, and conversely when the exposure is less than that specified by the manufacturer, the area covered by a square of shingles is decreased proportionately. Table 19–3 on pages 402–403 shows coverages of wood shingles for various exposures.

Starter Shingles

When shingling begins at the eave-line of a main roof or a dormer, *starting* shingles are laid first. The usual practice is to reverse the shingle, placing the butt end toward the ridge. Shingling is then begun with the first exposed course laid directly over the starters. This method is necessary because wood, and asphalt shingles are laid with a slight space between them, and the starters seal the roof under the spacing of the first course. Special starting material is available for asphalt shingles. Starter shingles should extend 1½ in. beyond the edge of the roof.

Hip and Ridge Caps

To seal the hips and ridge of a roof, and also to give a finished appearance, they are capped with the same shingles used on the roof, or with special caps provided by the manufacturer.

Waste

Waste in estimating roofing is the material lost by cutting and fitting around gables, ridges, hips, and valleys, plus starters and caps. The percentage to be added is a matter of an educated guess taking into consideration the shape, type, and size of the roof. More waste is added for cut-up roofs and for hip roofs versus gable roofs. On plain roofs, 10 to 15 percent is usually adequate, whereas on cut-up roofs, 15 to 20 percent may be needed.

WOOD SHINGLE ROOFING AND SIDING

Wood shingles are manufactured from several kinds of wood, the best of which are western red cedar, redwood, and cypress. All of these have a high decay resistance, low shrinkage, and a tendency to lie flat on the surface. The better grades of shingles are of clear vertical (edge) grain wood that consist entirely of heartwood with no defects. The intermediate and the economical grades are clear on the lower 6 inches but may contain defects such as knots, feather tips, flat grains, or some sapwood. These are recommended more for farm buildings, sidewalls of garages, summer homes, and secondary buildings. They are also used for undercourses where double coursing of side walls is contemplated. Grading rules of wood shingles, as with all lumber products, have been formulated by the Bureau of Standards of the U.S. Department of Commerce in cooperation with the industry.

The three standard lengths of wood shingles are 16-inch, 18-inch, and 24-inch. The standard exposure for 16-inch shingles is 5 inches; for 18-inch it is 5½ inches; and for 24-inch, 7½ inches. There are four bundles of shingles to a square (100 sq. ft.). The Red Cedar Shingle & Handsplit Shake Bureau recommends, on roof slopes less than 4/12 to a minimum of 3/12, that the shingle exposure be reduced to 3¾ inches, 4¼ inches, and 5¾ inches respectively for 16-, 18-, and 24-inch shingles. On all roofs, there should be a minimum of *three* layers of wood shingle at every point to prevent leakage in wind-driven rain storms. On sidewalls two layers of wood shingle at every point are considered adequate.

The standard thickness of four butts of 16- and 24-inch shingles is 2 inches. The standard thickness of four butts of 18-inch shingles is 2¼ inches.

Table 19–3 shows the covering capacities of Certigrade Red Cedar Shingles. Since the manufacturer specifies 5, 5½, and 7½ inches respectively for 16-, 18-, and 24-inch

TABLE 19–3
Covering Capacities of Certigrade Red Cedar Shingles

CERTIGRADE RED CEDAR SHINGLES

GRADE	Length	Thickness (at Butt)	No. of Courses Per Bundle	Bdls Cartons Per Square		Description
No. 1 BLUE LABEL	16" (Fivex) 18" (Perfections) 24" (Royals)	.40" .45" .50"	20 20 18 18 13 14	4 bdls. 4 bdls. 4 bdls.	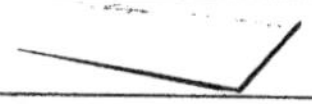	The premium grade of shingles for roofs and sidewalls. These top-grade shingles are 100% heartwood. 100% clear and 100% edge-grain.
No. 2 RED LABEL	16" (Fivex) 18" (Perfections) 24" (Royals)	.40" .45" .50"	20 20 18 18 13 14	4 bdls. 4 bdls. 4 bdls.		A good grade for many applications. Not less than 10" clear on 16" shingles, 11" clear on 18" shingles and 16" clear on 24" shingles. Flat grain and limited sapwood are permitted in this grade.
No. 3 BLACK LABEL	16" (Fivex) 18" (Perfections) 24" (Royals)	.40" .45" .50"	20 20 18 18 13 14	4 bdls. 4 bdls. 4 bdls.		A utility grade for economy applications and secondary buildings. Not less than 6" clear on 16" and 18" shingles, 10" clear on 24" shingles.
No. 4 UNDER-COURSING	16" (Fivex) 18" (Perfections)	.40" .45"	14 14 or 20 20 14 14 or 18 18	2 bdls. 4 bdls. 2 bdls. 4 bdls.		A utility grade for undercoursing on double-coursed sidewall applications or for interior accent walls.
No. 1 or No. 2 REBUTTED-REJOINTED	16" (Fivex) 18" (Perfections) 24" (Royals)	.40" .45" .50"	33 33 28 28 13 14	1 carton 1 carton 4 bdls.		Same specifications as above for No 1 and No 2 grades but machine trimmed for parallel edges with butts sawn at right angles. For sidewall application where tightly fitting joints are desired. Also available with smooth sanded face.

PITCH	Maximum exposure recommended for roofs:								
	NO. 1 BLUE LABEL			NO. 2 RED LABEL			NO. 3 BLACK LABEL		
	16"	18"	24"	16"	18"	24"	16"	18"	24"
3 IN 12 TO 4 IN 12	3¾"	4¼"	5¾"	3½"	4"	5½"	3"	3½"	5"
4 IN 12 AND STEEPER	5"	5½"	7½"	4"	4½"	6½"	3½"	4"	5½"

LENGTH AND THICKNESS	Approximate coverage of one square (4 bundles) of shingles based on following weather exposures																									
	3½"	4"	4½"	5"	5½"	6"	6½"	7"	7½"	8"	8½"	9"	9½"	10"	10½"	11"	11½"	12"	12½"	13"	13½"	14"	14½"	15"	15½"	16"
16" x 5 2"	70	80	90	100*	110	120	130	140	150‡	160	170	180	190	200	210	220	230	240†								
18" x 5 2¼"		72½	81½	90½	100*	109	118	127	136	145½	154½‡	163½	172½	181½	191	200	209	218	227	236	245½	254½†				
24" x 4 2"						80	86½	93	100*	106½	113	120	126½	133	140	146½	153‡	160	166½	173	180	186½	193	200	206½	213†

NOTES: * Maximum exposure recommended for roofs. ‡ Maximum exposure recommended for single-coursing No. 1 and No. 2 grades on sidewalls.
† Maximum exposure recommended for double-coursing No. 1 grades on sidewalls.

CERTIGROOVE GROOVED RED CEDAR SIDEWALL SHAKES

GRADE	Length	Thickness (at Butt)	No. Courses Per Carton	Cartons Per Square*		Description
No. 1 BLUE LABEL	16" 18" 24"	.40" .45" .50"	16 17 14 14 12 12	2 ctns. 2 ctns. 2 ctns.	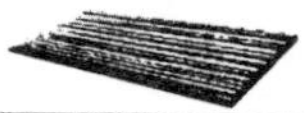	Machine-grooved shakes are manufactured from shingles and have striated faces and parallel edges. Used exclusively double-coursed on sidewalls.

NOTE: *Also marketed in one-carton squares.

CERTI-SPLIT RED CEDAR HANDSPLIT SHAKES

GRADE	Length and Thickness	18" Pack** Courses Per Bdl.	18" Pack** Bdls. Per Sq.		Description
No. 1 HANDSPLIT & RESAWN	15" Starter-Finish 18" x ½" Mediums 18" x ¾" Heavies 24" x ⅜" 24" x ½" Mediums 24" x ¾" Heavies	9 9 9 9 9 9 9 9 9 9 9 9	5 5 5 5 5 5		These shakes have split faces and sawn backs. Cedar logs are first cut into desired lengths. Blanks or boards of proper thickness are split and then run diagonally through a bandsaw to produce two tapered shakes from each blank.
No. 1 TAPERSAWN	24" x ⅝" 18" x ⅝"	9 9 9 9	5 5		These shakes are sawn both sides
No. 1 TAPERSPLIT	24" x ½"	9 9	5		Produced largely by hand, using a sharp-bladed steel froe and a wooden mallet. The natural shingle-like taper is achieved by reversing the block, end-for-end, with each split.
		20" Pack			
No. 1 STRAIGHT-SPLIT	18" x ⅜" True-Edge* 18" x ⅜" 24" x ⅜"	14 Straight 19 Straight 16 Straight	4 5 5		Produced in the same manner as tapersplit shakes except that by splitting from the same end of the block, the shakes acquire the same thickness throughout.

NOTE: * Exclusively sidewall product, with parallel edges.
** Pack used for majority of shakes.

SHAKE TYPE, LENGTH AND THICKNESS	Approximate coverage (in sq. ft.) of one square, when shakes are applied with ½" spacing, at following weather exposures, in inches (h): 5½	7½	8½	10	11½	16
18" x ½" Handsplit-and-Resawn Mediums (a)	55(b)	75(c)	85(d)	100	…	…
18" x ¾" Handsplit-and-Resawn Heavies (a)	55(b)	75(c)	85(d)	100	…	…
18" x ⅝" Tapersawn	55(b)	75(c)	85(d)			
24" x ⅜" Handsplit	…	75(e)	85	100(f)	115(d)	…
24" x ½" Handsplit-and-Resawn Mediums	…	75(b)	85	100(c)	115(d)	…
24" x ¾" Handsplit-and-Resawn Heavies	…	75(b)	85	100(c)	115(d)	…
24" x ⅝" Tapersawn		75b	85	100(c)	115(d)	
24" x ½" Tapersplit	…	75(b)	85	100(c)	115(d)	…
18" x ⅜" True-Edge Straight-Split	…	…	…	…	…	112(g)
18" x ⅜" Straight-Split	65(b)	90	100(d)	…	…	…
24" x ⅜" Straight-Split	…	75(b)	85	100	115(d)	…
15" Starter-Finish Course	Use supplementary with shakes applied not over 10" weather exposure.					

(a) 5 bundles will cover 100 sq. ft. roof area when used as starter-finish course at 10" weather exposure; 6 bundles will cover 100 sq. ft. wall area at 8½" exposure; 7 bundles will cover 100 sq. ft. roof area at 7½" weather exposure; see footnote (h).

(b) Maximum recommended weather exposure for 3-ply roof construction.

(c) Maximum recommended weather exposure for 2-ply roof construction.

(d) Maximum recommended weather exposure for single-coursed wall construction.

(e) Maximum recommended weather exposure for application on roof pitches between 4-in-12 and 8-in-12.

(f) Maximum recommended weather exposure for application on roof pitches of 8-in-12 and steeper.

(g) Maximum recommended weather exposure for double-coursed wall construction.

(h) All coverage based on ½" spacing between shakes.

For detailed information on all cedar products, write to: Red Cedar Shingle & Handsplit Shake Bureau, Suite 275, 515-116th Ave. N.E., Bellevue, WA 98004
(In Canada: Suite 1500, 1055 West Hastings Street, Vancouver, B.C., V6E 2H1)

BROCHURE NO. 34 PRINTED IN U.S.A. 4-80

Courtesy of Red Shingle & Handsplit Shake Bureau

exposures to apply one square of shingles, any variation of the exposure so specified will result in a square of shingles covering more, or less, than 100 square feet.

Applying Wood Shingles to Roofs

Wood shingles on roofs may be applied over spaced or solid sheathing. Spaced sheathing should be 1-by-4-inch or 1-by-6-inch boards. One method is to space 1-by-4-inch boards to coincide with the weather exposure. A second method of application, where 1-by-6-inch boards are used, two courses of shingles are nailed to each 1-by-6-inch board. Building paper is sometimes placed under wood shingles over the spaced or the solid wood sheathing. Saturated felt paper is undesirable as it retards the drying of the shingles when they have been wet. Shingles should be spaced at least 1/4 inch apart to allow for expansion.

Estimating Wood Shingles for Roofs

The quantity of wood shingles required for roofs is obtained by computing the square foot area to be covered, as discussed under Measuring Roof Areas. From Table 19–3, the coverage of one square of shingles of 16-, 18-, and 24-inch lengths, and of various exposures is shown.

Labor to Apply Wood Shingles

The labor to apply wood shingles to roofs is, in general, the same as outlined for sidewall application. Carpenters can apply the 16-inch and 18-inch shingles, at specified exposures and under average working conditions, at approximately 4 to 6 hours per square. For 24-inch shingles the rate of application will average approximately 3 to 5 hours per square. Where roof surfaces are irregular and cut up, the rate of laying wood shingles, as with other kinds of shingles, has to be adjusted accordingly. Also, where experienced shinglers are employed, their rate of productivity will increase 20 to 25 percent.

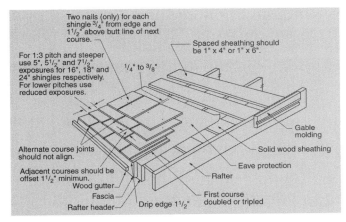

Two nails (only) for each shingle ³/₄" from edge and 1¹/₂" above butt line of next course.

Spaced sheathing should be 1" x 4" or 1" x 6".

For 1:3 pitch and steeper use 5", 5¹/₂" and 7¹/₂" exposures for 16", 18" and 24" shingles respectively. For lower pitches use reduced exposures.

¹/₄" to ³/₈"

Gable molding

Alternate course joints should not align.

Solid wood sheathing

Eave protection

Adjacent courses should be offset 1¹/₂" minimun.

Rafter

Wood gutter

Fascia

First course doubled or tripled

Rafter header

Drip edge 1¹/₂"

GENERAL
APPLICATION DETAILS

Shingle Application

Regardless of style, the following basic application details (Figure 19–9) must be observed.

1. Shingles must be doubled or tripled at all eaves.
2. Butts of first course shingles should project 1¹/₂" beyond the fascia.
3. Spacing between adjacent shingles (joints) should be a minimum of ¹/₄" and a maximum of ³/₈".
4. Joints in any one course should be separated not less than 1¹/₂" from joints in adjacent courses; and in any three courses, no two joints should be in direct alignment.
5. In lesser grade shingles containing both flat and vertical grain, joints should not be aligned with centerline of heart.
6. Flat grain shingles wider than 8" should be split in two before nailing. Knots and similar defects should be treated as the edge of the shingle and the joint in the course above placed 1¹/₂" from the edge of the defect.

Figure 19–9
Shingle Application to Roofs
Courtesy of Red Cedar Shingle & Handsplit Shake Bureau

It is important, therefore, when estimating the wood shingles required, to make certain of the exposure to be used for the particular length of shingle. *Both the quantity of material and the required labor to cover 100 square feet decreases as the exposure is increased and vice versa.*

The type and length of nail used to apply wood shingles is important, not more than two nails being used per shingle. Nails should be rust-resistant and may be zinc-coated or aluminum. For new work the 3d nail is used on 16- and 18-inch shingles, while the 4d nail is recommended for the 24-inch shingle. When new shingles are applied over old shingles, the 5d or 6d nails are used. For double coursing of shingles, the 5d nail is considered best. If shingles are to be applied directly over composition sheathing without nailing strips, a special self-clinching nail should be used in accordance with the specifications of the manufacturer.

Applying Wood Shingles to Sidewalls

Wood shingles may be applied on sidewalls over open sheathing, solid sheathing, or composition types of sheathing.

Open, or spaced sheathing, on centers equal to the exposure used, is common practice in warm climates and on seasonal bungalows or farm buildings. The strips may be 1 in. by 3 in., 1 in. by 4 in., or 1 in. by 6 in. nailed to the studs. Saturated building paper is first applied between the nailing strips and the studding.

When sidewall sheathing is solid boarding or plywood, a single layer of saturated felt paper is applied before the shingles are put in place.

There are two general methods of sidewall application: *single coursing* and *double coursing.* Because the sides of a building are usually less exposed to the direct action of rain, snow, and sun, the shingle exposure can be considerably greater than on roofs. In single coursing the first or lowest course is doubled, and the succeeding courses are applied above. Double coursing is principally used where a deep shadow line is desired for architectural effect.

Where shingles join at an outside corner, they may be mitered or they may be laced by having one overlap the other. Inside corners are finished by butting the shingles up against a vertical strip that has been installed prior to shingling. Mitering takes more time than lacing, making the labor cost higher.

Estimating Material for Sidewalls

The quantity of shingles needed for the sidewalls of a building is determined by measuring the area and deducting all window and door openings to obtain the net area to be covered in square feet. Table 19–3 shows the portion of a square of shingles required to cover 100 square feet for the corresponding exposure in inches. The usual waste allowance for starting courses and for cutting and fitting around windows and doors is 10 percent.

Labor Applying Wood Shingles to Sidewalls

When working under average conditions a *carpenter* should be able to apply 16- and 18-inch shingles at the rate of 4 to 6 hours per 100 square feet for exposures of 4 inches to 7 inches. On sidewalls that are difficult because of many openings and bays, the rate of application for *carpenters* will average 5 to 7 hours per 100 square feet for exposures of 4 inches to 7 inches. Experienced shinglers can apply shingles as much as 20 to 25 percent faster than carpenters who are not accustomed to that type of work.

When estimating labor productivity, working conditions, the experience of the workers, and other factors outlined and described in Chapter 10 must be taken into consideration.

HANDSPLIT SHAKES

Handsplit shakes are available in four grades: *handsplit and resawn, tapersplit, tapersawn,* and *straight split.* Table 19–3, courtesy of Red Cedar Shingle & Handsplit Shake

Bureau, shows lengths and thickness of each grade. It also describes each grade and shows the approximate coverage of one square when shakes are applied with 1/2-inch spacing at various weather exposures in inches.

Roof Decking for Shakes

Cedar shakes, like shingles, may be applied over spaced or solid roof decking. Spaced decking should be 1-by-6-inch boards on centers equal to the weather exposure that the shakes are to be laid (see weather exposure, Table 19–3) but not more than 7½ in. for 18-inch shakes, and 10 in. for 24-inch shakes on roof installations.

The minimum roof slope on which shakes are recommended is 1:4. They may be applied successfully, however, on solid decked roofs of lower slopes, providing a special method of application is followed. (For details, contact your nearest Red Cedar Shingle & Handsplit Shake Bureau.)

Handsplit Shake Coverage

Table 19–3 shows the type, length, and thickness of shakes, and the approximate square foot coverage at various weather exposures.

Hips and Ridges

Site-made or factory-assembled hip and ridge units may be used, but both types must have alternate overlaps and concealed nailing. Weather exposures should be the same as those on the main roof surface. Nails must be longer and sufficient to penetrate ½ inch into or completely through the roof decking.

Nailing Shakes

Two 5d or 6d nails, either aluminum or rust-resistant, should be driven at least 1-inch from each edge, and 1 or 2 inches above the butt line of the course to follow.

Applying Shakes to Roofs

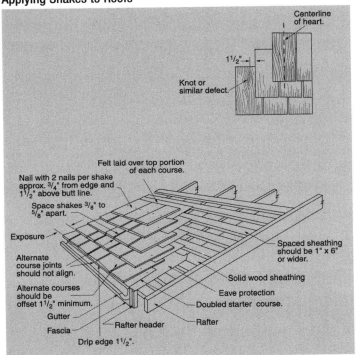

Shake Application

Shakes, like shingles, are normally applied in straight, single courses. The following application details (Figure 19–10) must be observed.

1. The starter course may be one or two layers of cedar shingles overlaid with the desired shake.
2. The first course or courses may be of shakes. A 15″ shake is made expressly for starter and finish courses.
3. An 18″ wide strip of No. 30 roofing felt (or No. 15 felt depending upon code requirements) should be laid over the top portion of the shakes and extend on to the sheathing. The bottom edge of the felt should be positioned above the butt of the shake at a distance equal to twice the weather exposure. For example, 24″ shakes laid with 10″ exposure would have felt applied 20″ above the butt. Thus the felt will cover the top 4″ of the shakes and extend up 14″ onto the sheathing. Note that the top edge of the felt must rest on the spaced sheathing.
4. Spacing between adjacent shakes should be a minimum of ⅜″ and a maximum of ⅝.″
5. Joints between shakes should be offset 1½″ over adjacent courses.
6. Straight-split shakes should be laid with the froe-end (the end from which the shake has been split and which is smoother) towards the ridge.

Figure 19–10
Shake Application to Roofs
Courtesy of Red Cedar Shingle & Handsplit Shake Bureau

410 *Roofing*

Longer nails should be used if the thickness of the shake indicates, to achieve proper penetration. It takes approximately 2 lb. of nails per square at standard exposures.

Sidewall Application of Shakes

Maximum exposure on sidewalls with single-course construction is 8½ inches for 18-inch shakes, and 11½ inches for 24-inch shakes. Double-course application requires an underlay of shakes or regular cedar shingles. Weather exposures of up to 14 inches are permissible with 18-inch handsplit-resawn or tapersplit shakes, and 20 inches with 24-inch shakes. If straight-split shakes are used, the dou-

Figure 19–11
Applying handsplit shakes, using
pneumatic nailing tools
Courtesy of Senco Products, Inc.

ble course exposure may be 16 inches for 18-inch shakes and 22 inches for 24-inch shakes.

Estimating Shakes for Roofs

Obtain the area of the roof to be covered and divide by 100 to get the number of squares required at specified exposures. Allow for double coursing at eaves and one square for each 100 lineal feet of valleys.

Labor Applying Shakes to Roofs

As with shingle application, the labor per square will vary with the pitch of the roof, how cut up the surface is with dormers or intersecting roofs, and with the experience of the workers. In general the rate at which shakes are applied is comparable to that for wood shingles. Figure 19–11 on page 410 shows workers applying handsplit shakes with Senco pneumatic nailing tools.

ASPHALT ROOFING PRODUCTS

The three principal types of asphalt roofing products are *saturated felts, roll roofing,* and *asphalt shingles.* All three are made by impregnating, with asphalt, felt paper made from organic fibers such as rag, wood, and jute. Fiberglass mats are also used in place of organic fibers.

Table 19–4 on page 413 shows typical asphalt rolls: mineral surface, smooth surface, and saturated felt. Table 19–5 on page 414 shows typical asphalt shingles, self-sealing strip shingles, and individual interlocking shingles. Both Tables, published in 1987 by the Asphalt Roofing Manufacturers Association of Rockville, Maryland, show specifications including configurations, approximate weights, widths, lengths, exposures, and Underwriters Laboratories listing.

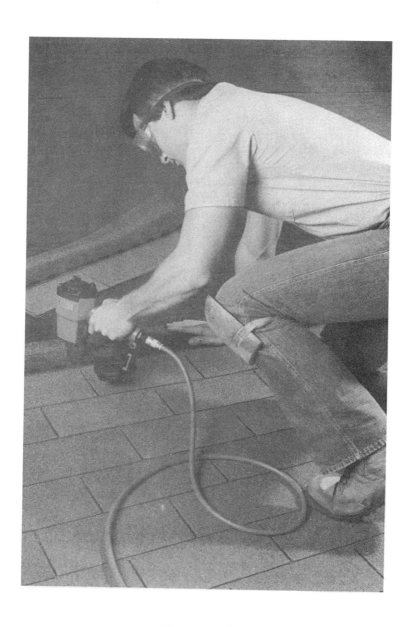

Figure 19–12
Applying shingles using
the Senco coil nailer
Courtesy of Senco Products, Inc.

TABLE 19–4

Typical Asphalt Rolls

PRODUCT	Approximate Shipping Weight		Squares Per Package	Length	Width	Selvage	Exposure	Underwriters Laboratories Listing*
	Per Roll	Per Square						
Mineral surface roll	75# to 90#	75# to 90#	1	36′ to 38′	36″	2″ to 4″	32″ to 34″	C
Mineral surface roll (double coverage)	55# to 70#	110# to 140#	½	36′	36″	19″	17″	C
Smooth surface roll	50# to 86#	40# to 65#	1 to 2	36′ to 72′	36″	2″	34″	None
Saturated felt (non-perforated)	45# to 60#	11# to 30#	2 to 4	72′ to 144′	36″	2″ to 19″	17″ to 34″	None

*UL rating at time of publication. Reference should be made to individual manufacturer's product at time of purchase.

Courtesy of Asphalt Roofing Manufacturers Association, Rockville, Maryland

TABLE 19–5

Typical Asphalt Shingles

| PRODUCT | Configuration | Per Square | | | Size | | Exposure | Underwriters Laboratories Listing |
		Approximate Shipping Weight	Shingles	Bundles	Width	Length		
Self-sealing random-tab strip shingle 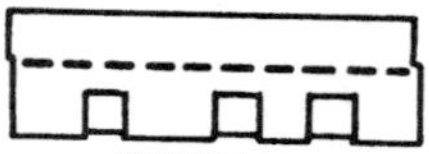 Multi-thickness	Various edge, surface texture and application treatments	240# to 360#	64 to 90	3, 4 or 5	11½" to 14"	36" to 40"	4" to 6"	A or C - Many wind resistant
Self-sealing random-tab strip shingle 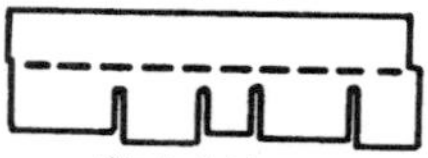 Single-thickness	Various edge, surface texture and application treatments	240# to 300#	65 to 80	3 or 4	12" to 13¼"	36" to 40"	5" to 5⅝"	A or C - Many wind resistant
Self-sealing square-tab strip shingle 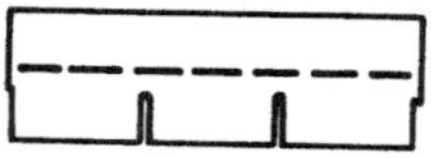 Three-tab	Two-tab or Four-tab	215# to 325#	65 to 80	3 or 4	12" to 13¼"	36" to 40"	5" to 5⅝"	A or C - All wind resistant
	Three-tab	215# to 300#	65 to 80	3 or 4	12" to 13¼"	36" to 40"	5" to 5⅝"	
Self-sealing square-tab strip shingle 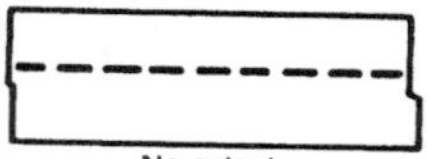 No-cutout	Various edge and surface texture treatments	215# to 290#	65 to 81	3 or 4	12" to 13¼"	36" to 40"	5" to 5⅝"	A or C - All wind resistant
Individual interlocking shingle 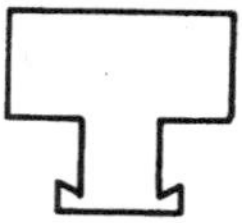 Basic design	Several design variations	180# to 250#	72 to 120	3 or 4	18" to 22¼"	20" to 22½"	—	A or C - Many wind resistant

Courtesy of Asphalt Roofing Manufacturers Association, Rockville, Maryland

414

ASPHALT SHINGLES

Asphalt shingles should be laid over 11-lb. to 30-lb. non-perforated saturated felt, depending on the pitch of the roof.

Estimating Material and Labor

Rust-resistant or aluminum nails, 10- to 12-gauge, 1¼ in. to 1½ in. long with 3/8-inch to 7/16-inch flat heads should be used over solid board decking. Shorter nails may be used over plywood decking, with threaded nails recommended for greater holding capacity. It takes approximately 1½ to 2 pounds of nails per square.

The quantity of shingles required is obtained by computing the total square foot area to be covered and adding the estimated waste for cutting and fitting, and for hips and caps. Usually 10 percent is added for simple gable roofs, with as much as 20 percent for more complicated roof surfaces.

Under average working conditions on roofs with moderate slope, experienced roofers can apply 3-tab strip shingles at a rate of approximately 1 to 1½ hours per square. On badly cut-up roofs with dormers, hips, and valleys, the rate will increase commensurably.

ROLL ROOFING

Asphalt roll roofing comes in rolls 36 inches wide and 36 feet long. A roll contains 108 square feet and, making an allowance for lapping, it will cover 100 square feet. The 65-lb. smooth surfaced and the 90-lb. mineral surfaced rolls are the most commonly used. An underlay of 15-lb. saturated felt is sometimes used, but is not required. Lapping cement and 7/8-inch galvanized nails are provided inside each roll. When laid over old roofing, longer nails should be used in order to penetrate the wood decking at least 3/4 inch.

Roll roofing is applied to roofs of every pitch. It is used on all types of structures with flat roofs, particularly where initial cost is a consideration. It is the least expensive roof to apply.

Roll roofing may be laid parallel to the eaves or the ridge of a pitched roof. Caps for ridges and hips are usually made by splitting a roll down the center. For better quality work, 9-inch starter strips are applied along the eaves and up the *rake* or gable ends.

The ends are lapped 6 inches and the edge laps are 2 inches. Roll roofing may be applied by nailing the upper edge of each strip and cementing the bottom edge. This is called the *concealed nail method*. The more common method of application is to nail the bottom edge of the overlapping strip, which is called the *exposed nail method*.

Estimating Material

The quantity of roll roofing required is obtained from the net roof area to be covered. Waste of about 10 percent is usually adequate to take care of cutting, lapping at edges of eaves and gables, and for caps.

Estimating Labor

Labor factors that affect the rate of laying other types of roofing should be considered when estimating the labor to lay roll roofing. The rolls are heavy and more cumbersome to manage as the pitch of the roof increases. On flat and moderately pitched roofs, a roofer or a carpenter should be able to cover a square in 1 to 1½ hours. On more steeply pitched and difficult roofs, it will take closer to 2 hours per square.

BUILT-UP ROOFING

A built-up roof is one that is applied by *building up* successive layers of saturated felt, lapped and cemented together with hot tar or asphalt. The layers of felt are called *plies*, and this type of roof is identified by the number of its plies, usually running 3 to 5 thick. The top may be finished off with a mopping of tar or asphalt and is referred to as a built-up smooth-top roof. Where slag or gravel is imbedded in the top coating, the roof is referred to as a tar and gravel, or a slag roof. When the slope of the roof exceeds 2 inches to the foot, the top ply is often mineral-surfaced 19-inch selvage, double coverage roll roofing. (See Table 19–4.)

The saturated felt used for built-up roofing is usually the 15-lb. type. It is 36 inches wide, 144 feet long, and contains 432 square feet. One roll weighs 60 pounds. On many roofs, particularly bonded roofs, the first ply is a 30-lb. felt which is available in rolls of 36 inches by 72 feet, containing 216 square feet.

On new roofs over wood decking, a layer of rosin-sized building paper or one layer of 15-lb. saturated felt is applied to prevent the pitch or asphalt from leaking through the cracks, knot holes, or joints in the decking. This layer is not counted as one of the plies. Successive layers of 15-lb. felt are then laid, lapping the sheets 19 inches and mopping between plies with hot pitch or asphalt. Approximately 30 pounds of asphalt or pitch per square is used for each mopping with at least double that quantity (60 pounds) applied to the surface if slag, gravel, or crushed marble is to be imbedded. About 300 pounds of slag or 400 pounds of gravel are required for each 100 square feet of surface covered. The quantity of pitch or asphalt applied to the surface when slag or gravel is not used is approximately 40 pounds per square.

Estimating Materials

The number of squares of built-up roofing required is obtained from the net roof area to be covered. Waste is taken care of in the rolls by the manufacturer. A roll of 15-lb. felt contains 432 square feet and is intended to cover 400 square feet or 4 squares. A roll of 30-lb. felt contains 216 square feet and will cover 200 square feet or 2 squares. Measurements of a roof are taken from the *outside* of the walls. Allowance should be made for flashing against bulkheads and parapet walls. Normally the flashing is carried up 1 foot, but in older buildings being re-roofed, the entire parapet may be flashed up to the coping.

Skylights, ventilators, and so forth need not be deducted unless the area they occupy is in excess of 100 square feet; then it is customary to deduct only one half of the area which allows for the additional work of cutting, flashing, and working around them.

The materials used in built-up roofing are saturated felt, asphalt or coal-tar pitch, and slag or gravel when so surfaced. Different manufacturers and roofers vary the specifications, but for estimating purposes, the quantities shown in Table 19–6 may be helpful as a guide.

Estimating Labor

Built-up roofing is usually applied by a crew ranging from three to five workers, depending on the size of the job and working conditions. Materials are raised by rope and pulley, by crane, and on multistoried buildings by outside elevators. Obviously roofs with numerous skylights, ventilators, bulkheads, and of irregular shape require more labor than plain flat roofs. Each job has to be studied for its individual characteristics before deciding upon the number of hours labor per square. As a guide in estimating, the approximate number of man hours per square for uncomplicated, one- to three-story roofs are:

3-ply and 4-coat . . . . 1½ to 2 hours
4-ply and 5-coat . . . . 2 to 2¼ hours
5-ply and 5-coat . . . . 2¼ to 2½ hours
Add for 1 coat
gravel or slag . . . . ¼ to ½ hour

Translating these guidelines into the number of squares laid by a crew of five workers in an 8-hour day (40 hours total):

3-ply and 4-coat. . . . . 20 to 26.6 squares
4-ply and 5-coat. . . . . 17.8 to 20 squares
5-ply and 5-coat . . . . 16 to 17.8 squares

TABLE 19–6

Materials Required to Apply 100 Square Feet Built-up Roofing Using Asphalt, and Surfacing with Slag or Gravel*

Number of Plies	Rolls of 15-Lb Saturated Felt (432 Square Feet)	Pounds of Asphalt	Pounds of Slag	Pounds of Gravel
3	¾	150	300	400
4	1	180	300	400
5	1¼	230	300	400
3(1 dry ply)	¾	120	300	400
4(2 dry plies)	1	120	300	400
5(2 dry plies)	1¼	150	300	400

* For smooth-top finish without slag or gravel, deduct 20 pounds of asphalt per 100 square feet of roofing. When 30-lb. saturated felt is used as a first ply, add ¼ of a roll of 15-lb. felt.

Fuel, Mops, Supplies

The fuel required to heat asphalt or pitch, the mops to apply it, brushes to spread gravel or slag, pails, nails, and so forth, are all expendable supplies. In computing the cost of built-up roofs, a flat allowance may be made to cover these items, or a percentage may be added to the total cost of labor and material for each job. The amount varies but ranges between 5 and 10 percent.

FIBERGLASS SHINGLES

Fiberglass shingles are actually asphalt shingles with a *fiberglass mat* in place of the *organic felt base*. This results in a stronger shingle with more asphalt for the same weight as the felt-base shingle; it has better wearing resistance and a longer life. Also, because the glass fiber base is noncombustible, the shingle is given a Class A label by the Underwriters Laboratory. (The organic felt base asphalt shingle has a Class C label.)

Fiberglass shingles are available in the conventional three-tab size. Hip and ridge units are available per lineal foot in bundles. Application is the same as for the organic-base shingles.

SLATE SHINGLES

Slate shingles vary in size from 6 inch to 14 inch in width, and from 20 inch to 24 inch in length. The standard thickness of slate shingles is 3/16 in., but they are available in thicknesses of 1 inch or more according to architectural requirements.

Slate shingles are usually laid over solid roof decking with a 30-lb. saturated felt underlay. However, they may be laid over sheathing that is spaced the same as the shingle exposure.

The holes in slate for nails are usually punched at the quarry. The *lap* of a slate shingle is the distance by which the upper slate overlaps the head of the second slate below. The standard lap is 3 inches. The exposure of slate shingles is 1/2 the length after deducting the lap. A 10-inch shingle with a 3-inch lap will have a 3½-inch exposure (10 – 3/2 = 3½ inches).

Hips and ridges are generally finished by alternately lapping shingles cut to proper size. The first course at the eaves is doubled for a starting course.

Waste from breakage and cutting, an allowance for the starter course and ridge and hip caps, varies between 10 and 20 percent.

Estimating Materials

The quantity of roofing slate is obtained by adding the appropriate waste to the net roof area and dividing by 100 to get the number of squares. Approximately one-half of a roll of saturated felt is needed per square.

Copper nails are recommended, 3d and 4d, for laying the 10-by-20-inch shingles. Approximately 3 lb. of nails per square of shingles are required.

Estimating Labor

Experienced slaters are difficult to find in most areas. Two slaters usually work as a team, and occasionally a helper is used to fill out a crew. The labor hours to apply a square of slate shingles will vary with the size of the shingles, the slope and shape of the roof, and the experience of the workers. On plain, uncomplicated roofs, a crew of two roofers and a helper can apply 10-by-20-inch slate shingles at approximately 2 to 2½ hours per square. On cut-up roof surfaces the rate of application may drop to 3 or 4 hours per square.

CORRUGATED STEEL AND ALUMINUM ROOFING

Corrugated aluminum and galvanized steel roofing and siding are used extensively on industrial-type buildings and also on farm and secondary buildings. The sheets are available with 1¼-inch and 2½-inch corrugations, and they come in different widths and lengths. The 26-inch wide sheets are the most popular, and lengths ranging from 6 to 12 feet should be selected according to the area to be covered to avoid as much cutting as possible.

Side laps should be a minimum of 1½ corrugations on sidewalls, and 2 to 3 corrugations on pitched roofs. End laps should be at least 2 to 6 inches depending upon the pitch of the roof.

Ridge and valley rolls, rubber end-seals, and end wall flashings may be obtained for both galvanized steel and aluminum corrugated sheets. Special rust-resistant nails and washers are used to apply this type roofing and siding. A 3/4-inch nail is used for new work; allow about 2 lb. per square.

Corrugated sheets may be applied to solid wood decking, or over open framing. When applied over steel framing, special types of fittings are required.

The price of corrugated roofing varies considerably with its thickness. Galvanized sheets are available in 12 gauge to 26 gauge. Aluminum sheet thickness is measured in inches and ranges from .019 inch (approximately 26 gauge) to .032 inch. The 26-gauge galvanized and .019-inch aluminum are often used on farm buildings, and heavier types are used on industrial buildings.

Estimating Material

The quantity of corrugated roofing is obtained from the net area to be covered. Waste to be added for lapping sides and ends and for cutting and fitting varies from 15 to 25 percent. Generally 20 percent added to the net area is adequate. Two pounds of nails should be included for each square of roofing.

Estimating Labor

The labor to apply corrugated metal roofing and siding varies with the job conditions, to some extent the size and weight of the sheets, and whether it is applied over wood or steel framing. Usually two roofers, or one roofer and a helper, work together.

As a guideline for estimating, assume that an experienced roofer and a helper make up a team. They should apply 26-by-96-inch corrugated steel or aluminum roofing on wood frame at an average rate of approximately 1 to 1½ hours per square. When applied as siding on wood frame, it takes a little longer and may average 1½ to 2 hours per square for the same team.

The labor to apply corrugated steel or aluminum roofing on steel frame will average 2½ to 3 hours per square for a roofer and help. When applied as siding on steel framing, the labor will average approximately 3 to 3½ hours per square for a roofer and helper. Scaffolding and hoisting of materials should be added.

ALUMINUM SHINGLES AND SHAKES

Aluminum shingles and shakes have a heavily textured surface which resembles the handsplit wood shake. They are available in 10-by-60-inch, 12-by-36-inch, and 12-by-48-inch sizes for easy handling. While basically marketed for roofing, they are often used as exterior siding. Aluminum shingles are relatively expensive, but they have good durability, light weight, and they need a minimum of maintenance.

Prices of aluminum shingles depend on the gauge (.020 ga. or .030 ga.) and on the finish. Ridge and hip caps, starter strips, trim, and so forth are available per lineal foot.

Labor to apply aluminum shingles, as with other roofing materials, is governed by the usual considerations such as how cut-up the roof shape is, the height and slope of the roof, the experience of the workers, and whether scaffolding is needed. Under average working conditions, it requires approximately 4 to 6 man-hours per square.

SQUARE FEET OF VARIOUS ROOFING MATERIALS DIRECT FROM TABLES

The number of square feet of various kinds of roofing required for *gable, hip,* and *intersecting roofs* may be read directly from Tables 19–7 through 19–12 (on pages 426–437). Quantities shown are for buildings with *ground-floor areas* from 200 to 3,000 square feet, and roof pitches of 1/8 to 1/2. The Tables apply regardless of story height as long as the roof covers the measured *ground-floor area.* Interpolation can be made for *ground-floor areas* that are between the 100 square foot increments.

Overhang

The Tables do not allow for overhang at the eaves. For each 12 inches of horizontal overhang at the eaves, add 5 percent to the Table quantities.

Dormers

The quantity of roofing for dormers is approximately the quantity for the area that the dormer takes up in the main roof. The only amount to be added is that of the dormer overhang.

Hip and Gable Roof Areas

The surface area of a hip roof is identical to that of a gable roof of the same pitch and covering the same horizontal area.

Nails Required per Square

Asphalt strip shingles	2 lb. 1¼-inch nail
Asphalt individual shingles	3 lb. 1¼-inch nail
Slate 10 in. by 20 in., 3-inch lap	3 lb. 3d or 4d nails
Clay tile	3 lb. 1¾ nail
Wood, 16-inch or 18-inch shingle	2 lb. 3d nail
Wood 24-inch	2 lb. 4d nail
Wood, handsplit shakes	2 lb. 6d nail

Formulas for Computing Gable and Hip Roof Areas

The square foot areas of *gable and hip* roofs can be quickly obtained by multiplying the horizontal area *covered by the roof* by the factor corresponding to the pitch of the roof in Table 19–1.

TABLE 19-7

Roofing for Gable and Hip Roofs, Including Intersecting Roofs, with Pitch of 1/8 (milling and cutting waste is included)

Ground Floor Area	Roof Surface Area	Asphalt Strip Shingles	Wood Shingles				Clay Roofing Tile	Number of Rolls of Saturated Felt Roofing Paper	
			18" Long		24" Long				
			4" Expo.	5½" Expo.	6" Expo.	7½" Expo.			
Sq. Ft.	Square Ft.	Square Ft.	Square Ft.	Square Ft.	Square Ft.	Square Ft.	Square Ft.	15-Lb.	30-Lb.
200	206	226	278	226	284	226	216	.57	1.14
300	309	339	417	339	426	339	324	.85	1.70
400	412	452	556	452	568	452	432	1.13	2.27
500	515	565	695	565	710	565	540	1.42	2.84
600	618	678	834	678	852	678	648	1.70	3.40
700	721	791	973	791	994	791	756	1.98	3.97
800	824	904	1112	904	1136	904	864	2.27	4.54
900	927	1010	1251	1010	1278	1010	972	2.55	5.10
1000	1030	1130	1390	1130	1420	1130	1080	2.83	5.66
1100	1133	1243	1529	1243	1562	1243	1188	3.11	6.22

426

1200	1236	1356	1668	1356	1704	1356	1296	3.40	6.80
1300	1339	1469	1810	1469	1846	1469	1404	3.68	7.36
1400	1442	1582	1946	1582	1990	1582	1512	3.96	7.92
1500	1545	1695	2085	1695	2130	1695	1620	4.25	8.50
1600	1648	1808	2224	1808	2272	1808	1728	4.53	9.06
1700	1751	1921	2363	1921	2414	1921	1836	4.81	9.62
1800	1854	2034	2500	2034	2556	2034	1944	5.09	10.18
1900	1957	2147	2640	2147	2700	2147	2052	5.38	10.76
2000	2060	2260	2780	2260	2840	2260	2160	5.66	11.32
2100	2163	2378	2920	2378	2982	2378	2268	5.94	11.88
2200	2266	2486	3058	2486	3124	2486	2376	6.23	12.46
2300	2369	2600	3200	2600	3266	2600	2484	6.51	13.02
2400	2472	2712	3336	2712	3410	2712	2592	6.80	13.58
2500	2575	2825	3475	2825	3550	2825	2700	7.08	14.16
2600	2678	2938	3615	2938	3692	2938	2808	7.36	14.72
2700	2781	3051	3753	3051	3834	3051	2916	7.64	15.28
2800	2884	3164	3892	3164	3976	3164	3024	7.92	15.84
2900	2987	3277	4030	3277	4118	3277	3132	8.21	16.42
3000	3090	3390	4170	3390	4260	3390	3240	8.49	16.98

Note: For types of roofing not shown, use Roof Surface Area column. Add the appropriate cutting and fitting waste. *Estimating Tables for Home Building*, Craftsman Book Co., 1986. Reprinted here with permission.

TABLE 19-8

Roofing for Gable and Hip Roofs, Including Intersecting Roofs, with Pitch of 1/6 (milling and cutting waste is included)

Ground Floor Area	Roof Surface Area	Asphalt Strip Shingles	Wood Shingles				Clay Roofing Tile	Number of Rolls of Saturated Felt Roofing Paper	
			18" Long		24" Long				
			4" Expo.	5½" Expo.	6" Expo.	7½" Expo.			
Sq. Ft.	Square Ft.	Square Ft.	Square Ft.	Square Ft.	Square Ft.	Square Ft.	Square Ft.	15-Lb.	30-Lb.
200	212	234	286	234	292	234	224	.59	1.17
300	318	351	429	351	438	351	336	.88	1.76
400	424	468	472	468	584	468	448	1.17	2.34
500	530	585	715	585	730	585	560	1.46	2.93
600	636	702	858	702	876	702	672	1.76	3.51
700	742	819	1000	819	1022	819	784	2.05	4.10
800	848	936	1144	936	1170	936	896	2.34	4.68
900	954	1053	1287	1050	1314	1050	1008	2.63	5.27
1000	1060	1170	1430	1170	1460	1170	1120	2.93	5.85
1100	1166	1287	1573	1287	1610	1287	1232	3.22	6.44

1200	1272	1404	1716	1404	1752	1404	1344	3.51	7.02
1300	1378	1521	1859	1521	1900	1521	1456	3.80	7.61
1400	1484	1638	2000	1638	2044	1638	1568	4.10	8.19
1500	1590	1755	2145	1755	2190	1755	1680	4.39	8.78
1600	1696	1872	2288	1872	2336	1872	1792	4.68	9.36
1700	1802	1989	2430	1989	2480	1989	1904	4.97	9.95
1800	1908	2106	2574	2106	2630	2106	2016	5.27	10.53
1900	2014	2223	2717	2223	2774	2223	2128	5.56	11.12
2000	2120	2340	2860	2340	2920	2340	2240	5.85	11.70
2100	2226	2457	3000	2457	3066	2457	2352	6.14	12.29
2200	2332	2574	3146	2574	3212	2574	2464	6.44	12.87
2300	2438	2691	3290	2691	3358	2691	2576	6.73	13.46
2400	2544	2808	3430	2808	3504	2808	2688	7.02	14.04
2500	2650	2925	3575	2925	3650	2925	2800	7.31	14.63
2600	2756	3042	3720	3042	3800	3042	2912	7.61	15.21
2700	2862	3159	3860	3159	3942	3159	3024	7.90	15.80
2800	2968	3276	4000	3276	4090	3276	3136	8.19	16.38
2900	3074	3393	4150	3393	4234	3393	3248	8.48	16.96
3000	3180	3510	4290	3510	4380	3510	3360	8.78	17.55

Note: For types of roofing not shown, use Roof Surface Area column. Add the appropriate cutting and fitting waste. *Estimating Tables for Home Building,* Craftsman Book Co., 1986. Reprinted here with permission.

TABLE 19-9

Roofing for Gable and Hip Roofs, Including Intersecting Roofs, with Pitch of 5/24 (milling and cutting waste is included)

Ground Floor Area	Roof Surface Area	Asphalt Strip Shingles	Wood Shingles				Clay Roofing Tile	Number of Rolls of Saturated Felt Roofing Paper	
			18" Long		24" Long				
			4" Expo.	5½" Expo.	6" Expo.	7½" Expo.			
Sq. Ft.	Square Ft.	Square Ft.	Square Ft.	Square Ft.	Square Ft.	Square Ft.	Square Ft.	15-Lb.	30-Lb.
200	218	240	294	240	300	240	228	.60	1.20
300	327	360	441	360	450	360	342	.90	1.80
400	426	480	588	480	600	480	456	1.20	2.40
500	545	600	735	600	750	600	570	1.50	3.00
600	654	720	882	720	900	720	684	1.80	3.60
700	763	840	1029	840	1050	840	798	2.10	4.20
800	872	960	1176	960	1200	960	912	2.40	4.80
900	981	1080	1323	1080	1350	1080	1026	2.70	5.40
1000	1090	1200	1470	1200	1500	1200	1140	3.00	6.00
1100	1199	1320	1617	1320	1650	1320	1254	3.30	6.60

1200	1308	1440	1764	1440	1800	1440	1368	3.60	7.20
1300	1417	1560	1911	1560	1950	1560	1482	3.90	7.80
1400	1526	1680	2058	1680	2100	1680	1596	4.20	8.40
1500	1635	1800	2205	1800	2250	1800	1710	4.50	9.00
1600	1744	1920	2352	1920	2400	1920	1824	4.80	9.60
1700	1853	2040	2500	2040	2550	2040	1938	5.10	10.20
1800	1962	2160	2646	2160	2700	2160	2052	5.40	10.80
1900	2071	2280	2793	2280	2850	2280	2166	5.70	11.40
2000	2180	2400	2940	2400	3000	2400	2280	6.00	12.00
2100	2289	2520	3087	2520	3150	2520	2394	6.30	12.60
2200	2398	2640	3234	2640	3300	2640	2508	6.60	13.20
2300	2507	2760	3381	2760	3450	2760	2622	6.90	13.80
2400	2616	2880	3528	2880	3600	2880	2736	7.20	14.40
2500	2725	3000	3675	3000	3750	3000	2850	7.50	15.00
2600	2834	3120	3822	3120	3900	3120	2964	7.80	15.60
2700	2943	3240	3969	3240	4050	3240	3078	8.10	16.20
2800	3052	3360	4116	3360	4200	3360	3192	8.40	16.80
2900	3161	3480	4263	3480	4350	3480	3306	8.70	17.40
3000	3270	3600	4410	3600	4500	3600	3420	9.00	18.00

Note: For types of roofing not shown, use Roof Surface Area column. Add the appropriate cutting and fitting waste. *Estimating Tables for Home Building,* Craftsman Book Co., 1986. Reprinted here with permission.

TABLE 19–10

Roofing for Gable and Hip Roofs, Including Intersecting Roofs, with Pitch of 1/4 (milling and cutting waste is included)

Ground Floor Area	Roof Surface Area	Asphalt Strip Shingles	Wood Shingles				Clay Roofing Tile	Number of Rolls of Saturated Felt Roofing Paper	
			18" Long		24" Long				
			4" Expo.	5½" Expo.	6" Expo.	7½" Expo.			
Sq. Ft.	Square Ft.	Square Ft.	Square Ft.	Square Ft.	Square Ft.	Square Ft.	Square Ft.	15-Lb.	30-Lb.
200	224	246	300	246	310	246	236	.62	1.24
300	336	369	450	369	465	369	354	.93	1.86
400	448	492	600	492	620	492	472	1.24	2.48
500	560	615	750	615	775	615	590	1.55	3.10
600	672	738	900	738	930	738	708	1.86	3.72
700	784	861	1050	861	1085	861	826	2.17	4.34
800	896	984	1200	984	1240	984	944	2.48	4.96
900	1008	1107	1350	1107	1395	1107	1062	2.79	5.58
1000	1120	1230	1500	1230	1550	1230	1180	3.10	6.20
1100	1232	1353	1650	1353	1705	1353	1298	3.41	6.82

1200	1344	1476	1800	1476	1860	1476	1416	3.72	7.44
1200	1344	1476	1800	1476	1860	1476	1416	3.72	7.44
1300	1456	1600	1950	1600	2015	1600	1534	4.03	8.06
1400	1568	1722	2100	1722	2170	1722	1652	4.34	8.68
1500	1680	1845	2250	1845	2325	1845	1770	4.65	9.30
1600	1793	1968	2400	1968	2480	1968	1888	4.96	9.92
1700	1904	2091	2550	2091	2635	2091	2006	5.27	10.54
1800	2016	2214	2700	2214	2790	2214	2124	5.58	11.16
1900	2128	2337	2850	2337	2945	2337	2242	5.89	11.78
2000	2240	2460	3000	2460	3100	2460	2360	6.20	12.40
2100	2352	2583	3150	2583	3255	2583	2478	6.51	13.02
2200	2464	2706	3300	2706	3410	2706	2596	6.82	13.64
2300	2576	2829	3450	2829	3565	2829	2714	7.13	14.26
2400	2688	2952	3600	2952	3720	2952	2832	7.44	14.88
2500	2800	3075	3750	3075	3875	3075	2950	7.75	15.50
2600	2912	3198	3900	3198	4030	3198	3068	8.06	16.12
2700	3024	3321	4050	3321	4185	3321	3186	8.37	16.74
2800	3136	3440	4200	3440	4340	3440	3304	8.68	17.36
2900	3248	3567	4350	3567	4495	3567	3422	9.00	18.00
3000	3360	3690	4500	3690	4650	3690	3540	9.30	18.60

Note: For types of roofing not shown, use Roof Surface Area column. Add the appropriate cutting and fitting waste. *Estimating Tables for Home Building,* Craftsman Book Co., 1986. Reprinted here with permission.

TABLE 19-11

Roofing for Gable and Hip Roofs, Including Intersecting Roofs, with Pitch of 1/3 (milling and cutting waste is included)

Ground Floor Area	Roof Surface Area	Asphalt Strip Shingles	Wood Shingles						Clay Roofing Tile	Number of Rolls of Saturated Felt Roofing Paper	
			18" Long		24" Long						
			4" Expo.	5½" Expo.	6" Expo.	7½" Expo.					
Sq. Ft.	Square Ft.	Square Ft.	Square Ft.	Square Ft.	Square Ft.	Square Ft.		Square Ft.		15-Lb.	30-Lb.
200	240	264	324	264	332	264		252		.66	1.32
300	360	396	486	396	498	396		378		.99	1.98
400	480	528	648	528	664	528		504		1.32	2.64
500	600	660	810	660	830	660		630		1.65	3.30
600	720	792	972	792	1000	792		756		1.98	3.96
700	840	924	1134	924	1162	924		882		2.31	4.62
800	960	1056	1296	1056	1328	1056		1008		2.64	5.28
900	1080	1188	1458	1188	1494	1188		1134		2.97	5.94
1000	1200	1320	1620	1320	1660	1320		1260		3.30	6.60
1100	1320	1452	1782	1452	1826	1452		1386		3.63	7.26

1200	1440	1584	1944	1584	1992	1584	1512	3.96	7.92
1300	1560	1716	2106	1716	2158	1716	1638	4.29	8.58
1400	1680	1848	2268	1848	2324	1848	1764	4.62	9.24
1500	1800	1980	2430	1980	2490	1980	1890	4.95	9.90
1600	1920	2112	2592	2112	2656	2112	2016	5.28	10.56
1700	2040	2244	2754	2244	2822	2244	2142	5.61	11.22
1800	2160	2376	2916	2376	2990	2376	2268	5.94	11.88
1900	2280	2508	3078	2508	3154	2508	2394	6.27	12.54
2000	2400	2640	3240	2640	3320	2640	2520	6.60	13.20
2100	2520	2772	3400	2772	3486	2772	2646	6.93	13.86
2200	2640	2904	3564	2904	3652	2904	2772	7.26	14.52
2300	2760	3036	2726	3036	3820	3036	2898	7.59	15.18
2400	2880	3168	3888	3168	3984	3168	3024	7.92	15.80
2500	3000	3300	4050	3300	4150	3300	3150	8.25	16.50
2600	3120	3432	4212	3432	4316	3432	3276	8.58	17.16
2700	3240	3564	4374	3564	4482	3564	3402	8.91	17.82
2800	3360	3696	4536	3696	4648	3696	3528	9.24	18.48
2900	3480	3828	4700	3828	4814	3828	3654	9.57	19.14
3000	3600	3960	4860	3960	4980	3960	3780	9.90	19.80

Note: For types of roofing not shown, use Roof Surface Area Column. Add the appropriate cutting and fitting waste. *Estimating Tables for Home Building,* Craftsman Book Co., 1986. Reprinted here with permission.

TABLE 19–12

Roofing for Gable and Hip Roofs, Including Intersecting Roofs, with Pitch of 1/2 (milling and cutting waste is included)

Ground Floor Area Sq. Ft.	Roof Surface Area Square Ft.	Asphalt Strip Shingles Square Ft.	Wood Shingles				Clay Roofing Tile Square Ft.	Number of Rolls of Saturated Felt Roofing Paper	
			18" Long		24" Long				
			4" Expo. Square Ft.	5½" Expo. Square Ft.	6" Expo. Square Ft.	7½" Expo. Square Ft.		15-Lb.	30-Lb.
200	284	312	384	312	392	312	298	.78	1.55
300	426	468	576	468	588	468	447	1.16	2.32
400	568	624	768	624	784	624	600	1.55	3.10
500	710	780	960	780	980	780	745	1.94	3.88
600	852	936	1152	936	1176	936	894	2.33	4.65
700	994	1092	1344	1092	1372	1092	1043	2.71	5.43
800	1136	1248	1536	1248	1568	1248	1192	3.10	6.20
900	1278	1404	1728	1404	1764	1404	1340	3.49	6.98
1000	1420	1560	1920	1560	1960	1560	1490	3.88	7.75
1100	1562	1716	2112	1716	2156	1716	1639	4.26	8.53

1200	1704	1872	2304	1872	2352	1872	1788	4.65	9.30
1300	1846	2028	2496	2028	2548	2028	1937	5.04	10.08
1400	1990	2184	2688	2184	2744	2184	2086	5.43	10.85
1500	2130	2340	2880	2340	2940	2340	2235	5.81	11.63
1600	2272	2500	3072	2500	3136	2500	2384	6.20	12.40
1700	2414	2652	3264	2652	3332	2652	2533	6.59	13.18
1800	2556	2810	3456	2810	3528	2810	2682	6.98	13.95
1900	2700	2964	3648	2964	3724	2964	2830	7.36	14.93
2000	2840	3120	3840	3120	3920	3120	2980	7.75	15.50
2100	2982	3276	4032	3276	4116	3276	3130	8.14	16.28
2200	3124	3432	4224	3432	4312	3432	3278	8.53	17.05
2300	3266	3590	4416	3590	4508	3590	3427	8.91	17.83
2400	3410	3744	4608	3744	4704	3744	3576	9.30	18.60
2500	3550	3900	4800	3900	4900	3900	3725	9.69	19.38
2600	3692	4056	4992	4056	5096	4056	3874	10.08	20.15
2700	3834	4212	5184	4212	5292	4212	4023	10.46	20.93
2800	3976	4368	5376	5488	4368	4368	4175	10.85	21.70
2900	4118	4524	5568	4524	5684	4524	4321	11.24	22.48
3000	4260	4680	5760	4680	5880	4680	4470	11.63	23.25

Note: For types of roofing not shown, use Roof Surface Area column. Add the appropriate cutting and fitting waste. *Estimating Tables for Home Building,* Craftsman Book Co., 1986. Reprinted here with permission.

CHAPTER 20

STRUCTURAL STEEL AND SHEET METAL

STRUCTURAL STEEL

Uniform Building Codes generally provide that:

> The design, fabrication, and erection of structural steel for building shall conform to the requirements of the "Specification for the Design, Fabrication, and Erection of Structural Steel for Buildings" of the American Institute of Steel Construction.

Estimating the materials, labor, and equipment for fabricating and erecting structural steel is a very detailed and at times complex process. Except where a relatively few beams, girders, and columns are required in a construction project, the estimating is usually subcontracted to firms that specialize in structural steel and have experienced workers and the necessary equipment. There are some general contractors who buy fabricated steel and either erect it with their own workers or sublet it to steel erection contractors.

In any case, general contractors should understand the methods and procedures that are customary in estimating materials and labor for structural steel.

439

Materials in Estimating

A materials estimate will usually include the following:

- Cost of the steel at the mill or warehouse
- Transportation to a fabrication shop
- Fabrication costs including shop drawings; handling steel shapes and fabricating them into specified columns; beams; and girders; painting; inspection; overhead; and profit
- Transportation costs of fabricated steel to the job site including all freight and trucking

Weights of Steel Shapes

Structural steel is priced and estimated per 100 lb., or per ton. A cubic foot of steel weighs 490 lb. A cubic inch weighs $\frac{490}{1728}$ = 0.28356 lb. A square inch of steel one foot long weighs 0.28356 × 12 = 3.4 lb. To obtain the weight per lineal foot of a structural steel member, multiply the area in square inches by 3.4. For example, a 4-by-4-by-¾-inch angle has an area of 5.44 sq. in. The weight is 5.44 × 3.4 = 18.5 lb. per lin. ft. Tables 20-1 through 20-8 (on pages 442–449) show the *depth, web* and *flange* dimensions, and the *weight per foot* of common structural steel shapes, courtesy of American Institute of Steel Construction, Inc.

Labor Factors in Estimating

An estimate of labor to erect structural steel depends, as with labor estimates of other building trades, on most of the factors considered in Chapter 10. Specifically, the labor costs will include, where relevant:

- The labor to set up and take down such heavy equipment as gin pole hoists, truck or crawler cranes, a power-operated steel tower with truss boom, and special scaffolding

- Actual erecting labor, plumbing, and temporarily or permanently bolting members
- Riveting or welding labor
- Labor to paint the steel

Permanent Erection Equipment

- Dollies and special trucks and trailers
- Winches, block and tackle
- Bolts, turnbuckles, and drift pins
- Rivets and riveting tools
- Air hammers, piping, and hose
- Forge, fuel, and tools
- Air compressor and fuel for power unit
- Welding rods, machines, fuel, and tools
- Bolts, washers, nuts, and tools
- Paint sprayers, brushes, and other tools

Table 20-9 on page 450 is a listing of various structural steel operations, and shows the average approximate man-hours of labor required to erect the steel. The cost of all erecting equipment previously listed must be added. This Table is presented as a guide for rough estimating only.

Welding Structural Steel

Arc welding has almost replaced riveting of structural steel in many parts of the country. Its principal advantages are economy and appearance. Uniform Building Codes provide:

> Details of design, workmanship, and technique for welding, inspection of welding and qualification of operators shall conform to the "Structural Welding Code" of American Welding Society, AWS D1.1.

TABLE 20–1

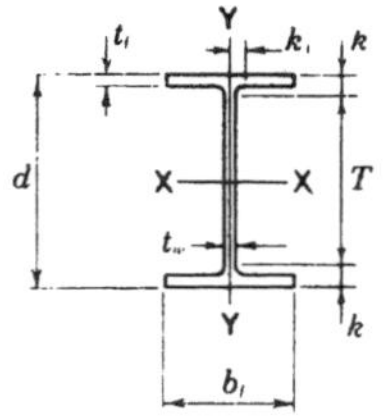

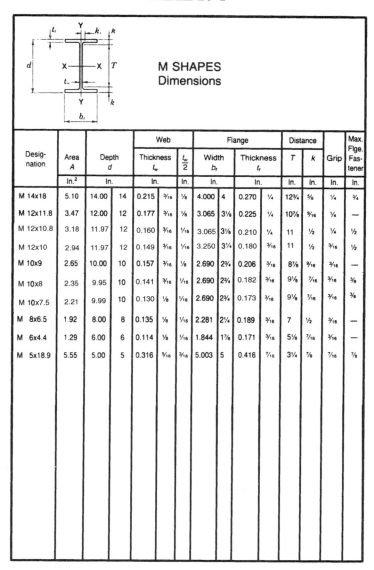

M SHAPES
Dimensions

Designation	Area A	Depth d		Web: Thickness t_w		Web: $\frac{t_w}{2}$	Flange: Width b_f		Flange: Thickness t_f		Distance: T	Distance: k	Grip	Max. Flge. Fastener
	In.^{2}	In.		In.		In.	In.		In.		In.	In.	In.	In.
M 14x18	5.10	14.00	14	0.215	3/16	1/8	4.000	4	0.270	1/4	12 3/4	5/8	1/4	3/4
M 12x11.8	3.47	12.00	12	0.177	3/16	1/8	3.065	3 1/8	0.225	1/4	10 7/8	9/16	1/4	—
M 12x10.8	3.18	11.97	12	0.160	3/16	1/16	3.065	3 1/8	0.210	1/4	11	1/2	1/4	1/2
M 12x10	2.94	11.97	12	0.149	3/16	1/16	3.250	3 1/4	0.180	3/16	11	1/2	3/16	1/2
M 10x9	2.65	10.00	10	0.157	3/16	1/8	2.690	2 3/4	0.206	3/16	8 7/8	9/16	3/16	—
M 10x8	2.35	9.95	10	0.141	3/16	1/16	2.690	2 3/4	0.182	3/16	9 1/8	7/16	3/16	3/8
M 10x7.5	2.21	9.99	10	0.130	1/8	1/16	2.690	2 3/4	0.173	3/16	9 1/8	7/16	3/16	3/8
M 8x6.5	1.92	8.00	8	0.135	1/8	1/16	2.281	2 1/4	0.189	3/16	7	1/2	3/16	—
M 6x4.4	1.29	6.00	6	0.114	1/8	1/16	1.844	1 7/8	0.171	3/16	5 1/8	7/16	3/16	—
M 5x18.9	5.55	5.00	5	0.316	5/16	3/16	5.003	5	0.416	7/16	3 1/4	7/8	7/16	7/8

TABLE 20-2

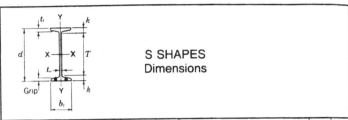

S SHAPES
Dimensions

Desig-nation	Area A	Depth d		Web Thickness t_w		$\dfrac{t_w}{2}$	Flange Width b_f		Flange Thickness t_f		Distance T	k	Grip	Max. Flge. Fas-tener
	In.²	In.		In.		In.	In.		In.		In.	In.	In.	In.
S 24x121	35.6	24.50	24½	0.800	¹³/₁₆	⁷/₁₆	8.050	8	1.090	1¹/₁₆	20½	2	1⅛	1
x106	31.2	24.50	24½	0.620	⅝	⁵/₁₆	7.870	7⅞	1.090	1¹/₁₆	20½	2	1⅛	1
S 24x100	29.3	24.00	24	0.745	¾	⅜	7.245	7¼	0.870	⅞	20½	1¾	⅞	1
x90	26.5	24.00	24	0.625	⅝	⁵/₁₆	7.125	7⅛	0.870	⅞	20½	1¾	⅞	1
x80	23.5	24.00	24	0.500	½	¼	7.000	7	0.870	⅞	20½	1¾	⅞	1
S 20x96	28.2	20.30	20¼	0.800	¹³/₁₆	⁷/₁₆	7.200	7¼	0.920	¹⁵/₁₆	16¾	1¾	¹⁵/₁₆	1
x86	25.3	20.30	20¼	0.660	¹¹/₁₆	⅜	7.060	7	0.920	¹⁵/₁₆	16¾	1¾	¹⁵/₁₆	1
S 20x75	22.0	20.00	20	0.635	⅝	⁵/₁₆	6.385	6⅜	0.795	¹³/₁₆	16¾	1⅝	¹³/₁₆	⅞
x66	19.4	20.00	20	0.505	½	¼	6.255	6¼	0.795	¹³/₁₆	16¾	1⅝	¹³/₁₆	⅞
S 18x70	20.6	18.00	18	0.711	¹¹/₁₆	⅜	6.251	6¼	0.691	¹¹/₁₆	15	1½	¹¹/₁₆	⅞
x54.7	16.1	18.00	18	0.461	⁷/₁₆	¼	6.001	6	0.691	¹¹/₁₆	15	1½	¹¹/₁₆	⅞
S 15x50	14.7	15.00	15	0.550	⁹/₁₆	⁵/₁₆	5.640	5⅝	0.622	⅝	12¼	1⅜	⁹/₁₆	¾
x42.9	12.6	15.00	15	0.411	⁷/₁₆	¼	5.501	5½	0.622	⅝	12¼	1⅜	⁹/₁₆	¾
S 12x50	14.7	12.00	12	0.687	¹¹/₁₆	⅜	5.477	5½	0.659	¹¹/₁₆	9⅛	1⁷/₁₆	¹¹/₁₆	¾
x40.8	12.0	12.00	12	0.462	⁷/₁₆	¼	5.252	5¼	0.659	¹¹/₁₆	9⅛	1⁷/₁₆	⅝	¾
S 12x35	10.3	12.00	12	0.428	⁷/₁₆	¼	5.078	5⅛	0.544	⁹/₁₆	9⅝	1³/₁₆	½	¾
x31.8	9.35	12.00	12	0.350	⅜	³/₁₆	5.000	5	0.544	⁹/₁₆	9⅝	1³/₁₆	½	¾
S 10x35	10.3	10.00	10	0.594	⅝	⁵/₁₆	4.944	5	0.491	½	7¾	1⅛	½	¾
x25.4	7.46	10.00	10	0.311	⁵/₁₆	³/₁₆	4.661	4⅝	0.491	½	7¾	1⅛	½	¾
S 8x23	6.77	8.00	8	0.441	⁷/₁₆	¼	4.171	4⅛	0.426	⁷/₁₆	6	1	⁷/₁₆	¾
x18.4	5.41	8.00	8	0.271	¼	⅛	4.001	4	0.426	⁷/₁₆	6	1	⁷/₁₆	¾
S 7x20	5.88	7.00	7	0.450	⁷/₁₆	¼	3.860	3⅞	0.392	⅜	5⅛	¹⁵/₁₆	⅜	⅝
x15.3	4.50	7.00	7	0.252	¼	⅛	3.662	3⅝	0.392	⅜	5⅛	¹⁵/₁₆	⅜	⅝
S 6x17.25	5.07	6.00	6	0.465	⁷/₁₆	¼	3.565	3⅝	0.359	⅜	4¼	⅞	⅜	⅝
x12.5	3.67	6.00	6	0.232	¼	⅛	3.332	3⅜	0.359	⅜	4¼	⅞	⅜	—
S 5x14.75	4.34	5.00	5	0.494	½	¼	3.284	3¼	0.326	⁵/₁₆	3⅜	¹³/₁₆	⁵/₁₆	—
x10	2.94	5.00	5	0.214	³/₁₆	⅛	3.004	3	0.326	⁵/₁₆	3⅜	¹³/₁₆	⁵/₁₆	—
S 4x 9.5	2.79	4.00	4	0.326	⁵/₁₆	³/₁₆	2.796	2¾	0.293	⁵/₁₆	2½	¾	⁵/₁₆	—
x 7.7	2.26	4.00	4	0.193	³/₁₆	⅛	2.663	2⅝	0.293	⁵/₁₆	2½	¾	⁵/₁₆	—
S 3x 7.5	2.21	3.00	3	0.349	⅜	³/₁₆	2.509	2½	0.260	¼	1⅝	¹¹/₁₆	¼	—
x 5.7	1.67	3.00	3	0.170	³/₁₆	⅛	2.330	2⅜	0.260	¼	1⅝	¹¹/₁₆	¼	—

Reprinted with permission of American Institute of Steel Construction

TABLE 20–3

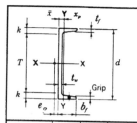

CHANNELS
AMERICAN STANDARD
Dimensions

Desig-nation	Area A	Depth d	Web Thickness t_w	Web $\frac{t_w}{2}$	Flange Width b_f	Flange Average thickness t_f	Distance T	Distance k	Grip	Max. Flge. Fas-tener
	In.²	In.	In.	In.	In.	In.	In.	In.	In.	In.
C 15x50	14.7	15.00	0.716	$^{11}/_{16}$ $^{3}/_{8}$	3.716 $3\frac{3}{4}$	0.650 $^{5}/_{8}$	12⅛	1$^{7}/_{16}$	$^{5}/_{8}$	1
x40	11.8	15.00	0.520	½ ¼	3.520 $3\frac{1}{2}$	0.650 $^{5}/_{8}$	12⅛	1$^{7}/_{16}$	$^{5}/_{8}$	1
x33.9	9.96	15.00	0.400	$^{3}/_{8}$ $^{3}/_{16}$	3.400 $3\frac{3}{8}$	0.650 $^{5}/_{8}$	12⅛	1$^{7}/_{16}$	$^{5}/_{8}$	1
C 12x30	8.82	12.00	0.510	½ ¼	3.170 $3\frac{1}{8}$	0.501 ½	9¾	1⅛	½	$^{7}/_{8}$
x25	7.35	12.00	0.387	$^{3}/_{8}$ $^{3}/_{16}$	3.047 3	0.501 ½	9¾	1⅛	½	$^{7}/_{8}$
x20.7	6.09	12.00	0.282	$^{5}/_{16}$ ⅛	2.942 3	0.501 ½	9¾	1⅛	½	$^{7}/_{8}$
C 10x30	8.82	10.00	0.673	$^{11}/_{16}$ $^{5}/_{16}$	3.033 3	0.436 $^{7}/_{16}$	8	1	$^{7}/_{16}$	¾
x25	7.35	10.00	0.526	½ ¼	2.886 $2\frac{7}{8}$	0.436 $^{7}/_{16}$	8	1	$^{7}/_{16}$	¾
x20	5.88	10.00	0.379	$^{3}/_{8}$ $^{3}/_{16}$	2.739 $2\frac{3}{4}$	0.436 $^{7}/_{16}$	8	1	$^{7}/_{16}$	¾
x15.3	4.49	10.00	0.240	¼ ⅛	2.600 $2\frac{5}{8}$	0.436 $^{7}/_{16}$	8	1	$^{7}/_{16}$	¾
C 9x20	5.88	9.00	0.448	$^{7}/_{16}$ ¼	2.648 $2\frac{5}{8}$	0.413 $^{7}/_{16}$	7⅛	$^{15}/_{16}$	$^{7}/_{16}$	¾
x15	4.41	9.00	0.285	$^{5}/_{16}$ ⅛	2.485 $2\frac{1}{2}$	0.413 $^{7}/_{16}$	7⅛	$^{15}/_{16}$	$^{7}/_{16}$	¾
x13.4	3.94	9.00	0.233	¼ ⅛	2.433 $2\frac{3}{8}$	0.413 $^{7}/_{16}$	7⅛	$^{15}/_{16}$	$^{7}/_{16}$	¾
C 8x18.75	5.51	8.00	0.487	½ ¼	2.527 $2\frac{1}{2}$	0.390 $^{3}/_{8}$	6⅛	$^{15}/_{16}$	$^{3}/_{8}$	¾
x13.75	4.04	8.00	0.303	$^{5}/_{16}$ ⅛	2.343 $2\frac{3}{8}$	0.390 $^{3}/_{8}$	6⅛	$^{15}/_{16}$	$^{3}/_{8}$	¾
x11.5	3.38	8.00	0.220	¼ ⅛	2.260 $2\frac{1}{4}$	0.390 $^{3}/_{8}$	6⅛	$^{15}/_{16}$	$^{3}/_{8}$	¾
C 7x14.75	4.33	7.00	0.419	$^{7}/_{16}$ $^{3}/_{16}$	2.299 $2\frac{1}{4}$	0.366 $^{3}/_{8}$	5¼	$^{7}/_{8}$	$^{3}/_{8}$	$^{5}/_{8}$
x12.25	3.60	7.00	0.314	$^{5}/_{16}$ $^{3}/_{16}$	2.194 $2\frac{1}{4}$	0.366 $^{3}/_{8}$	5¼	$^{7}/_{8}$	$^{3}/_{8}$	$^{5}/_{8}$
x 9.8	2.87	7.00	0.210	$^{3}/_{16}$ ⅛	2.090 $2\frac{1}{8}$	0.366 $^{3}/_{8}$	5¼	$^{7}/_{8}$	$^{3}/_{8}$	$^{5}/_{8}$
C 6x13	3.83	6.00	0.437	$^{7}/_{16}$ $^{3}/_{16}$	2.157 $2\frac{1}{8}$	0.343 $^{5}/_{16}$	4⅜	$^{13}/_{16}$	$^{5}/_{16}$	$^{5}/_{8}$
x10.5	3.09	6.00	0.314	$^{5}/_{16}$ $^{3}/_{16}$	2.034 2	0.343 $^{5}/_{16}$	4⅜	$^{13}/_{16}$	$^{3}/_{8}$	$^{5}/_{8}$
x 8.2	2.40	6.00	0.200	$^{3}/_{16}$ ⅛	1.920 $1\frac{7}{8}$	0.343 $^{5}/_{16}$	4⅜	$^{13}/_{16}$	$^{5}/_{16}$	$^{5}/_{8}$
C 5x 9	2.64	5.00	0.325	$^{5}/_{16}$ $^{3}/_{16}$	1.885 $1\frac{7}{8}$	0.320 $^{5}/_{16}$	3½	¾	$^{5}/_{16}$	$^{5}/_{8}$
x 6.7	1.97	5.00	0.190	$^{3}/_{16}$ ⅛	1.750 $1\frac{3}{4}$	0.320 $^{5}/_{16}$	3½	¾	—	—
C 4x 7.25	2.13	4.00	0.321	$^{5}/_{16}$ $^{3}/_{16}$	1.721 $1\frac{3}{4}$	0.296 $^{5}/_{16}$	2⅝	$^{11}/_{16}$	$^{5}/_{16}$	$^{5}/_{8}$
x 5.4	1.59	4.00	0.184	$^{3}/_{16}$ $^{1}/_{16}$	1.584 $1\frac{5}{8}$	0.296 $^{5}/_{16}$	2⅝	$^{11}/_{16}$	—	—
C 3x 6	1.76	3.00	0.356	$^{3}/_{8}$ $^{3}/_{16}$	1.596 $1\frac{5}{8}$	0.273 ¼	1⅝	$^{11}/_{16}$	—	—
x 5	1.47	3.00	0.258	¼ ⅛	1.498 $1\frac{1}{2}$	0.273 ¼	1⅝	$^{11}/_{16}$	—	—
x 4.1	1.21	3.00	0.170	$^{3}/_{16}$ $^{1}/_{16}$	1.410 $1\frac{3}{8}$	0.273 ¼	1⅝	$^{11}/_{16}$	—	—

Reprinted with permission of American Institute of Steel Construction

TABLE 20–4

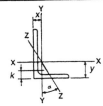

ANGLES
Equal legs and unequal legs
Properties for designing

Size and Thickness	k	Weight per Ft	Area	AXIS X-X				AXIS Y-Y				AXIS Z-Z	
				I	S	r	y	I	S	r	x	r	Tan
In.	In.	Lb.	In.²	In.⁴	In.³	In.	In.	In.⁴	In.³	In.	In.	In.	α
L 9×4× 5/8	1⅛	26.3	7.73	64.9	11.5	2.90	3.36	8.32	2.65	1.04	0.858	.847	0.216
9/16	1 1/16	23.8	7.00	59.1	10.4	2.91	3.33	7.63	2.41	1.04	0.834	.850	0.218
½	1	21.3	6.25	53.2	9.34	2.92	3.31	6.92	2.17	1.05	0.810	.854	0.220
L 8×8×1⅛	1¾	56.9	16.7	98.0	17.5	2.42	2.41	98.0	17.5	2.42	2.41	1.56	1.000
1	1⅝	51.0	15.0	89.0	15.8	2.44	2.37	89.0	15.8	2.44	2.37	1.56	1.000
⅞	1½	45.0	13.2	79.6	14.0	2.45	2.32	79.6	14.0	2.45	2.32	1.57	1.000
¾	1⅜	38.9	11.4	69.7	12.2	2.47	2.28	69.7	12.2	2.47	2.28	1.58	1.000
⅝	1¼	32.7	9.61	59.4	10.3	2.49	2.23	59.4	10.3	2.49	2.23	1.58	1.000
9/16	1 3/16	29.6	8.68	54.1	9.34	2.50	2.21	54.1	9.34	2.50	2.21	1.59	1.000
½	1⅛	26.4	7.75	48.6	8.36	2.50	2.19	48.6	8.36	2.50	2.19	1.59	1.000
L 8×6×1	1½	44.2	13.0	80.8	15.1	2.49	2.65	38.8	8.92	1.73	1.65	1.28	0.543
⅞	1⅜	39.1	11.5	72.3	13.4	2.51	2.61	34.9	7.94	1.74	1.61	1.28	0.547
¾	1¼	33.8	9.94	63.4	11.7	2.53	2.56	30.7	6.92	1.76	1.56	1.29	0.551
⅝	1⅛	28.5	8.36	54.1	9.87	2.54	2.52	26.3	5.88	1.77	1.52	1.29	0.554
9/16	1 1/16	25.7	7.56	49.3	8.95	2.55	2.50	24.0	5.34	1.78	1.50	1.30	0.556
½	1	23.0	6.75	44.3	8.02	2.56	2.47	21.7	4.79	1.79	1.47	1.30	0.558
7/16	15/16	20.2	5.93	39.2	7.07	2.57	2.45	19.3	4.23	1.80	1.45	1.31	0.560
L 8×4×1	1½	37.4	11.0	69.6	14.1	2.52	3.05	11.6	3.94	1.03	1.05	0.846	0.247
¾	1¼	28.7	8.44	54.9	10.9	2.55	2.95	9.36	3.07	1.05	0.953	0.852	0.258
9/16	1 1/16	21.9	6.43	42.8	8.35	2.58	2.88	7.43	2.38	1.07	0.882	0.861	0.265
½	1	19.6	5.75	38.5	7.49	2.59	2.86	6.74	2.15	1.08	0.859	0.865	0.267
L 7×4× ¾	1¼	26.2	7.69	37.8	8.42	2.22	2.51	9.05	3.03	1.09	1.01	0.860	0.324
⅝	1⅛	22.1	6.48	32.4	7.14	2.24	2.46	7.84	2.58	1.10	0.963	0.865	0.329
½	1	17.9	5.25	26.7	5.81	2.25	2.42	6.53	2.12	1.11	0.917	0.872	0.335
⅜	⅞	13.6	3.98	20.6	4.44	2.27	2.37	5.10	1.63	1.13	0.870	0.880	0.340

Reprinted with permission of American Institute of Steel Construction

TABLE 20–5

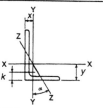

ANGLES
Equal legs and unequal legs
Properties for designing

Size and Thickness	k	Weight per Ft	Area	AXIS X-X				AXIS Y-Y				AXIS Z-Z	
				I	S	r	y	I	S	r	x	r	Tan
In.	In.	Lb.	In.²	In.⁴	In.³	In.	In.	In.⁴	In.³	In.	In.	In.	α
L 6×6 ×1	1½	37.4	11.0	35.5	8.57	1.80	1.86	35.5	8.57	1.80	1.86	1.17	1.000
⅞	1⅜	33.1	9.73	31.9	7.63	1.81	1.82	31.9	7.63	1.81	1.82	1.17	1.000
¾	1¼	28.7	8.44	28.2	6.66	1.83	1.78	28.2	6.66	1.83	1.78	1.17	1.000
⅝	1⅛	24.2	7.11	24.2	5.66	1.84	1.73	24.2	5.66	1.84	1.73	1.18	1.000
9/16	1 1/16	21.9	6.43	22.1	5.14	1.85	1.71	22.1	5.14	1.85	1.71	1.18	1.000
½	1	19.6	5.75	19.9	4.61	1.86	1.68	19.9	4.61	1.86	1.68	1.18	1.000
7/16	15/16	17.2	5.06	17.7	4.08	1.87	1.66	17.7	4.08	1.87	1.66	1.19	1.000
⅜	⅞	14.9	4.36	15.4	3.53	1.88	1.64	15.4	3.53	1.88	1.64	1.19	1.000
5/16	13/16	12.4	3.65	13.0	2.97	1.89	1.62	13.0	2.97	1.89	1.62	1.20	1.000
L 6×4 × ⅞	1⅜	27.2	7.98	27.7	7.15	1.86	2.12	9.75	3.39	1.11	1.12	0.857	0.421
¾	1¼	23.6	6.94	24.5	6.25	1.88	2.08	8.68	2.97	1.12	1.08	0.860	0.428
⅝	1⅛	20.0	5.86	21.1	5.31	1.90	2.03	7.52	2.54	1.13	1.03	0.864	0.435
9/16	1 1/16	18.1	5.31	19.3	4.83	1.90	2.01	6.91	2.31	1.14	1.01	0.866	0.438
½	1	16.2	4.75	17.4	4.33	1.91	1.99	6.27	2.08	1.15	0.987	0.870	0.440
7/16	15/16	14.3	4.18	15.5	3.83	1.92	1.96	5.60	1.85	1.16	0.964	0.873	0.443
⅜	⅞	12.3	3.61	13.5	3.32	1.93	1.94	4.90	1.60	1.17	0.941	0.877	0.446
5/16	13/16	10.3	3.03	11.4	2.79	1.94	1.92	4.18	1.35	1.17	0.918	0.882	0.448
L 6×3½× ½	1	15.3	4.50	16.6	4.24	1.92	2.08	4.25	1.59	0.972	0.833	0.759	0.344
⅜	⅞	11.7	3.42	12.9	3.24	1.94	2.04	3.34	1.23	0.988	0.787	0.767	0.350
5/16	13/16	9.8	2.87	10.9	2.73	1.95	2.01	2.85	1.04	0.996	0.763	0.772	0.352
L 5×5 × ⅞	1⅜	27.2	7.98	17.8	5.17	1.49	1.57	17.8	5.17	1.49	1.57	0.973	1.000
¾	1¼	23.6	6.94	15.7	4.53	1.51	1.52	15.7	4.53	1.51	1.52	0.975	1.000
⅝	1⅛	20.0	5.86	13.6	3.86	1.52	1.48	13.6	3.86	1.52	1.48	0.978	1.000
½	1	16.2	4.75	11.3	3.16	1.54	1.43	11.3	3.16	1.54	1.43	0.983	1.000
7/16	15/16	14.3	4.18	10.0	2.79	1.55	1.41	10.0	2.79	1.55	1.41	0.986	1.000
⅜	⅞	12.3	3.61	8.74	2.42	1.56	1.39	8.74	2.42	1.56	1.39	0.990	1.000
5/16	13/16	10.3	3.03	7.42	2.04	1.57	1.37	7.42	2.04	1.57	1.37	0.994	1.000

TABLE 20-6

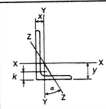

ANGLES
Equal legs and unequal legs
Properties for designing

Size and Thickness	k	Weight per Ft	Area	AXIS X-X				AXIS Y-Y				AXIS Z-Z	
				I	S	r	y	I	S	r	x	r	Tan
In.	In.	Lb.	In.^{2}	In.^{4}	In.^{3}	In.	In.	In.^{4}	In.^{3}	In.	In.	In.	α
L 5×3½× ¾	1¼	19.8	5.81	13.9	4.28	1.55	1.75	5.55	2.22	0.977	0.996	0.748	0.464
⅝	1⅛	16.8	4.92	12.0	3.65	1.56	1.70	4.83	1.90	0.991	0.951	0.751	0.472
½	1	13.6	4.00	9.99	2.99	1.58	1.66	4.05	1.56	1.01	0.906	0.755	0.479
⁷⁄₁₆	¹⁵⁄₁₆	12.0	3.53	8.90	2.64	1.59	1.63	3.63	1.39	1.01	0.883	0.758	0.482
⅜	⅞	10.4	3.05	7.78	2.29	1.60	1.61	3.18	1.21	1.02	0.861	0.762	0.486
⁵⁄₁₆	¹³⁄₁₆	8.7	2.56	6.60	1.94	1.61	1.59	2.72	1.02	1.03	0.838	0.766	0.489
¼	¾	7.0	2.06	5.39	1.57	1.62	1.56	2.23	0.830	1.04	0.814	0.770	0.492
L 5×3 × ⅝	1	15.7	4.61	11.4	3.55	1.57	1.80	3.06	1.39	0.815	0.796	0.644	0.349
½	1	12.8	3.75	9.45	2.91	1.59	1.75	2.58	1.15	0.829	0.750	0.648	0.357
⁷⁄₁₆	¹⁵⁄₁₆	11.3	3.31	8.43	2.58	1.60	1.73	2.32	1.02	0.837	0.727	0.651	0.361
⅜	⅞	9.8	2.86	7.37	2.24	1.61	1.70	2.04	0.888	0.845	0.704	0.654	0.364
⁵⁄₁₆	¹³⁄₁₆	8.2	2.40	6.26	1.89	1.61	1.68	1.75	0.753	0.853	0.681	0.658	0.368
¼	¾	6.6	1.94	5.11	1.53	1.62	1.66	1.44	0.614	0.861	0.657	0.663	0.371
L 4×4 × ¾	1⅛	18.5	5.44	7.67	2.81	1.19	1.27	7.67	2.81	1.19	1.27	0.778	1.000
⅝	1	15.7	4.61	6.66	2.40	1.20	1.23	6.66	2.40	1.20	1.23	0.779	1.000
½	⅞	12.8	3.75	5.56	1.97	1.22	1.18	5.56	1.97	1.22	1.18	0.782	1.000
⁷⁄₁₆	¹³⁄₁₆	11.3	3.31	4.97	1.75	1.23	1.16	4.97	1.75	1.23	1.16	0.785	1.000
⅜	¾	9.8	2.86	4.36	1.52	1.23	1.14	4.36	1.52	1.23	1.14	0.788	1.000
⁵⁄₁₆	¹¹⁄₁₆	8.2	2.40	3.71	1.29	1.24	1.12	3.71	1.29	1.24	1.12	0.791	1.000
¼	⅝	6.6	1.94	3.04	1.05	1.25	1.09	3.04	1.05	1.25	1.09	0.795	1.000
L 4×3½× ½	¹⁵⁄₁₆	11.9	3.50	5.32	1.94	1.23	1.25	3.79	1.52	1.04	1.00	0.722	0.750
⁷⁄₁₆	⅞	10.6	3.09	4.76	1.72	1.24	1.23	3.40	1.35	1.05	0.978	0.724	0.753
⅜	¹³⁄₁₆	9.1	2.67	4.18	1.49	1.25	1.21	2.95	1.17	1.06	0.955	0.727	0.755
⁵⁄₁₆	¾	7.7	2.25	3.56	1.26	1.26	1.18	2.55	0.994	1.07	0.932	0.730	0.757
¼	¹¹⁄₁₆	6.2	1.81	2.91	1.03	1.27	1.16	2.09	0.808	1.07	0.909	0.734	0.759

TABLE 20-7

ANGLES
Equal legs and unequal legs
Properties for designing

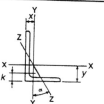

Size and Thickness	k	Weight per Ft	Area	AXIS X-X				AXIS Y-Y				AXIS Z-Z	
In.	In.	Lb.	In.²	I In.⁴	S In.³	r In.	y In.	I In.⁴	S In.³	r In.	x In.	r In.	Tan α
L 4 ×3 × ½	15/16	11.1	3.25	5.05	1.89	1.25	1.33	2.42	1.12	0.864	0.827	0.639	0.543
7/16	7/8	9.8	2.87	4.52	1.68	1.25	1.30	2.18	0.992	0.871	0.804	0.641	0.547
3/8	13/16	8.5	2.48	3.96	1.46	1.26	1.28	1.92	0.866	0.879	0.782	0.644	0.551
5/16	3/4	7.2	2.09	3.38	1.23	1.27	1.26	1.65	0.734	0.887	0.759	0.647	0.554
1/4	11/16	5.8	1.69	2.77	1.00	1.28	1.24	1.36	0.599	0.896	0.736	0.651	0.558
L 3½×3½× ½	7/8	11.1	3.25	3.64	1.49	1.06	1.06	3.64	1.49	1.06	1.06	0.683	1.000
7/16	13/16	9.8	2.87	3.26	1.32	1.07	1.04	3.26	1.32	1.07	1.04	0.684	1.000
3/8	3/4	8.5	2.48	2.87	1.15	1.07	1.01	2.87	1.15	1.07	1.01	0.687	1.000
5/16	11/16	7.2	2.09	2.45	0.976	1.08	0.990	2.45	0.976	1.08	0.990	0.690	1.000
1/4	5/8	5.8	1.69	2.01	0.794	1.09	0.968	2.01	0.794	1.09	0.968	0.694	1.000
L 3½×3 × ½	15/16	10.2	3.00	3.45	1.45	1.07	1.13	2.33	1.10	0.881	0.875	0.621	0.714
7/16	7/8	9.1	2.65	3.10	1.29	1.08	1.10	2.09	0.975	0.889	0.853	0.622	0.718
3/8	13/16	7.9	2.30	2.72	1.13	1.09	1.08	1.85	0.851	0.897	0.830	0.625	0.721
5/16	3/4	6.6	1.93	2.33	0.954	1.10	1.06	1.58	0.722	0.905	0.808	0.627	0.724
1/4	11/16	5.4	1.56	1.91	0.776	1.11	1.04	1.30	0.589	0.914	0.785	0.631	0.727
L 3½×2½× ½	15/16	9.4	2.75	3.24	1.41	1.09	1.20	1.36	0.760	0.704	0.705	0.534	0.486
7/16	7/8	8.3	2.43	2.91	1.26	1.09	1.18	1.23	0.677	0.711	0.682	0.535	0.491
3/8	13/16	7.2	2.11	2.56	1.09	1.10	1.16	1.09	0.592	0.719	0.660	0.537	0.496
5/16	3/4	6.1	1.78	2.19	0.927	1.11	1.14	0.939	0.504	0.727	0.637	0.540	0.501
1/4	11/16	4.9	1.44	1.80	0.755	1.12	1.11	0.777	0.412	0.735	0.614	0.544	0.506
L 3 ×3 × ½	13/16	9.4	2.75	2.22	1.07	0.898	0.932	2.22	1.07	0.898	0.932	0.584	1.000
7/16	3/4	8.3	2.43	1.99	0.954	0.905	0.910	1.99	0.954	0.905	0.910	0.585	1.000
3/8	11/16	7.2	2.11	1.76	0.833	0.913	0.888	1.76	0.833	0.913	0.888	0.587	1.000
5/16	5/8	6.1	1.78	1.51	0.707	0.922	0.865	1.51	0.707	0.922	0.865	0.589	1.000
1/4	9/16	4.9	1.44	1.24	0.577	0.930	0.842	1.24	0.577	0.930	0.842	0.592	1.000
3/16	1/2	3.71	1.09	0.962	0.441	0.939	0.820	0.962	0.441	0.939	0.820	0.596	1.000

TABLE 20–8

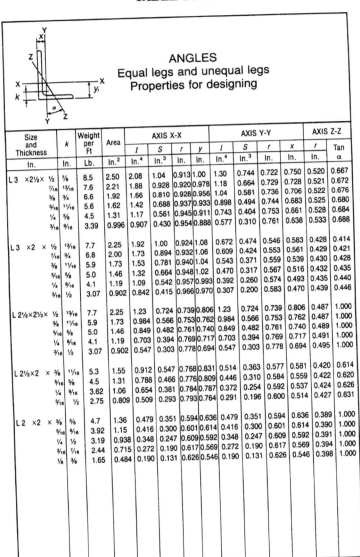

ANGLES
Equal legs and unequal legs
Properties for designing

Size and Thickness	k	Weight per Ft	Area	AXIS X-X				AXIS Y-Y				AXIS Z-Z	
				I	S	r	y	I	S	r	x	r	Tan
In.	In.	Lb.	In.²	In.⁴	In.³	In.	In.	In.⁴	In.³	In.	In.	In.	α
L 3 ×2½× ½	⅞	8.5	2.50	2.08	1.04	0.913	1.00	1.30	0.744	0.722	0.750	0.520	0.667
7/16	13/16	7.6	2.21	1.88	0.928	0.920	0.978	1.18	0.664	0.729	0.728	0.521	0.672
⅜	¾	6.6	1.92	1.66	0.810	0.928	0.956	1.04	0.581	0.736	0.706	0.522	0.676
5/16	11/16	5.6	1.62	1.42	0.688	0.937	0.933	0.898	0.494	0.744	0.683	0.525	0.680
¼	⅝	4.5	1.31	1.17	0.561	0.945	0.911	0.743	0.404	0.753	0.661	0.528	0.684
3/16	9/16	3.39	0.996	0.907	0.430	0.954	0.888	0.577	0.310	0.761	0.638	0.533	0.688
L 3 ×2 × ½	13/16	7.7	2.25	1.92	1.00	0.924	1.08	0.672	0.474	0.546	0.583	0.428	0.414
7/16	¾	6.8	2.00	1.73	0.894	0.932	1.06	0.609	0.424	0.553	0.561	0.429	0.421
⅜	11/16	5.9	1.73	1.53	0.781	0.940	1.04	0.543	0.371	0.567	0.539	0.430	0.428
5/16	⅝	5.0	1.46	1.32	0.664	0.948	1.02	0.470	0.317	0.567	0.516	0.432	0.435
¼	9/16	4.1	1.19	1.09	0.542	0.957	0.993	0.392	0.260	0.574	0.493	0.435	0.440
3/16	½	3.07	0.902	0.842	0.415	0.966	0.970	0.307	0.200	0.583	0.470	0.439	0.446
L 2½×2½× ½	13/16	7.7	2.25	1.23	0.724	0.739	0.806	1.23	0.724	0.739	0.806	0.487	1.000
⅜	11/16	5.9	1.73	0.984	0.566	0.753	0.762	0.984	0.566	0.753	0.762	0.487	1.000
5/16	⅝	5.0	1.46	0.849	0.482	0.761	0.740	0.849	0.482	0.761	0.740	0.489	1.000
¼	9/16	4.1	1.19	0.703	0.394	0.769	0.717	0.703	0.394	0.769	0.717	0.491	1.000
3/16	½	3.07	0.902	0.547	0.303	0.778	0.694	0.547	0.303	0.778	0.694	0.495	1.000
L 2½×2 × ⅜	11/16	5.3	1.55	0.912	0.547	0.768	0.831	0.514	0.363	0.577	0.581	0.420	0.614
5/16	⅝	4.5	1.31	0.788	0.466	0.776	0.809	0.446	0.310	0.584	0.559	0.422	0.620
¼	9/16	3.62	1.06	0.654	0.381	0.784	0.787	0.372	0.254	0.592	0.537	0.424	0.626
3/16	½	2.75	0.809	0.509	0.293	0.793	0.764	0.291	0.196	0.600	0.514	0.427	0.631
L 2 ×2 × ⅜	⅝	4.7	1.36	0.479	0.351	0.594	0.636	0.479	0.351	0.594	0.636	0.389	1.000
5/16	9/16	3.92	1.15	0.416	0.300	0.601	0.614	0.416	0.300	0.601	0.614	0.390	1.000
¼	½	3.19	0.938	0.348	0.247	0.609	0.592	0.348	0.247	0.609	0.592	0.391	1.000
3/16	7/16	2.44	0.715	0.272	0.190	0.617	0.569	0.272	0.190	0.617	0.569	0.394	1.000
⅛	⅜	1.65	0.484	0.190	0.131	0.626	0.546	0.190	0.131	0.626	0.546	0.398	1.000

TABLE 20–9

Approximate Hours Labor Erecting Structural Steel

Item of Work		Man-Hours
Erecting, plumbing and temporary bolting		
Steel mill buildings, one and two stories,	per ton	7–12
Skeleton framed, multistory buildings	" "	6–10
Columns, beams and channels	" "	5–8
Roof trusses, light	" "	8–12
Roof trusses, heavy	" "	6–10
Riveting, air driven, 30 to 40 rivets per ton		
Steel mill buildings	per 100 rivets	10–14
Skeleton framed multistory buildings	" " "	10–16
Columns, beams and channels	" " "	16–24
Roof trusses, light	" " "	8–12
Roof trusses, heavy	" " "	12–16
Bolting in place of riveting, 20 to 30 bolts per ton		
	per 100 bolts	4–5
Painting steel work, two coats		
Light weight framing members,	per ton	1–2
Heavy weight members	" "	0.5–1

FLASHING AND SHEET METAL

Sheet-metal work is a specialized building trade encompassing the fabrication and installation of building units made of sheet metal, including but not limited to such things as gutters and downspouts, roofing, flashing and counterflashing, ventilators, and skylights.

Duct systems for heating, air conditioning, and manufacturing processes, while falling into the classification of sheet-metal work, will not be taken up here. The majority of these items are best referred to subcontractors.

Flashing and Counterflashing

Flashing is used in building construction to prevent water from entering through joints created by intersecting roofs (valleys); chimneys, stack pipes, and ventilators that project through a roof; over windows and doors; at skylight curbings; and where roofs, flat or pitched, join a vertical wall. Another and more limited use for flashing is a shield at foundation sills and post bases against insect damage, particularly by termites.

The metals most commonly used for flashing are galvanized sheet metal, lead, copper, tin, and aluminum. A bituminous material frequently used in flashing is the 90-lb. mineral-surfaced roll roofing.

Copper flashing is commonly 16-ounce, but 20- and 24-ounce rolls and sheets are also used. *Galvanized sheet metal* is usually the 26-gauge and comes in 8-inch, 14-inch, and 20-inch widths. A common thickness of *aluminum flashing* is .019 inches and is furnished in rolls 6 inches to 28 inches wide and 50 lineal feet long.

The labor to apply flashing should be estimated by the individual job considering the physical working conditions, the type of flashing, and the experience of the workers. As a guide, a worker can place roll metal flashing in valleys at approximately 10 to 20 lineal feet per hour depending on the slope of the roof. Placing flashing over windows and doors, where narrow widths are used, workers should average approximately 15 lineal feet per hour. Ridge and hip flashing can be applied at an average rate of 15 to 20 lineal feet per hour.

Gutters and Downspouts

Gutters, installed at the cornice line of a roof to collect water runoff, are sometimes called eave troughs. The vertical downspout can also be called a leader pipe or conductor. Fittings include hangers, end caps, slip joints, funnels, elbows, leader straps or hooks, and wire strainers. Most of these terms are self-descriptive. Elbows are used to connect the gutter outlet to the downspout, which is located against the building.

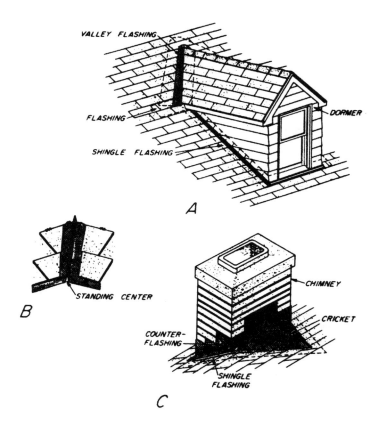

Flashing at center of rooflines. *A,* Valley and shingle flashing; *B,* standing-center valley flashing; *C,* chimney and shingle flashing.

Figure 20–1
Wood-Frame House Construction . . .
Agricultural Handbook No. 73

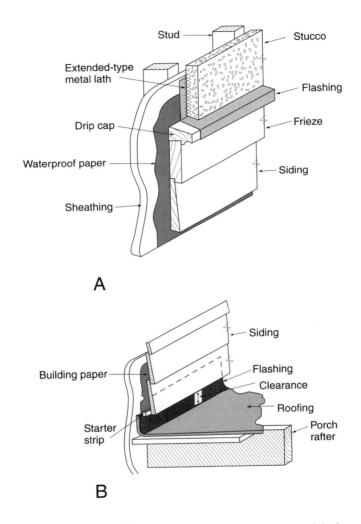

A

B

Location of flashing. *A,* at material change; *B,* at roof deck.

Figure 20–2
Wood-Frame House Construction . . .
Agricultural Handbook No. 73

Gutters and leaders are made in several styles such as half-round and box type. Downspouts are shaped round and rectangular, either of which may be smooth or corrugated. The materials may be copper, galvanized steel, vinyl, fiberglass, and, less frequently, wood.

In estimating material costs, the shape of the gutter, its material, and its thickness or gauge are important considerations.

Gutters and leaders are installed by two journeymen or one journeyman and a helper. On one- or two-story buildings, working from ladders, two workers can hang gutters and leaders at approximately 100 lineal feet in 8 to 10 hours.

Metal Skylights and Ventilators

Skylights and ventilators are generally manufactured in the shop and delivered to the job, where they are installed by the fabricator. The curbing is usually prepared by carpenters and should be figured separately. Prices for skylights, including installation, should be quoted by subcontractors.

SELECTED REFERENCE

Manual of Steel Construction, 9th Edition, American Institute of Steel Construction, 400 North Michigan Avenue, Chicago, IL 60611-4185, (AISC Manuals and Textbooks Catalog available), 1989.

CHAPTER 21

INTERIOR AND EXTERIOR PAINTING

JOB CONSIDERATIONS

There are different methods used to estimate the approximate number of gallons of painting material and the hours required to apply it. Some of these are discussed in the following pages. Whatever method is used, the objective should be to compute, as accurately as possible, the amount of material and to determine a reasonable number of hours to apply it considering all of the conditions that pertain to the particular job. The following conditions need to be given special attention.

Shifting and Protecting Contents

Painters work faster in a vacant room than in one that requires shifting and covering the contents to conveniently reach the areas that are to be painted. It takes time to cover and move furnishings or take down wall hangings. In mercantile buildings it should be determined who will have the responsibility for moving stocks off shelves and away from walls that are to be painted.

Preparation of Surfaces When Repainting

Walls or ceilings that are badly soiled have to be washed down to remove grease and dirt before applying paint. Blistered paint on exterior surfaces has to be sanded or burned off prior to repainting. Frequently, minor cracks in plaster, or in the joints of wallboard, have to be filled with patch plaster or spackle, and nail holes in woodwork must be puttied.

Kind of Surface

The quantity of painting material and the labor to apply it will vary with the kind of surface. Rough plastered masonry or cement walls require more paint than smooth plastered walls and woodwork. Also, a painter will cover more square feet per hour on smooth surfaces than on porous ones.

Shape and Form of Surface

Stairways, trim, cabinets, windows, and doors take longer to paint per square foot than smooth uninterrupted areas such as walls, ceilings, wood siding, and wood floors.

Type of Painting Material

The covering capacities of some painting materials are much greater than others regardless of the kind of surface. Cold water-mix paints and calcimine cover less square feet per gallon than oil, rubber, or alkyd resin base paints; shellac and varnishes generally have greater covering capacities than any of those named. When estimating the quantity of material it is important to know the probable covering capacity of that particular material. See Table 21–1 on page 458 for the approximate covering capacities of paint materials.

Number of Colors

When doors, windows, and trim in a room or on the exterior are to be painted different colors from other areas, the painter requires more time than if only one color material is to be used.

Number of Coats

The number of coats of paint material required naturally will affect the cost both as to material and labor.

Method of Application

Brush work is slower than using a roller on walls and ceilings. In some jurisdictions a painter who uses a roller is paid a slightly higher rate of wages, which may offset the advantage of speed. Where spray painting is permissible, the labor cost is considerably reduced although more paint is usually needed.

Layout of Area

Where painters are working in a large area, for example, a loft building, the rate of applying paint on walls is higher than when working in a confined area or in a cut-up room.

Degree of Workers' Skill

Some painters have considerably more experience and skill than others. Painting contractors generally put workers where their talents and capabilities are most productive.

Working from Ladders or Scaffolds

Painters who have to work from ladders or scaffolding are naturally less productive than when working at ground or floor level. It also takes extra time to set up and move ladders or scaffolds from one position to another.

COVERING CAPACITY OF PAINTS

The number of square feet that a gallon of paint material will cover varies with the type and the condition of the surface, the particular paint material used, and whether

TABLE 21–1

Approximate Covering Capacity of Painting Materials Using Brush Application

Approximate Covering Capacity of Painting Materials Using Brush
Application

USE OF PAINT MATERIALS	Sq Ft Per Gallon
Interior	
Flat paint on smooth walls	450 – 500
Flat paint on sand finish plaster	300 – 350
Flat paint on texture finish plaster	300 – 350
Gloss or semi-gloss on smooth plaster	400 – 450
Gloss or semi-gloss on sand plaster	300 – 400
Gloss or semi-gloss on texture plaster	250 – 350
Calcimine-Watersize	720
Concrete block - Resin Emulsion	200
Wood veneer-Lacquer	500 – 540
Wood veneer-Synthetic resin	600 – 675
Penetrating wax	600 – 675
Enamel on wood floors	400 – 450
Stainwax on floors or trim	500 – 600
Shellac (4 lb cut)	500 – 550
Varnish over shellac	500
Varnish remover	150
Exterior	
Wood siding and trim-flat or semi-gloss	450 – 550
Staining shingle roofs and siding	200 – 300
Oil Paint-asbestos shingles	200 – 300
Cedar shakes-Latex	200 – 300
Brick -Latex	200 – 300
Concrete floors and steps-enamel	300 – 400
Link fences	750
Gutters and leaders	200–250 lin ft per coat

Estimating Tables for House Building, Craftsman Book Co., 1986. Reprinted with permission.

it is applied with a brush, roller, or sprayer. Other factors
include the color or *hiding* capacity of the paint material, and whether the individual who applies it uses excessive and unnecessary amounts or brushes it out well. The
first coat generally covers less square feet per gallon than
second and third coats. This is because of the absorption

of the paint by the surface to which it is applied, and the fact that the first coat acts as a sealer.

While good judgment is essential in estimating coverage capacities, along with the latest information on products from manufacturers, Table 21–1 may provide useful guidelines.

MEASUREMENT OF AREAS

The unit of measurement for painting is the square yard or square foot. All measuring is done in lineal feet and areas calculated in square feet. When estimating the number of square feet, the actual surface to be painted should be measured as accurately as possible. The flat surface areas, such as walls and ceilings, present no special problem in measurement, but questions often arise as to measuring the areas of windows, trimmed openings, stairs and balustrades, doors, exterior cornices, lattices, and so forth. Rules-of-thumb are used by most painters and, while there are differences in the rules used, each one strives to reasonably approximate the actual surface area to be covered. The following methods for measuring flat and irregular surface areas are recommended.

Wall Areas

The square foot area of walls is obtained by measuring the distance around the room (the perimeter) and multiplying by the height from the floor to the ceiling. Table 21–2 shows perimeters of rooms 2 ft. by 3 ft. to 20 ft. by 30 ft. for convenient calculating. Walls that are to be painted only above or below a wainscoting should be measured on that basis.

Whether window and door openings are to be deducted will depend on the method being used to estimate. When openings are deducted, the cost of material

TABLE 21-2

Lineal Feet of Perimeter of Rooms Measuring 2' × 3' to 20' × 30'

Feet	2	3	4	5	6	7	8	9	10	11	12	13	14	15	16	17	18	19	20
3	10	12	14	16	18	20	22	24	26	28	30	32	34	36	38	40	42	44	46
4	12	14	16	18	20	22	24	26	28	30	32	34	36	38	40	42	44	46	48
5	14	16	18	20	22	24	26	28	30	32	34	36	38	40	42	44	46	48	50
6	16	18	20	22	24	26	28	30	32	34	36	38	40	42	44	46	48	50	52
7	18	20	22	24	26	28	30	32	34	36	38	40	42	44	46	48	50	52	54
8	20	22	24	26	28	30	32	34	36	38	40	42	44	46	48	50	52	54	56
9	22	24	26	28	30	32	34	36	38	40	42	44	46	48	50	52	54	56	58
10	24	26	28	30	32	34	36	38	40	42	44	46	48	50	52	54	56	58	60
11	26	28	30	32	34	36	38	40	42	44	46	48	50	52	54	56	58	60	62
12	28	30	32	34	36	38	40	42	44	46	48	50	52	54	56	58	60	62	64
13	30	32	34	36	38	40	42	44	46	48	50	52	54	56	58	60	62	64	66
14	32	34	36	38	40	42	44	46	48	50	52	54	56	58	60	62	64	66	68
15	34	36	38	40	42	44	46	48	50	52	54	56	58	60	62	64	66	68	70
16	36	38	40	42	44	46	48	50	52	54	56	58	60	62	64	66	68	70	72
17	38	40	42	44	46	48	50	52	54	56	58	60	62	64	66	68	70	72	74
18	40	42	44	46	48	50	52	54	56	58	60	62	64	66	68	70	72	74	76
19	42	44	46	48	50	52	54	56	58	60	62	64	66	68	70	72	74	76	78
20	44	46	48	50	52	54	56	58	60	62	64	66	68	70	72	74	76	78	80
21	46	48	50	52	54	56	58	60	62	64	66	68	70	72	74	76	78	80	82
22	48	50	52	54	56	58	60	62	64	66	68	70	72	74	76	78	80	82	84
23	50	52	54	56	58	60	62	64	66	68	70	72	74	76	78	80	82	84	86
24	52	54	56	58	60	62	64	66	68	70	72	74	76	78	80	82	84	86	88
25	54	56	58	60	62	64	66	68	70	72	74	76	78	80	82	84	86	88	90
26	56	58	60	62	64	66	68	70	72	74	76	78	80	82	84	86	88	90	92
27	58	60	62	64	66	68	70	72	74	76	78	80	82	84	86	88	90	92	94
28	60	62	64	66	68	70	72	74	76	78	80	82	84	86	88	90	92	94	96
29	62	64	66	68	70	72	74	76	78	80	82	84	86	88	90	92	94	96	98
30	64	66	68	70	72	74	76	78	80	82	84	86	88	90	92	94	96	98	100

and labor to paint the openings is added separately. If window and door openings are not deducted, the additional time to cut in those openings is included in the labor allowance for the room, or per square foot area. Actually it requires no additional paint, even when different colors are used for the finish coat. It does require more time than painting a flat surface.

Wall Areas from Table 17-1 (Page 358)

Table 17-1 shows the gross square foot area of the sidewalls of rooms 2 ft. by 3 ft. to 20 ft. to 30 ft., having a floor-to-ceiling height of 10 ft. The sidewall areas of rooms with higher or lower ceiling heights can be readily determined from this Table using the conversion factors shown. There has been no deduction for door or window openings, and the rooms are assumed to be rectangular in shape. For rooms that are irregular in shape, Table 21–2 shows the lineal feet of perimeter of rooms 2 ft. by 3 ft. to 20 ft. by 30 ft. Wall areas are obtained by multiplying the perimeter by the ceiling height.

Windows and Doors

Measure the width and height from the outside trim of the opening. Add 2 feet to the width and the height of a window. Add 2 feet to the width and 1 foot to the height of a door. This allows for moldings, edges, and frames.

A window that measures 3 ft. by 6 ft. would be figured as 5 ft. by 8 ft., or 40 square feet. A door measuring 3 ft. by 7 ft. would be figured 5 ft. by 8 ft. or 40 square feet. Some painters use a unit of 40 square feet for all average-sized openings when figuring windows and doors.

Windows or doors with sash having more than one light of glass, are figured at 2 square feet for each additional light. A 6-light sash would add 10 square feet to the window area.

The proper measurement of window and door areas is very important when estimating brick, masonry, or other surfaces where painting of such openings is strictly an individual operation.

Floor and Ceiling Areas

The number of square feet in a floor is obtained by multiplying the length by the width. Openings less than 10 ft. by 10 ft. in the floor are generally not deducted.

Floor and Ceiling Areas from Table 17-2 (Page 360)

Table 17-2 shows the gross square foot ceiling and floor areas in rooms 2 ft. by 3 ft. to 20 ft. by 30 ft. The rooms are assumed to be rectangular.

Wainscot Areas

If the wainscoting is painted on a flat surface such as plaster, the area is obtained by multiplying the length by the width. If the wainscoting is paneled, the area should be multiplied by 1½ or 2, depending on the form of the surface.

Stair and Balustrade Areas

To determine the surface area of stairs that are to be painted, allow 1 foot for the width of the tread, and 1 foot for the width of the riser to take care of the nosing and molding. Multiply the widths by the length of the treads or risers, plus 2 feet for stair strings. A stairway with treads and risers 3 ft. 6 in. wide would have 3 ft. 6 in + 2 ft. = 5½ square feet for each tread and riser. Multiply the area of the tread and the riser by the number of each.

The area calculation for a balustrade is obtained by measuring the height from the top of the tread to the handrail. Multiply that height by the length of the balustrade. The area of square edge and relatively simply designed balustrades is computed by multiplying the height of the balustrade by its length by 4. Turned and fancy balustrade areas should be multiplied by 5.

For example, a balustrade that measures 2 ft. 6 in. by 18 ft. has a plane area of 45 square feet. The paint area for a simple balustrade would be 45 square feet × 4, or 180 square feet. For a turned and fancy balustrade, the paint area would be 45 square feet × 5 or 225 square feet.

Baseboard, Chair Rail, Picture Moldings

The paint area of baseboards, chair rails, plate rails, picture moldings, and other individual trim items should be estimated on the basis of square feet per lineal foot. Add 50 percent to the width of the particular trim item. For example, the paint area of a 4-inch chair rail would be 4 in. + 2 in. = 6 in., or 1/2 square foot per lineal foot.

Trimmed Openings Areas

The paint area of two sides of a *trimmed* opening (one with no door) is obtained by measuring the lineal feet around the two sides and top of the opening and multiplying by 3. A trimmed opening 8 ft. wide and 7 ft. high would have a paint area of:

8 ft. + 7 ft. + 7 ft. – 22 ft. × 3 = 66 square feet

This area is for both sides of the opening.

Built-in Shelving, Cupboards, and Cases

To obtain the paint area for built-in shelves, cupboards, bookcases and cabinets, measure the front area (width × height). Multiply the front area by 3 for open-front units.

Multiply the front area by 5 for units that have doors. The total area obtained includes the finishing of both interior and exterior.

Exterior Shingles and Siding Areas (Wood, Metal, and Composition)

To obtain the paint area of exterior shingle and bevel or drop siding, measure the actual wall area including any gables. Do not deduct window or door openings less than 10 ft. by 10 ft. For narrow clapboards and for shingles, add 20 percent to the actual area. For wide clapboards (over 5 inches), add 10 percent to the actual area.

Exterior Cornices and Eaves Areas

Measure the length and the width of the cornice to obtain the area. Multiply this area by 2 for relatively simple cornices, and multiply it by 3 for fancy or ornate cornices.

Where cornices are open, with exposed rafters, multiply the cornice area (length × overhang) by 3 to obtain the paint area.

Exterior Masonry Wall Areas

The paint area of brick, stucco, cement block, and other masonry walls is the actual square foot area including any gables. Window and door openings should not be deducted if less than 10 ft. by 10 ft.

Fence Areas

The paint area of solid board fences is the length times the height for one side. Multiply by 2 when both sides are to be painted.

The paint area of *one side* of a picket fence is obtained by multiplying the length times the height by 2. For two sides, multiply by 4. Paint area of chain link fences is the area of *one* side multiplied by 3.

Lattice Areas

The paint area of lattice work is obtained by multiplying the length times the width. To paint one side, multiply that area by 3.

ESTIMATING MATERIALS

When the paint area has been measured and calculated, the *material* required per coat is obtained by dividing that area by the number of square feet a gallon of paint material will cover. (See Table 21–1.)

ESTIMATING LABOR

The conditions that have an influence on painting labor specifically have been discussed earlier in this chapter. The factors that affect labor in general should be reviewed under "Estimating Labor" in Chapter 10.

The labor-hours required to apply paint materials vary but may be estimated reasonably accurately when based on average approximate rates. Table 21–3 shows the approximate number of square feet per hour required to apply paint materials for various types of interior work. Table 21–4 shows approximate number of square feet per hour to apply paint materials for various types of exterior work.

TABLE 21-3

Approximate Number of Hours to Apply Painting Materials for Type of Interior Work Shown—Premises Unoccupied

TYPE OF WORK	Sq Ft Per Hour Per Coat	
	Brush	Roller
Washing down stained walls and ceilings	100–150	
Sanding woodwork, preparatory	150–250	
Sizing plastered walls	300–400	
Burning paint from trim	25–40	
Painting smooth plaster incl openings	150	
Painting smooth plaster excl openings	200–250	300–350
Painting sandfinish plaster excl openings	175–200	350
Painting sandfinish plaster incl openings	125	
Painting texture plaster-semi-gloss	100–125	350
Painting texture plaster-flat paint	180	350
Painting Cement block or Brick	150	250
Spray painting		350–400
Painting Concrete floors or walls	200–250	300–350
Floors-remove finish with liquid	50	
Floors shellac, paint, stain or varnish	200–300	400–500
Floors wax and polish	200	
Paneling-Staining, or finishing	250	
Trim-back priming-paint	350	
*Windows–one side	125–150	
**Doors–one side	125–250	
Molding etc–1 to 6 inches wide	175 lin ft	
Note: No scaffolding included in above hour rates.		

See pages 459 to 469 Measurement of Areas.
*Average 12–14 openings per day, one coat, one side per 8 hours. 6/6 lites.
**Flush-average 16–18 openings per 8 hours, one coat, one side.
  Panel–average 12–14 openings per 8 hours, one coat, one side.
There are some painters who estimate 1/16 gal. of paint per side per opening,
and 1/3 hours labor per side per opening.

Estimating Tables for House Building, Craftsman Book Co., 1986. Reprinted here with permission.

TABLE 21–4

Approximate Number of Hours to Apply Painting Materials for Type of Exterior Work Shown—Average from Ground and Ladders

	Sq Ft Per Hour Per Coat	
TYPE OF WORK	Brush	Roller
Sanding woodwork lightly Sanding blistered woodwork Burning off paint from woodwork	250 100–200 25–40	
Wood siding, incl window and door openings " " excl window and door openings *Exterior trim only incl windows and doors	100–125 200–225 80–100	300–350
Shingle staining-walls Brick walls–Oil paint Stucco walls–Oil Paint Concrete block	150 125–150 100–150 100–125	200 200–250 200–250 250–300
Floors–wood Floors–concrete	200 200–250	250 300–350
Roofs–Wood shingle stain Spraying	200–225	250–300 500
Shutters Screens Storm sash	4–5 units 5–6 " 4–5 "	
Fences–solid board " –picket " –picket (spray) " –link Leaders and gutters	250 150–200 125 100 lin ft	400 250–300 450–500 300

*Windows–painter will paint (one-coat) 12–14 openings per day, one side only.
Flush doors–Painter will paint (one-coat) 16–18 openings per day, one side only.
Panel doors–painter will paint (one-coat) 12–14 openings per day, one side only.
Note: See pages 459 to 469. Measurement of Areas.

METHODS OF ESTIMATING PAINTING

There are five basic methods or systems for estimating painting which are commonly used by contractors. In some sections of the country one particular method may become established by custom and usage. It therefore is important to understand each system, especially when analyzing painting estimates or bids. When properly used, there should be very little difference in the results obtained.

Method 1—Material and Labor Estimated Separately

In this method the total gallons of paint and the total hours of labor are determined from the *gross* wall and ceiling area. The number of gallons of paint material is based on the average square foot covering capacity of one gallon. The number of hours of labor is based on the number of square feet a painter can average per hour.

Method 2—Unit Cost Applied to Gross Area

A method that is frequently used by many painters is one in which a unit cost per square foot or square yard is developed and applied to the gross area without deductions for window and door openings. It employs the same principle as method 1 in that the rate of application allows for the time to cut in window, door, and other wood trim where different color paints are used.

Method 3—Separate Unit Costs for Flat Wall and Ceiling Surfaces, and for Openings and Trim

A method that is quite common in midwestern and western states is one where individual unit costs are applied to walls and ceilings, window and door openings, and to trim items. The unit costs take into consideration the conditions applicable to each kind or type of surface, on the principle that paint coverage and the rate of application vary with the kind of paint and the form of the surface being painted. This method is taught in a number of trade schools for painters. Because of the refinement in the estimate, it is considered by many to be more accurate than other methods. It has particular merit where the treatment of the wall or ceiling surfaces and that given to openings and trim are substantially different in the cost per square foot.

Method 4—Material and Labor Per Gallon of Material

Occasionally, a rough estimate for painting is figured on the cost of labor and paint material *per gallon*. The unit cost of labor and material per gallon is multiplied by the estimated number of gallons required.

Method 5—Material and Labor Per Hour

Another rather rough method used by some painters is to develop a unit cost per hour for both material and labor. For example, an average covering capacity of the particular paint might be figured at 450 square feet per gallon, and the rate of application 150 square feet per hour. On this basis, the unit cost of material and labor *per hour* would be the cost of 1/3 gallon of paint (150/450 = 1.3), plus 1 hour's wages for the painter. Some estimators prefer this method, especially for rough estimating, because they can look at a room, a house, or a complete job and approximate the number of hours of labor to do the work.

Summary of Methods

The choice of method by which painting is estimated is one of individual preference. Whatever system is used, it should consider all of the conditions of the particular job.

SELECTED REFERENCES

Paint Handbook, McGraw-Hill Publishing Company, New York.

Painter's Handbook, William McElroy, Craftsman Book Company, Carlsbad, CA, 1987.

CHAPTER 22

INTERIOR WALLCOVERINGS

Wallpaper is sold by the single or the double roll. A single roll is a strip 18 inches wide, 24 feet long, and containing 36 square feet. A double roll is 48 feet long and contains 72 square feet. While other widths and lengths are available, the foregoing are the most commonly used. Wallpaper borders are sold by the lineal yard. The quality and the price of wallpaper have wide ranges. It is, therefore, important to make certain that the type ordered meets specifications.

PREPARATION OF SURFACE

Before walls are papered they should be given a coat of glue size. Old wallpaper must be removed, the walls washed, cracks patched, and a coat of sizing applied. Wallpaper can be removed by wetting it down thoroughly with water using brush or spray and scraping it off with wide putty knives. An electric-steam appliance is available which makes removal of old paper quicker.

Old wallpaper can be removed by wetting and scraping at the rate of 150 square feet per hour for single

471

layers, and approximately 100 square feet per hour where there are two layers. When permissible, new wallpaper can be applied directly over old paper provided the surface is smooth (not embossed) and the old paper is butt-jointed and tight on the wall. Under these conditions papering over old paper is usually satisfactory.

ESTIMATING MATERIALS

When hanging wallpaper, some allowance should be made for the waste that results from cutting, fitting around openings and built-in units, and for matching patterns. Generally, the waste will average between 15 and 20 percent. A single roll of paper contains 36 square feet. A practical and acceptable method for allowing waste is to figure single rolls at 30 square feet and double rolls at 60 square feet.

To obtain the number of rolls required, the distance around the room (perimeter) is multiplied by the height from the top of the baseboard to the ceiling. Some estimators deduct window and door openings in full; others deduct one single roll of paper. The area of built-ins is also deducted. The result is the wall area to be papered. That area is divided by 30 to obtain the number of single rolls. Wallpaper is sold in full rolls and the number of rolls needed must be carried out to the next whole roll. If double rolls are used, the area should be divided by 60. Table 21-2 (page 460) shows the perimeter of rooms in feet from 2 ft. by 3 ft. to 20 ft. by 30 ft. Table 17–1 shows the square foot area of sidewalls in rooms (page 358).

Where the walls are to be papered above the chair rail only, or above or below a wainscoting, the height measurement should be taken accordingly. If a single wall is to be papered, the length of that wall is measured and multiplied by the height.

Where ceilings are to be papered, the area is obtained by multiplying the length of the ceiling by the width. See Table 17–2 (page 360) for ceiling areas of rooms.

If a border is required, the distance around the room divided by 3 will give the number of yards needed.

Table 22–1 shows the approximate single rolls for paperhanging walls and ceilings, including yards of border.

Paste to Hang Wall Coverings

The quantity of paste needed to hang wallpaper varies with the type and weight of the paper. Heavy or rough-textured papers require more paste than the lighter types. Prepared dry pastes that are on the market are mixed with water and are ready for use. One pound will make about 2 gallons of paste, which should hang about 18 to 20 single rolls of lightweight paper or 8 to 10 single rolls of heavy or rough-textured paper.

ESTIMATING LABOR

An experienced paperhanger should hang 3 to 4 rolls per hour under average working conditions and where the rooms are vacant. If the joints are lapped instead of butted, the rate may increase to about 4 to 4½ rolls per hour. Where work and materials are of the highest grade, a paperhanger will hang paper at an average rate of 2 single rolls per hour.

Paper borders can be hung, under average working conditions, at the rate of approximately 30 lineal yards per hour. Table 22–2 shows the approximate hours of labor required for various paperhanging operations.

TABLE 22-1

Estimating Quantities for Paperhanging Walls and Ceilings:

Measure straight area to secure number of square feet and divide by 30. (This allows 6 sq. ft. for waste.) Deduct ½ roll for every opening. Or, for the following standard-size rooms use this handy chart:

Size of Room	Single Rolls of Side Wall Height of Ceiling			Yards of Border	Rolls of Ceiling
	8 feet	9 feet	10 feet		
4 × 8	6	7	8	9	2
4 × 10	7	8	9	11	2
4 × 12	8	9	10	12	2
6 × 10	8	9	10	12	2
6 × 12	9	10	11	13	3
8 × 12	10	11	13	15	4
8 × 14	11	12	14	16	4
10 × 14	12	14	15	18	5
10 × 16	13	15	16	19	6
12 × 16	14	16	17	20	7
12 × 18	15	17	19	22	8
14 × 18	16	18	20	23	8
14 × 22	18	20	22	26	10
15 × 16	15	17	19	23	8
15 × 18	16	18	20	24	9
15 × 20	17	20	22	25	10
15 × 23	19	21	23	28	11
16 × 18	17	19	21	25	10
16 × 20	18	20	22	26	10
16 × 22	19	21	23	28	11
16 × 24	20	22	25	29	12
16 × 26	21	23	26	31	13
17 × 22	19	22	24	28	12
17 × 25	21	23	26	31	13
17 × 28	22	25	28	32	15
17 × 32	24	27	30	35	17
17 × 35	26	29	32	37	18
18 × 22	20	22	25	29	12
18 × 25	21	24	27	31	14
18 × 28	23	26	28	33	16
20 × 26	23	28	28	33	17
20 × 28	24	27	30	34	18
20 × 34	27	30	33	39	21

Deduct one single roll of side wall for every two ordinary sized doors or windows or every 36 square feet of opening. Yard goods with no match such as wide vinyls, measure area, take out for openings and allow 10" for waste.
Courtesy of Painting and Decorating Contractors of America, Falls Church, Va. 22046

TABLE 22–2

Approximate Hours Labor Applying Wall Coverings

Type of Covering	Number of 30 Sq. Ft. Rolls Per Hours
Wallpaper, butt joint	
Lightweight, medium grade work	3–4 rolls
Lightweight, high grade work	2–3 rolls
Heavy, rough-textured	1½–2 rolls
Scenic paper	1–1½ rolls
Vinyl, lightweight	1½–2 rolls
Vinyl, textured, medium weight	1–1½ rolls
Borders	(25–35 yds per hour)
Sanatas™/Walltex™	2–2½ rolls
Canvas on plastered walls	1½–2 rolls

SANATAS™ AND WALLTEX™

Sanatas™ and Walltex™ come in rolls containing 9 square yards. They are customarily used on walls of bathrooms and kitchens for their water-resistant and washable qualities. The quantity of these wall coverings required is equal to the area to be covered plus 10 percent waste. A paperhanger can hang approximately 2 rolls of Sanatas™ and Walltex™ per hour under average working conditions.

CHAPTER 23

GLASS AND GLAZING

TYPES OF GLASS

Float Glass

The method of manufacturing flat window and plate glass changed radically with the discovery of the *float glass process*. It was developed in the United Kingdom by Pilkingtons Ltd. in 1959, and by the late 1970s most glass manufacturers had converted over to the new process.

Float glass is made by passing molten glass onto a float bath of molten tin where it spreads out to produce a uniformly flat sheet. It is gradually cooled to emerge with a firm surface that is not deformed by the rollers that take it into an annealing chamber where it is further cooled before passing over more rollers to a computer-controlled cutting and stacking operation.

Regular plate or float glass, used for residential and commercial glazing, is 1/8-inch and 1/4-inch thick. *Heavy* plate or float glass for residential and commercial glazing comes in thicknesses of 5/16 inch to 7/8 inch.

Window Glass

Window or sheet glass comes in two thicknesses, single strength (SS) which is 3/32-inch thick, and double strength (DS) which is 1/8-inch thick. Qualities, as defined by Federal Specification DD-G-451d, are quality AA specially selected for the highest grade of work; quality A selected for superior glazing; and quality B for general glazing.

Heavy window or sheet glass is 3/16-inch to 7/32-inch thick and comes in the same three grades.

Plate Glass

Plate glass, like float glass, is manufactured in two types, depending on the thickness. *Regular plate* is 1/8-inch to 1/4-inch thick; *heavy plate* is 5/16-inch to 7/8-inch thick. Plate glass is ground and polished on both sides.

Patterned Glass

Patterned glass is manufactured in a variety of textures and designs on only one side or both sides. It is translucent and is produced by the same method as plate glass, though it is not ground and polished.

Patterned glass is usually made in thicknesses of 1/8 inch and 7/32 inch, but other thicknesses are available up to 3/8 inch. It is used where privacy is desired, yet it allows diffused light transmission.

Laminated Glass

Laminated or safety glass is made up of two or more layers of high quality sheet or polished plate glass with a layer or layers of transparent tough vinyl sandwiched in between the layers.

Laminated glass is called safety glass because when it breaks, the pieces are held by the vinyl, which prevents shattering. Glass with a minimum of two layers of glass and one layer of vinyl is often required, under many building codes, for bathtub doors and enclosures, shower doors and enclosures, storm doors, and patio sliding doors.

Wired Glass

Wired glass has a wire mesh (vertical, hexagonal, or diamond-shaped) embedded in the middle of the glass thickness, generally 1/4 inch. The wire mesh is not larger than 7/8 inch, and the gauge of wire not less than no. 24. This type of glass is used where fire resistance is required, as in fire doors, windows, and skylights (size limited to 5 square feet), and where impact and abuse occur, such as entrance doors and partitions. It is available in clear glass and patterned glass.

Corrugated Glass

Corrugated glass is 3/8-inch thick, translucent with a pattern on both sides, and is available wired or unwired. Standard lengths are between 10 and 12 feet. The unwired type is used mainly in partitions where both light and privacy are desired. The wired type is used for roofing and in skylights. It is also used in partitions, doors, and other places where breakage is to be avoided.

Tempered Glass

Tempered glass, made from plate and certain types of patterned glass, is heat treated in a process that leaves the exterior surface in compression and the interior in tension. It is three or four times stronger than plate-annealed glass of the same thickness.

Stock sizes, usually 1/4-inch to 1/2-inch thick, are available up to 6 feet by 10 feet.

Tempered glass, because of its nature, has to be cut and drilled for hardware prior to being heat treated.

ESTIMATING MATERIALS

Where glazing is to be done on the job, the plans, working drawings, and specifications must be reviewed carefully to establish the type or types of sash (wood, steel, etc.), the type, quality and thickness of glass, size of windows and doors, and where any plate glass is to be installed.

The amount of putty needed for glazing is a relatively minor item. As a guide for estimating purposes, the following amounts are suggested:

Wood sash (1⅛-inch frames) 1 lb. per 8 lin. ft. putty
Wood sash (1¼-inch frames) 1 lb. per 7 lin. ft. putty
Steel sash 1 lb. per 5 lin. ft. putty

(Some estimators use ½ lb. putty per sq. ft.)

Plate glass in doors, storefronts, transoms, and so forth should be measured to the next *even inch*.

ESTIMATING LABOR

When estimating the time to glaze sash, consideration has to be given to the physical problems that confront the person or persons who will be doing the work. On new construction projects or ones that require a large number of lights of glass, an average hourly rate per light can be estimated fairly accurately.

Table 23-1 shows the approximate number of hours of labor to install various sizes and kinds of glass. The hourly rates should be used as guidelines and adjusted according to the working conditions.

ACRYLIC PLASTIC GLAZING

Acrylic plastics are used in place of glass in flat, curved, or domed skylights. They are known for superior optical properties, resistance to breakage, and excellent weathering properties. A polycarbonate plastic sheet, used for glazing windows in public buildings where the breakage hazard is high, is available in transparent sheets in colors, or in a glare-reducing shade known as *Lexan*™. Acrylic plastics scratch easily.

TABLE 23–1
Approximate Hours of Labor to Install
Glass

Glazing Wood Sash	Hours/Light	Hours/100
8″ × 10″ to 12″ × 14″ Lights	.10–.14	10–14
14″ × 20″ to 20″ × 24″	.20–.22	20–22
30″ × 36″ to 32″ × 40″ ″	.25–.30	25–30
40″ × 48″ to 48″ × 60″ ″	.40–.45	40–45
Glazing Steel Sash		
9″ × 12″ to 12″ × 16″	.12–.14	12–14
16″ × 20″ to 16″ × 30″	.25–.30	25–30
Skylight Sash		
24″ × 36″	.40–.50	40–50
24″ × 60″	.60–.65	60–65
Store Fronts—Per 100 Sq. Ft.		
Setting plate glass 4′ × 6′ to 8′ × 10′	10 man-hours	
″ ″ ″ 8′ × 10′ to 10′ × 15′	15 man-hours	

CHAPTER 24

ELECTRICAL WIRING

Electrical power reaches a building through service wires extending from the power company's distribution system to the building, passing through a meter to the service equipment. This generally consists of a meter, a main entrance switch, followed by a distribution panel with fuses, circuit breakers or other protective devices for each *branch* circuit. Branch circuits bring current to points within the premises by way of junction boxes, to switches, outlet receptacles, lights, motors, and other equipment.

Adequate wiring contemplates:

1. Service entrance of sufficient capacity.

2. Wires of sufficient capacity throughout.

3. Sufficient number of circuits.

4. Adequate number of outlets, switches, etc., for livable flexibility.

ELECTRICAL TERMS

To understand the language of electrical wiring, whether in manuals, on plans, or in a discussion with electrical contractors, one should be familiar with the various terms used. Those terms most frequently encountered, and their definitions, are the following:

Alternating current (A.C.). Electrical current that reverses direction in a circuit at regular intervals.

Ampacity. The maximum number of amperes a wire of specified diameter can safely carry continuously is called the ampacity of that wire.

Ampere. The rate of flow of electric current through a conductor. More technically, the amount of current that will flow through one ohm under pressure of one volt.

Circuit. A continuous path for current to travel over two or more conductors (wires) from source to point of use and back to source.

Circuit breaker. A device designed to open automatically on a predetermined overload of current, and close, either automatically or manually, all without injury to the device itself.

Conductor. Any material that will permit current to flow through it. Wire in a circuit is a conductor.

Current. A flow of electrical charge.

Cycle. The flow of alternating current in one direction, then in the other direction, is one cycle. In a 60-cycle circuit this occurs 60 times a second.

Direct current (D.C.). Current in which electricity flows in only one direction. Current from any type of battery is always direct current.

Fuse. A protective device inserted in a circuit, containing a short length of alloy wire with a low melting point and which will carry a given amperage indefinitely. Any overload of current melts the alloy wire and opens the circuit. (Plug fuses, when burned out, cannot be reused.)

Fusetron. A time-delay fuse designed to tolerate temporary overloads for several seconds without blowing out.

Fustat. A nontamperable fuse designed to prevent changing fuses to a higher amperage. For example, No. 14 wire has an ampacity of 15 amps. If larger amp fuses are used, there is danger of the wires in the circuit heating sufficiently to cause a fire on the premises before the fuse blows.

Grounding. A connection from the wiring system to the ground.

Horsepower. Watts measure the total energy flowing in a circuit at a given moment; 746 watts equal one horsepower.

Insulation. Any material that will not permit current to flow is an insulator. When such material is used to isolate a charged conductor, it is called insulation.

Kilowatt. One thousand watts.

Kilowatt hours (k.w.h.). A kilowatt of power used for one hour.

National Electrical Code (N.E.C.). A set of standards for the design and manufacture of electrical materials and devices and the manner of their installation.

Ohm. The unit of electrical resistance. It is the resistance through which one volt will force one ampere.

Outlets. Any point in a circuit where electrical power can be taken and consumed.

Parallel wiring. This type of circuit wiring is commonly used in electrical construction wiring. Outlets and lights are wired across the two source wires rather than in series along one of the source wires.

Phase. The type of power available: single-phase or three-phase. The former is used in residence wiring, while three-phase is used in commercial and manufacturing properties where heavy power is required.

Resistance. Different materials offer resistance to the flow of current in different degrees. Aluminum wire of the same diameter as copper wire has a greater resistance than the copper wire. Iron wire has a greater resistance to the flow of current than aluminum wire. Resistance depends on the kind of material, the size of the conductor, and its length.

Receptacle. An outlet into which appliances can be plugged.

Series wiring. A scheme of wiring, seldom used, where the current path is along one conductor or wire. Example: Christmas tree lights, when one lamp is removed, all others go out because the circuit is broken.

Service drop. The overhead service conductors (wires) at the building.

Source. Wherever the current comes from; may be a generator, battery, or where it enters the premises at the meter.

Switch. A device to open and close an electrical circuit or divert current from one conductor to another.

Transformer. An apparatus to increase or decrease voltage. The former are step-up; the latter, step-down transformers.

Underwriters Laboratories (UL). A "not-for-profit" organization whose cost of operation is supported by manufacturers who submit merchandise for testing and labels. (See Figure 24–1.)Many cities require UL listing on all electrical materials used or sold in their jurisdiction. Underwriters Laboratories, Inc. was created in 1894 by the insurance industry, but today is independent and self-supporting.

Volt. The pressure to force one ampere through a resistance of one ohm. Voltage is electrical pressure.

Voltage drop. The voltage loss due to overcoming resistance in wires and devices. The Code recommends wire sizes be used so that voltage drop will not exceed three percent in any branch circuit, measured at the furthest outlet.

Watt. The unit of electrical power. Amperes × Volts = Watts.

ELECTRICAL SYMBOLS

Electrical installation plans, whether for new work or repairs, consist of sketches or outline drawings of the

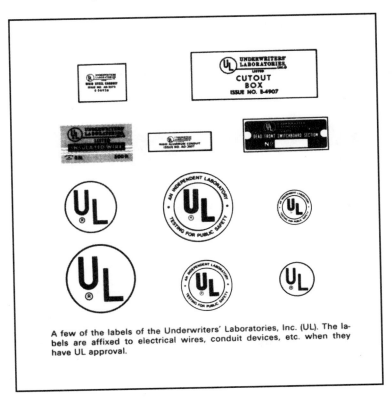

A few of the labels of the Underwriters' Laboratories, Inc. (UL). The labels are affixed to electrical wires, conduit devices, etc. when they have UL approval.

Figure 24–1

room or area involved. On these drawings are shown, by *symbols,** the location of equipment, fixtures, outlets, etc. It is as important to understand these symbols as it is to understand the language or terms used in electrical construction wiring.

Figure 24–2 is a floor plan of a residence showing how symbols are used in actual practice. The lines show the location of the wiring system.

---

*See page 36, Chapter 2, Drawings and Specifications.

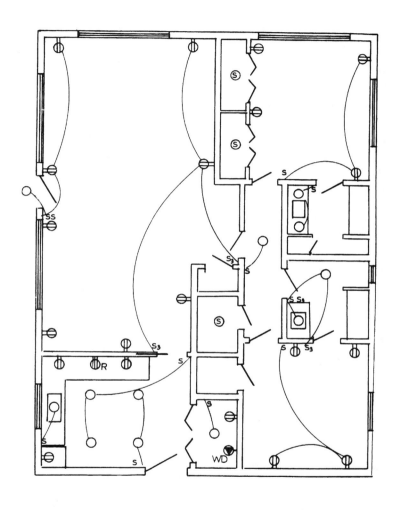

Figure 24–2
Use of symbols in the wiring layout for
a residence.

SIZES OF WIRE

Electrical wire sizes are referred to by *number* in accord with the gauge of the American wire gauge (AWG) which is the same as that of Brown and Sharpe (B & S gauge) for nonferrous metals. For example, No. 14 wire, a minimum size for general building wiring, is 0.064 inches (64 mils) in diameter. Smaller wires range from No. 16 to No. 36 on the standard wire gauge. Larger wires range from No. 12 to No. 0. There are larger and also smaller wires made but we are not concerned with them. Note that as the number gets larger, the wire gets smaller.

Many years ago No. 14 wire was considered adequate for wiring residential and farm buildings. With the increase in appliances that have motors and heating elements, this size wire is not approved in most cities of any size. No. 12 wire is specified and in some cases No. 10. Table 24–1 shows the ampacity of various sized copper wires. Note also that these are the maximum ratings for protective devices (fuses) for the size wire shown.

TABLE 24–1

Ampacity of Various Size Wires

No. of Wire	Ampacity	No. of Wire	Ampacity
14	15	4	70
12	20	2	95
10	30	1/0	125
8	40	2/0	145
6	50	3/0	165

COLORS OF WIRE

The insulation on building wires is colored for a specific purpose—to assure that "hot" wires will be connected only to "hot" wires and neutral wires will run uninterrupted back to ground terminal. The white may be used as a grounding neutral wire only.

The color of wires, established by the N.E.C., are:

2-wire circuit . . . . . . white, black
3-wire circuit . . . . . . white, black, red
4-wire circuit . . . . . . white, black, red, blue
5-wire circuit . . . . white, black, red, blue, yellow

Sometimes another grounding wire is used, but it must be green with a yellow stripe or completely bare.

METHODS OF WIRING

There are many different methods of wiring, but the most commonly used systems for residential, farm, and noncommercial or nonmanufacturing buildings are the following:

Rigid conduit   Nonmetallic sheathed cable
Thin-wall conduit   Armored cable
Flexible conduit   Surface raceways

Rigid Conduit

Much like water-pipe but annealed for ease of bending and treated to prevent corrosion, this conduit comes in lengths of 10 feet with UL label. It is sold in nominal inside diameter sizes from 1/2 to 6 inches, although actual measurements are slightly larger. The pipe is threaded for coupling and connecting to boxes. Conduit systems are installed before the wire is pulled through.

Galvanized rigid conduit can be used inside, outside, and also underground. The black enamel conduit may only be used indoors away from moisture.

The N.E.C. limits the number of wires permitted in the conduit, depending on the size of the pipe and the size and type of wire. Both rigid and thin-wall conduit are more costly to install than other systems.

Thin-Wall Conduit

This is the common name for *electrical metallic tubing* (EMT) as designated by the N.E.C. In many respects this conduit is similar to rigid conduit; it is sold in 10-foot lengths, carries the UL label, the internal diameters are the same, and the method of installation is similar. Its advantage over rigid conduit, obviously, is its ease of cutting, bending, and handling. Fittings are compression types as it cannot be threaded. EMT must be supported within 3 ft. of a box, fitting, or cabinet, and at least every 10 ft.

Flexible Conduit

This is to be distinguished from *armored cable* (BX) which contains wires. The wires have to be pulled through *flexible conduit* in the same manner as rigid conduit or EMT. Its use is somewhat limited to situations requiring flexibility, or in conjunction with rigid conduit where freakish bends are required.

Nonmetallic Sheathed Cable (originally called Romex)

This system of wiring is used where economy is a major influence due to lower material costs and ease of installation. The N.E.C. recognizes two types. In Type NM the wires are enclosed in an outer fabric braid or plastic sheath that is flame-retardant and moisture resistant. It may be used for exposed and concealed work where excessive moisture is not present. There should be a strap every 4½ feet and within 12 inches of every outlet box.

Type NMC has the wires imbedded in solid plastic, and may be used in damp locations but not in water or underground.

Both Types NM and NMC are available with an uninsulated ground wire. Connections may only be made in junction boxes.

Local codes will govern the use of this system of wiring.

Armored Cable

This type of wiring was originally called BX, which was a trade name. It is still referred to as BX regardless of brand. It consists of two or more insulated wires enclosed in a flexible, spiral, galvanized steel sheath to protect the wires from mechanical injury. It is not a watertight armor, and nails can be driven through it. Its use should be confined to interiors, never underground, and in relatively dry surroundings. It is the second easiest system to install, nonmetallic sheathed cable being first. It has a small bond wire assuring a continuous circuit through the metal covering. Only steel junction boxes, switches, and outlet boxes should be used.

Surface Raceways

These are channels of metal or plastic for holding and actually concealing wires on the surface of a wall or ceiling. Special elbows, junction boxes, switches, receptacles, fittings, and connectors are available. Usually the raceways and fittings are mounted and then the wires are pulled through. This system of wiring is often used in making repairs where it is undesirable to break through walls, ceilings, and decorations.

NUMBER OF BRANCH CIRCUITS

In determining the number of circuits required, there are four considerations to be explored:

1. Lighting circuits

2. Small appliance circuits

3. Special laundry circuits

4. Circuits serving individual equipment

Lighting Circuits

The Code recommends, in Section 220-3(a), one 20-ampere circuit for every 500 square feet of livable floor area for *lighting* (or a 15-ampere circuit for every 375 square feet). A dwelling, for example, with 1,500 square feet of livable area would require 3 circuits; one with 2,500 square feet would need 5 circuits. The minimum required is slightly less and is based on 3 watts for every square foot of area under Section 220-2. In practice, one circuit for each 400 feet of floor area would be better. Open basements that are not used for living area may be ignored in computing floor area.

Small Appliance Circuits

Section 220-3(b) of the Code requires, in addition to the previous lighting recommendation, that two or more 20-ampere *appliance branch circuits* shall be provided for the outlets servicing small appliances in the kitchen, pantry, family room, dining room, and breakfast room of dwelling occupancies. Both circuits must run to the kitchen; other rooms may be served by either of the circuits.

Special Laundry Circuit

At least one 20-ampere branch circuit should be provided for *laundry* appliance receptacle(s).

No lighting outlets may be connected to the foregoing circuits.

A three-wire 115/230-volt branch circuit is equal to two 115-volt receptable branch circuits.

Circuits Serving Individual Equipment

Appliances or equipment that use large amounts of current are usually provided with an individual circuit. While the Code does not require that all of the following appliances have a separate circuit, it is good practice to provide one.

- Electric range
- Water heater
- Electric clothes dryer
- Automatic clothes washer
- Dishwasher
- Disposal
- Appliances rated over 1,000 watts
- Electric motors rated over 1/8 hp

These circuits may be 115- or 230-volt depending on their rating. Table 24–2 on page 496 shows the watts consumed by many electrical units.

EXAMPLE

1. Lighting Circuits 1,500 sq. ft. ÷ 500	3 circuits
2. Small Appliances (minimum)	2 "
Laundry circuit outlets for small appliances	1 "
3. Individual Equipment	
¼ hp motor, oil burner	1 "
Range	1 "
Water heater	1 "
½ hp motor water pump	1 "
Clothes washer	1 "
Clothes dryer	1 "
Bathroom heater	1 "
	Total 13 circuits

From the foregoing three considerations, the total number of circuits may be determined. Assume a dwelling with a total of 1,500 square feet of livable floor area. Also assume it has the individual equipment noted.

SERVICE ENTRANCE

Electrical current from the power company comes into the premises through lead-in cables called a *service drop* if it is from overhead. It passes through a watt-hour meter to measure consumption and to the main power switch or *service disconnect,* and into the *distribution panel.* This panel *(service panel, panel board)* contains the protective devices for each branch circuit. These may be fuses or circuit breakers. The branch circuits are designed to handle, safely, subdivisions of the total anticipated electrical requirements.

60-Ampere Service. Standard for 1930–1940, but considered minimum in 1975 by NEC to "get by." Provides for lights, appliances including range, washer–dryer, *or* hot water heater. Considered adequate where gas is used to heat water and for clothes dryers. Entrance service is over three No. 6 gauge wires.

TABLE 24-2

Watts Consumed by Electrical Units

Lighter Units		
Blankets (electric)	150–	200
Blender (food)	250–	1,000
Fans (portable)	50–	200
Heat lamp (infrared)	250–up	
Knife	100–	
Lights, incandescent-noted on lamp	10–up	
Lights, fluorescent	20–up	
Mixer (food)	150–	
Radio (variable)	30–	150
Refrigerator	200–	300
Shaver	10–	
Stereo	300–	
Sump pump	300–	
Sun lamp (ultraviolet)	250–	400
T-V	300–	
Water pump	300–	
Intermediate Units		
Electric iron	800–	1,200
Freezer	300–	500
Power saw (depends on motor)	300–	600
Vacuum cleaner	300–	600
Waffle iron	500–	1,000
Heavier Units		
Air-conditioner (window type)	600–	1,500
Coffeemaker	600–	1,000
Deep fryer	1,300–	1,500
Dishwasher	1,000–	1,500
Clothes dryer	4,000–	8,000
Fry pan	1,000–	1,200
Garbage disposal	600–	1,000
Grill (outdoor type)	1,300–	
Heater (portable)	1,000–	1,600
Range-burners and oven on	8,000–	14,000
" -oven on only	4,000–	6,000
" -burners on only	4,000–	8,000
Rotisserie	1,200–	1,500
Washing machine	600–	800
Water heater	2,000–	5,000

100-Ampere Service. Standard for 1950–1960 when all-electric homes became popular. Now considered *minimum* for lighting, hot water, dryer, freezer, irons, refrigeration, air conditioner, electric range, and small appliances. Entrance service is over three No. 2 or No. 3 gauge wires of RH-RW insulated type.

200-Ampere Service. The coming of electric heat, plus 12,000 watt ranges, dryers, central air conditioning, roasters and rotisseries, other appliances with heating elements, and numerous small appliances made the 100-ampere service inadequate. Entrance service is over three No. 1/0 or No. 3/0 wires of RH-RW insulated type. Some local codes now require 200-ampere minimum service.

Determining Total Service Entrance Load

The Code requires a minimum service of 100 amperes, 115/230 volts, if there are more than five 2-wire branch circuits.

It is possible to fairly well approximate the size service entrance needed by totaling the load carried. Returning to the four considerations in determining the number of branch circuits, the wattage (load) is computed.

1. The Code permits an alternate method of computing lighting and general purpose circuits. That is to figure 3 watts for each square foot area, the same area used to determine the number of circuits on page 493.

2. Small appliance load under Section 220-3(b) was two 20-ampere 115-volt circuits. This works out to $20 \times 115 = 2,300$ watts ($A \times V = W$); but it is recommended 3,000 watts be used.

3. The special laundry outlet required by the Code in 1968 is rounded out at 1,500 watts.

4. Individual equipment (heavy appliances) are rated individually. In Table 24–2 there is a listing of the most common units and the range of wattage for each. Include each at full rating in watts.

By adding the wattage from these four considerations and dividing by 230 volts, service entrance amperage is determined.

Special Notes

The Code has provided a "demand factor" for safety. If the total for the above *items 1, 2, and 3* exceeds 3,000 watts, 35 percent of the excess over 3,000 is added, *not the entire excess.*

Electric ranges get special attention. Since not all burners will be in use at once with the oven turned on, the range's full rating in watts is not used.

If the range is rated at 12,000 watts, use an arbitrary 8,000 watts.

If the range is rated over 12,000 watts, use 8,000 watts plus 400 for each one thousand watts in excess of 12,000. Add wall-mounted ovens and counter-mounted (table-top) burners together to get the equivalent range rating.

Electric heating treatment under the Code can be complicated, but, in general, the *total* wattage consumed by all heating elements is used.

Air-conditioning requires that the rating in watts of the motor or motors be used.

In the following example, the total watts, 27,800, divided by 230 volts equals 120.87-amp entrance service needed. A 100-amp service would not be adequate but a 150-amp service would be ideal as it would have a good safety factor for any expansion.

EXAMPLE

(Dwelling 1,500 sq ft)	Estimated Watts	Demand Factor %	Net Watts
1. Lighting 1,500 sq ft @ 3 watts	4,500		
2. Small appliances	3,000		
3. Special laundry outlet	1,500		
First 3,000 watts Total	9,000	100	3,000
Balance, 6,000 watts		35	2,100
4. ¼ hp motor Oil burner			300
½ hp motor Water pump			500
Range rated 15,000 watts (arbitrary)			8,000
Balance 7 × 400			2,800
Water heater			3,500
Clothes washer			600
Clothes dryer			6,000
Bathroom heater			1,000
		Total	27,800

ESTIMATING ELECTRICAL WIRING

All of the basic principles discussed in Chapter 10, Estimating Labor, are applicable to estimating the cost of electrical wiring. The major factors in forming judgment and making decisions will be *specifications, material, labor, overhead,* and *profit.*

Materials

It is best to make up a *branch circuit schedule* in which the locations of lighting outlets, switch outlets, and convenience receptacles are shown. Figure 24–3 is a sample of one type of branch circuit schedule.

BRANCH CIRCUIT SCHEDULE

Circuit No.	Location on Premises	Light Outlets		Switch Outlets			Receptacles Outlets General	Special Circuit
		Clg	Wall	S₁	S₃	S₄		
1	Kitchen	4			2			
2	Laundry	1			1			
3	Kitchen						1	Oven & Range
4	Dining Rm.		2		1		3	
5	Living Rm.		2	2			2	
	Master B.R.	1		1			6	
	Master Bath	2		1			4	
6	Guest Rm.	1		1			1	Heater
	Guest Bath	1		1			4	
7	Garage	2			1		1	Furnace

Figure 24–3
Branch Circuit Schedule

From this schedule the number and type of fixtures are listed for later pricing.

The next step for estimating materials can be quite complicated on jobs other than minor repair and replacement. It will require meticulous attention to detail. A typical *Branch Circuit Material Schedule* is shown in Figure 24–4. There are other methods of take-off of materials that are less detailed and used by estimators of long experience. Some use a notebook in which they write down the materials as they inspect and examine the premises or study a floor plan. (See page 488.)

Labor

The number of hours for an electrician to install new electrical wiring, outlets, boxes, and so forth, calls for some experience and judgment. Much will depend on the type of wiring as to whether it is exposed, concealed, or in raceways; also, how much cutting or boring through partitions, floors, or joists is needed to "fish" wires through. Much thought and consideration must be given to those factors that affect labor as outlined in the Tables.

The *labor hours* given in Table 24–3 on page 503 for some of the operations encountered in run-of-the-mill electrical wiring should be used as a starting point in estimating the hours required for the particular job being considered. In many cases, pure judgment will supplement or even supersede the hours shown.

Pricing the Estimate

The principles for pricing the labor and material in an electrical wiring estimate follow those outlined in Chapter 10.

BRANCH CIRCUIT MATERIAL SCHEDULE

No.	Cable		Boxes			Plaster Rings			Switches			Plug Recep	Plates			Wire Nuts-Connectors Hangers-Grd. Straps			
	No. 12 2-wire	No. 12 3-wire	4" Oct.	4" Sq	SR	4" R	4" Sq	4" SS	S_1	S_3	S_4		SS	2G	PR	W.N.	C	H	G.S.
1	120'	20'	2	2	6	2			2			4	2		4	6	4	4	
2	160'	35'	1	3	8		2	1	2	4		3	2	4	3	10	2	2	2
3	40'	14'			5	1	1	1				8		1	8	4	2	6	
4	18'	70'		3	2		1		1	3	1	4	1	3	4	4		2	4
ETC																			
Total																			

SR Standard receptable box   2G Two gang switch plate
SS Single pole switch plate   PR Plug receptacle plate

Figure 24–4
Branch Circuit Material Schedule

TABLE 24-3

Approximate Hours to Install Various Electric Wiring and Units

Type of Work		Hours
Conduit, Cable, and Wire		Hours Per 100 Lin. Ft.
Rigid conduit-galvanized	½" to 1"	10–14
" " "	1¼" to 2"	16–20
Rigid conduit-Thin wall	½"	6–10
" " " "	1"	10–12
" " " "	1½"	12–16
" " " "	2"	14–18
Flexible conduit (Greenfield)	½" to 1"	6–8
" " "	1½" to 2"	10–14
Non-metallic sheathed cable	#12 and #14	3–4
Armored cable (BX)	#10 wire	5–6
" "	#12 and #14 wire	4–5
Pulling wire through conduit	#10, #12 and #14	2–4

Service Entrance Installation	
Meter	1–2 each
Main switch	1½–2 "
Panel or circuit-breaker box-200 amp	2–3 "

Setting Outlet Boxes	
4" × 3" and octagonal	0.5 "
2 gang	0.6 "
4 gang	1.0 "

Installing Switches and Receptables	.3 to .5 "

Overhead and Profit

An electrician is entitled to a reasonable charge for his overhead and profit. The same factors will apply in arriving at these charges as are outlined in detail under Overhead, Chapter 10. As previously pointed out, the general contractor or builder adds only a supervisory charge to the electrical contractor's estimate.

Alternate Estimating Method

Frequently electrical contractors are able to estimate the cost of wiring in buildings on a unit cost for labor and material per outlet. This method is based on knowledge and experience in similar jobs.

Under this method each lighting outlet and each convenience outlet is counted and multiplied by the unit cost. To the total must be added any special circuits to heavy-duty appliances or equipment, and also any lighting fixtures, door chimes, and so forth.

SELECTED REFERENCES

Electrical Wiring Commercial, Ray C. Mullin and Robert L. Smith, Delmar, New York, 1999.

National Electrical Estimator, Edward J. Tyler, Craftsman Book Company, Carlsbad, CA, 1999.

CHAPTER 25

PLUMBING

The work of the plumbing contractor is roughly separated into two categories: *rough* plumbing, called *roughing-in*, and *finish* plumbing which is the installation of fixtures.

Roughing-in involves the labor and materials to place piping for supplying water and to provide for drainage and sewerage. Figure 25–1 on page 507 shows a schematic drawing of a typical plumbing system in a two-story house.

WATER SUPPLY MATERIALS

Galvanized pipe for potable water supply, while still used to some extent, is rapidly being displaced by copper tubing which is easier to install, takes less time, and requires no threading or threaded fittings. Two types of copper tubing are available: *rigid* (hard-drawn) and *flexible* (soft-drawn). The latter can be put in place with sweeping bends. For common household fixtures such as sinks, tubs, water closets, showers, and dishwashers, the ½-inch pipe is minimum. Automatic clothes washers should have ¾-inch pipe.

DRAINAGE PIPING MATERIALS

A drainage system in a building includes the piping that carries off sewage and other liquid wastes to the sewer or septic tank. Piping materials may be cast iron, galvanized wrought iron, copper, brass, or plastic. Cast iron is generally used for underground drains or those buried under concrete floors.

Plastic Piping

Carefully selected and properly installed plastic pipe offers several advantages over conventional metallic piping materials. There are no perfect plumbing piping materials and all must be installed with knowledge of their physical properties and limitations. Plastic pipe and fittings are often the most economical and are nearly immune to the attack of aggressive waters. At the present time PE (polyethylene) pipe is used most commonly for underground service. Since it is available in long coils, it requires a minimum of fittings for long pipe runs. For short runs, the friction loss in the insert fittings can be a disadvantage.

PVC (polyvinyl chloride) pipe is available in nearly twice the pressure rating for the same cost as PE pipe. PVC pipe is most often assembled with solvent-welded fittings. Heavy-wall PVC Schedule 80 pipe may be threaded. CPVC (chlorinated-polyvinyl-chloride) pipe is available for hot water service.

ABS (acrylonitrile-butadiene-styrene) pipe was once primarily used in potable water distribution in a size known as SWP (solvent welded pipe). Today ABS is used in DWV (drainage-waste-vent) systems. PVC is also used in DWV systems.

To be sure of getting quality plastic pipe and fittings, see that the material is marked with the manufacturer's

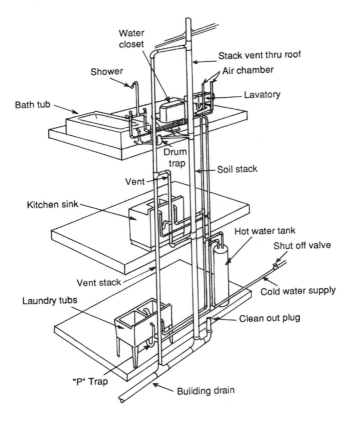

Figure 25–1
Plumbing arrangement in a two-story
house with basement

name and trademark, pipe size, the plastic material type
or class code, pressure rating, standard to which the pipe
is manufactured (usually ASTM standard), and the seal
of approval of an accredited testing laboratory (usually
N.S.F., the National Sanitation Foundation).

ESTIMATING MATERIALS

Estimates of plumbing materials are usually prepared according to the type or class of work involved. A typical situation might be:

1. *Soil, sewer* and *outside drainage* piping and fittings,
2. *Cold water supply* piping and fittings, and
3. *Interior fixture drainage* piping, stacks, sump pumps, and so forth.

The lineal feet of piping, sizes, and fittings are listed for each class of work, and then priced locally. The various plumbing fixtures that are to be installed, including such items as hot water heaters, are listed and priced for the area.

ESTIMATING LABOR

Some plumbing contractors estimate the hours of labor by applying an experienced percentage to the cost of roughing-in materials, and another percentage to the cost of the fixtures to be installed. Most contractors, however, estimate the labor for each operation, considering all factors that could influence productivity.

While it is neither feasible nor practical to set fixed hourly rates of labor to rough-in plumbing or install fixtures, Table 25–1 shows the approximate average *man hours* as a guide in estimating labor.

Additional Charges, Overhead, and Profit

Table 25–1 does not include the costs of trucking, any needed excavating or trenching, special equipment, and hoisting or moving materials and fixtures above the second or third stories. These charges must be estimated and added.

When plumbing is subcontracted, the plumber adds the customary overhead and profit. The general contractor usually adds a percentage for supervision.

TABLE 25–1

Approximate Hours Labor Roughing-In Plumbing Lines and Installing Fixtures

Roughing-In	Unit	Man Hours
4" Soil pipe, C.I. or PVC, below ground	10 lin. ft.	2.5– 3.0
6" " " " " " "	10 lin. ft.	3.5– 4.0
Water closets, Supply, waste and vent	Each	8.0– 9.0
Lavatories " " " "	"	6.0– 8.0
Bath tub and shower" " " "	"	8.0– 9.0
Stall showers " " " "	"	7.0– 9.0
Kitchen sink " " " "	"	10.0–12.0
Laundry tubs, double" " " "	"	10.0–12
Installing Fixtures and Accessories		
Water closets		3.0–4.0
Lavatories		3.0–4.0
Bath tub and shower		4.0–6.0
Stall showers		7.0–8.0
Kitchen sinks		4.0–6.0
Laundry tubs, double		3.0–5.0
Dishwasher		4.0–6.0
Water heaters, 40 to 60 gals.		4.0–5.0

SELECTED REFERENCE

Plumbing, L. V. Ripka, American Technical Publishers, Inc., 1155 W. 175th Street, Homewood, IL 60430, 1994.

CHAPTER 26

FLOOR COVERINGS

RESILIENT FLOORING

Resilient-type floor coverings are made in two forms: *floor tile* and *sheet*. The common kinds of floor tile are asphalt, vinyl, solid vinyl, cork, and rubber. Vinyl sheet flooring has replaced linoleum, which is no longer made in this country.

Underlayments are used to provide a smooth flat base for resilient floors when the existing surface is irregular or is unsuited for proper bonding. Plywood, hardboard, and, sometimes, particleboard are used over wood floors. A mastic underlayment is available for uneven and irregular surfaces of concrete. It should be mixed and applied according to the manufacturer's specifications and directions.

Adhesives are used to apply resilient flooring of all types. They are designed for their adaptability to the specific flooring that will be used, the particular underlayment, and whether the flooring is below grade, at grade level, or above grade. In all cases the manufacturer's recommendations should be followed.

Table 26-1 lists various types of resilient flooring materials, common thicknesses, sizes, and their approximate weight per square foot.

TABLE 26-1

Thickness, Sizes, and Approximate Weights of Resilient Flooring

Type of Flooring	Thickness in Inches	Common Sizes	Weight Per Square Foot
Asphalt tile, greaseproof	1/8 3/16	9" × 9", 12" × 12"	1.17 1.74
Vinyl Plastic tile	3/32 3/16	9" × 9", 12" × 12"	0.93 1.86
Rubber tile	1/8 3/16	9" × 9", 12" × 12"	1.24 1.70
Cork tile	1/8 3/16 5/16	6" × 12", 9" × 9"	0.68 1.05 1.70
	Gauge	Size	
Sheet Vinyl	.065 to .160	6' and 12' wide	

Note: All types except *cork* are suitable for grade level and below grade level application.

Estimating Materials

To obtain the quantity of materials, compute the square feet of floor to be covered. Add a percentage for waste from cutting and fitting. Waste is a matter of judgment. For small areas of a few hundred square feet, waste may run 10 percent; for large areas of several thousand square feet, a waste of 3 or 4 percent may be adequate. The exact number of tiles of various sizes may be figured from the following simple formulas:

- For 6-by-6-inch tile, divide the square foot floor area by 0.25
- For 6-by-12-inch tile, divide the square foot floor area by 0.50

- For 9-by-9-inch tile, divide the square foot floor area by 0.5625
- For 9-by-12-inch tile, divide the square foot floor area by 0.75
- For 12-by-12-inch tile, divide the square foot floor area by 1.00

Estimating Labor

Table 26–2 lists various types and sizes of resilient floor tile and vinyl sheet floor covering with the approximate hours of labor per 100 square feet. Careful consideration in estimating should be given to the size of tile and area to be covered and also the working conditions at the job site.

TABLE 26–2

Approximate Hours Labor to Install Resilient Flooring in Large Rooms

Type of Flooring	Size of Squares	Man-Hours Per 100 S.F.
Asphalt tile, Greasless	9" × 9"	2.0–2.5
Vinyl plastic tile	12" × 12"	1.5–2.0
Rubber tile	9" × 9"	2.0–3.0
Rubber tile	12" × 12"	1.5–2.0
Cork tile	9" × 9"	2.0–3.0
Cork tile	12" × 12"	1.5–2.5
Vinyl sheet flooring	.065" gauge	2.0–2.5
Vinyl sheet flooring	.160" gauge	2.5–3.0

Notes: Add the time to apply felt underlayment
Adjust hours of labor for small or cut-up rooms.

CERAMIC FLOOR TILE

Ceramic mosaic floor tile is covered in Chapter 16, page 348.

TERRAZZO FLOORS

There are two basic methods of laying terrazzo floors over concrete. One is to bond the terrazzo directly to the concrete. A second method, which is used where cracking may occur from structural movement, is to separate the terrazzo from the concrete with a very thin layer of sand cushion.

In the bonding method, the terrazzo contractor installs an underbed of mortar, 1 part portland cement and 4 to 5 parts sand, bringing it to 3/4 inch below the surface of the finish floor.

In the second method, the concrete base is covered with a layer of sand about 1/4-inch thick. Saturated felt or some other suitable membrane is placed over the sand and lapped about 4 inches at all of the ends. Metal reinforcing mesh is then placed over the membrane. A mortar underbed, like that installed for bonded terrazzo, is laid over the wire mesh.

Whenever terrazzo is laid over wood floors, a layer of saturated roofing felt is placed over the flooring, and galvanized wire mesh is nailed over the felt. A mortar underbed of a minimum of 2 inches is laid over the wire mesh.

Divider strips are installed while the underbed is still plastic and before the terrazzo topping is poured. Standard divider strips, 1¼-inch or 1½-inch, of brass, zinc or stainless steel, are set in the desired pattern with the tops at the finish floor level.

Terrazzo topping is customarily composed of 200 lb. of marble chips per (94-lb.) bag of white or gray portland cement mixed in a power-driven mortar mixing machine. The topping is spread 1/2-inch to 3/4-inch thick and trowled even with the divider strips, then rolled with a heavy roller to bring up any excess water and create a hard surface.

After proper curing, wet grinding and grouting operations are performed followed by final polishing. The entire process may take several days.

Estimating Materials

Terrazzo floors are estimated by the square feet to be covered, deducting any openings such as stairwells. Base and narrow boarders are estimated by the lineal foot. The estimate of materials required should include the mortar underbedding and, where specified, any membrane, reinforcing, and dividers; also, marble granules, coloring, and portland cement for the terrazzo topping.

Estimating Labor

Terrazzo flooring is usually contracted to experienced specialists who have both the workers and the equipment to do the job. As a guide, Table 26–3 lists several types of terrazzo installations showing approximate hours of labor for a team of one journeyman and a helper. Installation of 1¼-inch to 1½-inch underbed is included in the labor.

TABLE 26–3

Approximate Hours of Labor
for Cast-in-Place Terrazzo

Type of Work Including Underbed	Hours Per 100 Sq. Ft.	
	Journeyman	Helper
Floors		
4′ × 4′ squares, metal dividers	5	6
2′ × 2′	5	7
Base	**Hours Per 100 Lin. Ft.**	
Cove, 6″ high	14	14
Hand finishing base		10

CHAPTER 27

THERMAL INSULATION AND VAPOR BARRIERS

Thermal insulation materials reduce the transmission of heat through ceilings, walls, and floors of buildings. Most building materials possess some degree of insulating quality. Even "dead air space" in the walls, floors, and ceilings provides a certain amount of insulation. However, the term "dead air space" is misleading, because temperature differentials on opposite sides of the air space can cause air movement.

CONDUCTIVITY

Heat is *conducted* toward the cold. If you heat one end of a metal rod, the heat will be *conducted* toward the cooler end. On the same principle, if the interior of a building with masonry walls is warmer than the exterior, the heat from the interior will be conducted through the masonry toward the cooler exterior.

The rate at which heat is conducted through building materials is a measure of their *thermal conductivity*, their thickness, and the temperature differential between one side and the other. This *thermal conductivity* is tested with samples according to methods prescribed by the American Society for Testing and Materials (ASTM). The results are translated into a *k-value* that represents the heat loss. This is defined as the number of BTUs transmitted through one square foot of material 1-inch thick, per one-degree Fahrenheit difference in temperature across the sample of material. The lower the k-value, the better the insulating qualities.

INSULATING VALUES

The effectiveness of a material to *resist* or retard heat transmission is called its *R-value*. The R-value is the reciprocal of *k* and is used as a method of rating the insulating values of building materials and of thermal insulation. The higher the R-values, the better the insulating qualities. For example, if the k-value of face brick is 9.0 per square inch of thickness, the R-value would be its reciprocal, 1/9k = 0.11. Table 27–1 lists the R-values of typical building materials.

The term *U-value* is the total insulating effect of materials in a building *section,* such as an insulated wall or roof, glass windows, and storm windows.

INSULATING MATERIALS

Insulating materials come in a variety of forms including *loose-fill, flexible, foamed-in-place, rigid,* and *reflective foil.* Table 27–2 lists the R-values of various insulating materials.

Loose-fill Insulation

Most loose-fill insulating materials are composed of rock or slag wool, fiberglass, vermiculite, and cellulosic fibers of recycled newsprint, wood chips, ground cork, or other organic fibers. Loose-fill materials are used more in attics and between floor and ceiling joists than in walls. The material is either poured or blown in by machine.

TABLE 27-1

R-values of Common Building Materials

Material	R-Value	Material	R-Value
Concrete		Wood and composition	
Poured concrete, sand and gravel	0.08	Bevel lap siding, 1/2″ × 8″	0.81
Concrete slab, 4 in. thick	0.32	Bevel lap siding, 3/4″ × 10″	1.05
Concrete wall, 8 in. thick	0.64	Drop siding 1″ × 8″	0.79
Masonry Units		Wood shingle siding, 16 in.	0.87
Cinder block, 8 in. thick	1.72	Vertical tongue & groove 3/4″	1.00
Cinder block, 12 in. thick	1.89	Wood subfloor, 3/4 in.	0.94
Concrete block, 8 in. thick	1.11	Hardwood finish floor, 25/32 in.	0.70
Lightweight concrete block, 8 in.	2.00	Plywood, 1/4 in.	0.31
Cement mortar, per in. thick	0.20	Plywood, 3/8 in.	0.47
Common brick, per in. thick	0.20	Plywood, 1/2 in.	0.62
Face brick, per in. thick	0.11	Plywood, 5/8 in.	0.79
Roofing		Hardboard, 1/4 in.	0.18
Asphalt shingles	0.44	Gypsum board, 3/8 in.	0.32
Wood cedar shingles, 16 in.	0.90	Gypsum board, 1/2 in.	0.45
Built-up roofing, 3/8 in.	0.33	Gypsum board, 5/8 in.	0.56
Plaster		Floor Coverings	
Gypsum plaster on gypsum lath 1/2 in.	0.32	Linoleum or rubber tile	0.08
Metal lath and plaster 3/4 in.	0.13	Carpeting, foam rubber pad	1.23

TABLE 27–2
R-Values of Insulating Materials

Material	R-Value 1 Inch Thick
Loose Fill	
Mineral fiber (rock or glass)	2.2 –3.0
Mineral fiber (blown in)	2.8 –3.7
Cellulose (blown in)	2.8 –3.7
Perlite	2.7
Vermiculite	2.08–2.20
Flexible	
Mineral wool batts or blankets	2.8 –3.7
Fiberglass batts	3.1 –3.5
Rigid	
Cellular glass	2.63
Expanded polystyrene	3.6 –4.4
Extruded polystyrene (Styrofoam)	5.0
Expanded polyurethane	6.0 –8.0
Mineral fiberboard	3.45
Foamed in Place	
Polystyrene	3.5 –4.0
Urethane	5.9 –6.7
Reflective Foil	
1-inch Air Space/2 Reflective Surfaces	1.39

The advantages of loose-fill insulation are that it can be blown into areas that are inaccessible and can be put in place faster than batts and blanket type. Fiberglass and rock wool loose-fill have good fire-resistance qualities. A disadvantage of loose-fill is that its R-value may be lower than batts.

Estimating Material

The quantity of loose-fill needed depends on the type of material, how well it is packed down (density), and the depth. A density of 6 to 8 lb. per cu. ft. is usually recommended.

The material comes in bags weighing 40 lb. and containing 4 cubic feet. If it is packed to a density of 6 lb. for each cubic foot, a 40-lb. bag will cover 40/6 or 6.67 cubic feet. Table 27–3 lists the number of 40-lb. bags of loose-fill required for the depth and density shown.

Estimating Labor

Pouring or packing in loose-fill insulation is a relatively fast operation. Pouring in between joists takes less time than packing between studding. Physical conditions and the thickness must be considered. Ordinary working conditions permit a worker to pour granule insulating wool 3 to 5 inches thick at the rate of about 100 to 125 sq. ft. per hour. Packing between studding, a worker should cover about 40 to 60 sq. ft. per hour. Granulated vermiculite, a dry and slippery material, pours easily and will cover floor areas much faster than insulating wool.

Flexible Insulation

A popular type of insulation is wool batts and wool blanket. The material consists of mineral or glass wool or cellulosic fibers. It is available unfaced, faced on one side only, or faced on both sides with paper, aluminum foil, or plastic. Batts and blankets come in thicknesses 1 to 6 inches, and in widths of 15 inches to go between framing 16 inches on center; 19 inches to go between framing of 20 inches on center; and 23 inches to go between framing that is 24 inches on center. Batts come in lengths 4 to 8 feet; blankets come in rolls 24 feet and longer. The facing materials have flanges on the long sides for stapling to studs and joists.

Flexible insulation is light and easy to cut and install; it is fairly inexpensive and has good fire resistance. A disadvantage is the difficulty of placing it properly in confined areas.

TABLE 27-3

Loose Fill Insulation for the Ceilings or Floors of Dwellings (mineral wool or fiberglass) Number of 40-lb. Bags Required for the Depth and Density Shown (No openings have been deducted.)

Ground Floor Area Sq. Ft.	Density 6 Lb./Cubic Foot — Depth of Fill						Density 8 Lb./Cubic Foot — Depth of Fill						Density 10 Lb./Cubic Foot — Depth of Fill					
	1"	2"	3"	3½"	4"	6"	1"	2"	3"	3½"	4"	6"	1"	2"	3"	3½"	4"	6"
200	2.4	4.7	7.0	8.2	9.4	14.1	3.1	6.3	9.4	10.9	12.5	18.9	3.9	7.8	11.8	13.8	15.7	23.5
300	3.5	7.1	10.6	12.3	14.2	21.1	4.7	9.4	14.1	16.4	18.8	28.3	5.9	11.8	17.6	20.7	23.6	35.3
400	4.7	9.4	14.0	16.5	18.9	28.2	6.3	12.5	18.8	21.9	25.0	37.7	7.8	15.7	23.5	27.6	31.5	47.0
500	5.9	11.8	17.6	20.6	23.6	35.2	7.8	15.7	23.5	27.3	31.3	47.2	9.8	19.6	29.4	34.5	39.4	58.8
600	7.1	14.1	21.1	24.7	28.3	42.3	9.4	18.8	28.2	32.8	37.5	56.6	11.8	23.5	35.3	41.4	47.2	70.6
700	8.2	16.5	24.7	28.8	33.0	49.3	11.0	21.9	32.9	38.3	43.8	66.0	13.7	27.5	41.2	48.3	55.0	82.4
800	9.4	18.8	28.2	32.9	37.7	56.3	12.5	25.1	37.6	43.7	50.0	75.5	15.7	31.4	47.0	55.2	63.0	94.0
900	10.6	21.2	31.7	37.0	42.5	63.4	14.1	28.2	42.3	49.2	56.3	84.9	17.6	35.3	53.0	62.0	71.0	106
1000	11.8	23.5	35.2	41.2	47.2	70.4	15.7	31.3	47.0	54.6	62.5	94.3	19.6	39.2	58.8	69.0	78.7	118
1100	13.0	25.9	38.7	45.3	51.9	77.5	17.2	34.5	51.6	60.1	68.8	104	21.6	43.0	64.7	76.0	86.6	129
1200	14.1	28.2	42.3	49.4	56.6	84.5	18.8	37.6	56.3	65.6	75.0	113	23.5	47.0	70.6	83.0	94.5	141

Ground Floor Area Sq. Ft.	Density 6 Lb./Cubic Foot Depth of Fill						Density 8 Lb./Cubic Foot Depth of Fill						Density 10 Lb./Cubic Foot Depth of Fill					
	1"	2"	3"	3½"	4"	6"	1"	2"	3"	3½"	4"	6"	1"	2"	3"	3½"	4"	6"
1300	15.3	30.6	45.8	53.5	61.3	91.5	20.4	40.8	61.0	71.0	81.3	122	25.5	51.0	76.5	89.7	102	153
1400	16.5	33.0	49.3	57.6	66.0	98.6	22.0	43.9	65.7	76.5	87.5	132	27.5	55.0	82.4	96.6	110	165
1500	17.6	35.3	52.8	61.7	70.8	106	23.5	47.0	70.4	82.0	93.8	141	29.4	59.0	88.2	104	118	177
1600	18.8	37.6	56.3	65.8	75.5	113	25.1	50.2	75.1	87.4	100	151	31.4	63.0	94.0	110	126	188
1700	20.0	40.0	60.0	70.0	80.2	120	26.6	53.3	79.8	92.9	106	160	33.3	66.7	100	117	134	200
1800	21.2	42.4	63.4	74.1	84.9	127	28.2	56.4	84.5	98.4	113	170	35.3	70.6	106	124	142	212
1900	22.4	44.7	66.9	78.2	89.6	134	29.8	59.6	89.2	104	119	179	37.3	74.5	112	131	150	224
2000	23.5	47.0	70.4	82.3	94.3	141	31.3	62.7	93.9	109	125	189	39.2	78.4	118	138	158	235
2100	24.7	49.4	73.9	86.4	99.1	148	32.9	65.8	98.6	115	131	198	41.2	82.4	124	145	165	247
2200	25.9	51.8	77.5	90.5	104	155	34.5	69.0	103	120	138	208	43.0	86.3	129	152	173	259
2300	27.1	54.1	81.0	94.7	109	162	36.1	72.1	108	126	144	217	45.0	90.2	135	158	181	271
2400	28.2	56.5	84.5	98.8	113	169	37.6	75.2	113	131	150	226	47.0	94.0	141	166	189	282
2500	29.4	58.8	88.0	103	118	176	39.2	78.4	117	137	156	236	49.0	98.0	147	172	197	294
2600	30.6	61.2	91.5	107	123	183	40.8	81.5	122	142	163	245	51.0	102	153	179	205	306
2700	31.8	63.5	95.1	111	127	190	42.3	84.6	127	148	169	255	53.0	106	159	186	213	318
2800	33.0	65.9	98.6	115	132	197	43.9	87.8	132	153	175	264	55.0	110	165	193	221	329
2900	34.1	68.2	102	119	137	204	45.5	90.9	136	159	181	274	57.0	114	171	200	228	341
3000	35.3	70.6	106	124	142	211	47.0	94.0	141	164	187	283	59.0	118	177	207	236	353

Estimating Material

The square foot area that batts will cover, *including the area occupied by studs, joists, and rafters,* is easily determined. Multiply the length in feet of a 15-inch batt by 1⅓; the length of a 19-inch batt by 1⅔; and the length of a 23-inch batt by 2.

EXAMPLE
A 15-inch batt 4 ft. long covers 4 × 1⅓ = 5.33 sq. ft.
A 19-inch batt 4 ft. long covers 4 × 1⅔ = 6.67 sq. ft.
A 23-inch batt 4 ft. long covers 4 × 2   = 8.00 sq. ft.

Table 27–4 shows the number of batts of different sizes needed to cover 100 sq. ft. including the area taken up by studs or joists.

Estimating Labor

The number of hours of labor required to install batts and blanket insulation depends on several important variables. How long are the batts or blankets? Are they to be applied between studding, ceiling or floor joists, or between rafters? Are they friction (loose fitting),

TABLE 27–4

**Batts Needed to Cover an Area of 100 sq. ft.
Including Studs and Joists**

Batt Size Width and Length	Square Feet Per Batt	Batts Per 100 Sq. Ft.	Batts per Sq. Ft.
15″×4′	5.33	18.75	.188
15″×8′	10.67	9.37	.094
19″×4′	6.67	15.00	.150
19″×8′	13.34	7.50	.075
23″×4′	8.00	12.50	.125
23″×8′	16.00	6.25	.063

Note: The 100 sq. ft. area of walls, floor, or ceiling includes the area taken up by studs and joists.

or stapled? Good judgment and experience are necessary to estimate accurately.

Foamed-in-Place Insulation

This type insulation can be poured or sprayed in place to form a continuous airtight seal. The commonly used material is urethane, which has a high R-value, 6 to 7 per inch. It is highly suited to irregular surfaces and must be applied by a qualified subcontractor using special equipment. Compared to flexible insulation, it is very expensive.

Rigid Insulation Board

Commonly used materials to make rigid insulating board are expanded polystyrene, urethane, extruded polystyrene, and some forms of glass, wood, and vegetable fibers.

Expanded polystyrene is called "beadboard" because it is composed of compressed pellets or beads. *Extruded polystyrene,* manufactured by Dow Chemical as Styrofoam, and by U.C. Industries as Foamular, is excellent for exterior insulative sheathing and decking.

Urethane insulation boards have the highest R-value per inch of thickness.

Polystyrene and urethane boards must be covered with gypsum wallboard or some other fireproof material. There are many thicknesses and sizes of rigid insulating board.

Reflective Insulation

Reflective insulation is made up of highly reflective metallic foil, such as aluminum foil, with air spaces of at least 3/4 inch between the sheets. Its effectiveness depends largely on how long the reflective surface remains undulled. It has also been determined that reflective foils are better suited for keeping heat out of a building than preventing or reducing heat loss in a building. For this reason reflective insulation serves best in warm climates.

THE FEDERAL TRADE COMMISSION RULE

Building contractors have specific responsibilities and obligations with regard to recommending, specifying, and installing insulation. One of the important areas that may be applicable will be found in the Federal Trade Commission Rule, Labeling and Advertising of Home Insulation. This rule, which is mandatory and enforceable, requires that a potential buyer be given information about the insulation in the home which includes the R-value and thickness of the different kinds of insulation used in the home.

If you say or imply that a combination of products can cut fuel bills or use, you must have a reasonable basis for such claim. The information must be provided prior to selling the house, and it also must be included in the contract of sale.

UNIFORM BUILDING CODE REGULATIONS

Another area of importance regarding the contractor's responsibilities for labeling and installing insulation will be found in local building codes which have considerable uniformity countrywide.

SECTION 36—Efficient Energy Utilization in New
   Buildings. (Effective April 1, 1982)*

1.1—General Requirements.

    (a)   These insulation requirements apply to all new dwellings that are heated and/or cooled regardless of the type of fuel used (electric, oil, gas, or wood) as follows:

    (1)   All one- and two-family dwellings specified in Section I of Volume I-B.

---

* Uniform Building Code Requirements

(2) All new multiple-family dwellings (apartments and condominiums) three (3) stories and less in height specified in Section 400 of Volume I.

(b) It is imperative that close attention be paid to workmanship in the installation of the materials specified if the full benefits of these requirements are to be realized.

(1) The vapor-resistant facing furnished on blanket and roll-type insulation shall always face the interior of the structure. Insulation shall be wedged between pipes and electrical outlets and the external surface of the wall.

(2) If unfaced blankets or rolls are used, a vapor barrier of at least 4-mil polyethylene or its equivalent shall be stapled to the studs or foil-backed gypsum board shall be used on the interior wall.

(3) Voids shall not exist at the top or bottom of the stud cavity.

(4) All cracks around windows and doors shall be filled with insulation with a vapor barrier properly installed.

(5) Vapor barriers shall be carefully checked to assure that no tears exist and any tear shall be patched.

(6) The manufacturers' installation procedures for all shall be strictly adhered to.

(c) On blanket and roll-type insulation furnished with a vapor-resistant facing, the R-value of the insulation shall be marked at three (3) foot intervals on the exposed facing.

(d) For unfaced blankets and rolls, the manufacturer shall furnish sufficient identifying markings to indicate the insulation R-Value.

(e) When the exterior sheathing or exterior siding of any insulated stud cavity wall has a permeance of less than 0.6 perm (ASTM Dry Cup Method), the interior vapor barrier shall be a minimum of 4-mil polyethylene or its equivalent with all penetrations sealed by either taping or caulking. Unless the sheathing manufacturer specifically requires moisture relief vents, no moisture vents are required.

(f)   The required thermal value of any one assembly, such as roof/ceiling, wall or floor, may be increased and the thermal value for other components decreased, provided the overall heat loss from the entire building envelope does not exceed the total resulting from conformance to the required thermal values.

1.2—Maximum "U" Values for Exterior Walls and Ceilings

(a)   All buildings that are heated or mechanically cooled shall have sections exposed to the exterior or unheated spaces constructed to comply with the maximum "U" value shown in Tables A and B.

(b)   Blown or poured-type loose-fill insulation may be used in attic spaces where the slope of the roof is a minimum of 2½ feet in 12 feet and there is at least 30 inches clear headroom at the roof ridge. (Clear headroom is defined as the distance from the top of the bottom chord of the truss or ceiling joists to the underside of the roof sheathing.)

(1)   When soffit vents are installed, adequate baffling of the vent opening shall be provided to deflect the incoming air above the surface of the material and shall be installed at the soffit on a 60-degree angle from horizontal.

(2)   Baffles shall be in place at the time of inspection.

(c)   When loose-fill insulation is proposed, the R-value of the material shall be determined in accordance with ASTM Standards C-687, C-236, and C-518.

TABLE A—MAXIMUM "U" VALUES FOR CEILING AND WALL SECTIONS

Flat Roof Deck[1]	Masonry Wall Construction		Frame Wall Construction		Doors and Windows
"U"	Ceilings "U"	Walls "U"	Ceilings "U"	Walls "U"	
.09	.05	.10	.05	.08	1.13[2]

[1]Indicates construction with rigid roof insulation and exposed structural system. Where ceiling cavity exists, use value for ceilings.

[2]In any room where 20% or more of the exterior wall is composed of windows and doors their maximum "U" Value shall be 0.65 (This will require insulating glass and doors or storm windows or doors). An exterior wall is any wall that faces to the outside of or is adjacent to any unconditioned space, such as: garages, car-ports, storage rooms, or porch areas. In any room that has two or more exterior walls, the total percentage of window and door area may be combined and used in any one of these walls.

(1) The "R" value shall be shown on the building plans together with the total number of bags required and net coverage per bag.

(2) Upon completion of the installation of insulation, an insulation certification card shall be furnished by the insulation applicator and posted at a conspicuous location within the structure.

(3) This certification shall indicate the R-value, minimum thickness, maximum net coverage, and weight per square foot of the insulation installed.

(d) Minimum ventilation for roof-ceiling cavities shall conform to the following requirements. The required net free vent area may be reduced 50% if an approved vapor barrier is installed behind the ceiling finish material.

(1) Gabled Roofs—Screened louvers having a net free area of 1 square foot for each 300 square feet of ceiling area shall be provided at each gabled end.

(2)   Hip-Roofs—Screened soffit vents having a net free area of 1 square foot for each 900 square feet of ceiling area and screened outlet vents located near the roof peak having a net free area of 1 square foot for each 1,600 square feet of ceiling shall be provided.

(3)   Flat Roofs—Screened openings having a net free area of 1 square foot for each 250 square feet of ceiling area shall be provided along with the overhanging eaves. Blocking and bridging shall be arranged so as not to interfere with the movement of air.

(4)   Cathedral ceilings with joist cavities shall have a screened soffit intake and a screened outlet at the roof ridge or at the intersection of the roof with a vertical surface. The intake and outlet openings shall each have a net free area of 1 square foot for each 250 square feet of roof surface. There shall be 1 inch minimum clearance between the bottom of the roof deck and the insulation.

1.3—Maximum "U" Values for Floors

(a)   For floors over unheated basements, unheated garages, breezeways or ventilated crawl spaces with operable vents, the thermal value of the floor section shall not exceed the values shown in Table B. (A basement is considered unheated unless it is provided with a positive heat supply equivalent to at least 15% of the total

TABLE B—FLOOR SECTION MAXIMUM "U" VALUES[1]

Structural Slab U	Wood and Steel Framing U
.12	.08

[1]U Value for heat flow down

---

calculated heat loss of the structure or is provided a positive heat supply to maintain a minimum temperature of 50° F.)

(b) Insulation may be omitted from floors over unheated areas if the crawl space foundation walls are insulated.

(1) The U value of insulated foundation walls from above a point 12 inches below grade or top of footing shall not exceed 0.17 (R=5.88). ("Foundation wall insulation for underfloor supply plenums shall have a minimum of R-11. See Volume III, Section 609-L.")

(2) A minimum of 75 to 80% of the crawl space ground area shall be covered with a 6-mil polyethylene vapor barrier or its equivalent.

(c) Crawl spaces under buildings without basements shall be ventilated by approved mechanical means or by openings in the foundation walls. Openings shall be arranged to provide cross-ventilation and shall be covered with corrosion-resistant wire mesh of not less than 1/4 inch nor more than 1/2 inch in any dimension. Such wall openings shall have a net free area of not less than 2 square feet for each 100 linear feet of exterior wall plus 1/3 square feet net free area for each 100 square feet of crawl space area. Where at least 75 to 80% of the crawl space ground surface is covered with a 6-mil polyethylene vapor barrier or its equivalent, the areas specified above may be reduced 50 percent. It is recommended that a maximum of 80% of the crawl space ground surface be covered to prevent excessive drying of the flooring. Vents shall be so placed as to provide ventilation at all points and to prevent dead air pockets.

(d) When used, crawl space ventilation openings shall not be covered with insulation. Vents shall be of the closeable type and insulation shall be attached to the closing device. When fuel-burning equipment is located in crawl space, adequate means for combustion air shall be provided.

(e)   Basement walls below a point 12 inches below grade need not be insulated. Walls above a point 12 inches below grade shall be insulated in accordance with Table A.

(f)   Slab-on-grade floors shall be insulated around the perimeter of the floor exposed to the outside with rigid insulation having a minimum "R" value of 3.75 and specifically designed and recommended by the manufacturer for this type application.

(1)   The insulation may be installed vertically on the interior or the exterior of the foundation wall with the insulation extending 24 inches below the top of the slab. In areas where the frost line is deeper than 24 inches, the insulation shall extend to the frost line.

(2)   Insulation may be installed downward to the bottom of the slab, then horizontally beneath the slab for a total distance of 24 inches.

(3)   Insulation extending above grade shall be protected from physical damage.

(4)   With either method, the entire slab edge thickness exposed to the outside shall be insulated.

Uniform Building Code Requirements

THE THERMAL ENVELOPE

The term *thermal envelope* refers to the insulating of the shell or boundary of a room or of an entire building. Figure 27–1 is a map of the continental U.S. showing the climatic regions with recommended R-values for ceilings, walls, and floors.

Also given are insulation recommendations, in terms of R-values, for ceilings, walls, and floors applicable to each region.

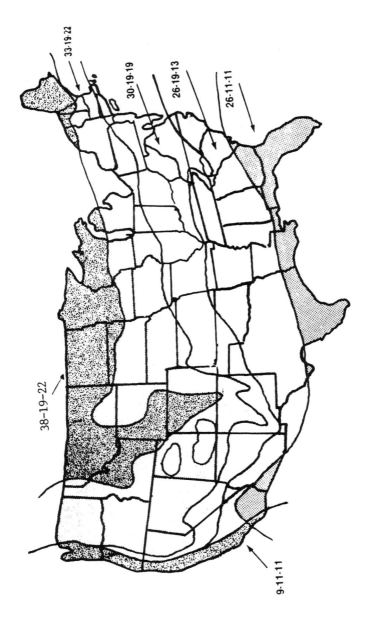

Figure 27-1
Recommended Insulation R-Values for Ceilings, Walls, and Floors

33-19-22

30-19-19

26-19-13

26-11-11

38-19-22

9-11-11

Ceilings, using double layers of batts

R-38: Two layers of R-19 (6") mineral fiber

R-33: One layer of R-22 (6½") and one layer of R-11 (3½") mineral fiber

R-30: One layer of R-19 (6") and one layer of R-11 (3½") mineral fiber

R-26: Two layers of R-13 (3⅝") mineral fiber

Ceilings, using loose-fill mineral wool batts

R-38: R-19 (6") mineral fiber and 20 bags of wool per 1,000 sq. ft. (8¾")

R-33: R-22 (6½") mineral fiber and 11 bags of wool per 1,000 sq. ft. (5")

R-30: R-19 (6") mineral fiber and 11 bags of wool per 1,000 sq. ft. (5")

R-26: R-19 (6") mineral fiber and 8 bags of wool per 1,000 sq. ft. (3¼")

Walls, 2" × 6" framing

R-19: R-19 (6") mineral fiber batts

Walls, 2" × 4" framing

R-19: R-11 (3½") mineral fiber batts and 1" polystyrene

R-13: R-13 (3⅝") mineral fiber batts

R-11: R-11 (3½") mineral fiber batts

Floors

R-22: R-22 (6½") mineral fiber

R-19: R-19 (6") mineral fiber

R-13: R-13 (3⅝") mineral fiber

R-11: R-11 (3½") mineral fiber

VAPOR BARRIERS

The insulating of buildings introduced a new problem. It required applying *vapor barriers* to walls and floors. When the temperature inside the air space between studs, joists, or rafters is much lower than that within the building, the water vapor condenses and moisture collects in the wall space. If the wall or floor spaces are completely filled with insulating material, the problem is greatly reduced.

Concrete slabs laid on fill require a vapor or moisture barrier between the fill and the concrete. The most widely used material is 6-mil polyethylene.

Most uniform building codes require that vapor-resistant facing on blanket and roll type insulation shall face the interior of the structure. Where unfaced blanket and roll type insulation is used, a vapor barrier of 4-mil polyethylene or its equivalent will be stapled to the studs, or foil-backed gypsum board shall be used on exterior walls. These building codes also require that vapor barriers be checked to assure that no tears exist and that any tears shall be patched.

A minimum of 75 to 80 percent of the crawl space ground area shall be covered with a 6-mil polyethylene vapor barrier or its equivalent.

CHAPTER 28

HEATING AND AIR CONDITIONING

Estimating and installing heating and air conditioning systems is the complex business of highly trained and experienced personnel. The purpose of this chapter is to briefly outline the commonly used systems. The final choice of a particular system will depend upon several important factors, including the size, type, and construction of the building; its location climate-wise; availability of the fuel to be used; initial cost; operating and maintenance costs; adaptability of heating and cooling within one system; and preference of the owner.

Chapter 27, Thermal Insulation and Vapor Barriers, is directly related to the subject of heating and air conditioning. Effective insulation and weather stripping can reduce the size and cost of any central heating and cooling unit. A review of that chapter is recommended.

HEATING SYSTEMS

There are three major types of central heating systems: warm air, hot water, and steam. Central heating systems distribute heat by ducts for warm air or pipes for steam and hot water. Electric radiant heating is a major heating system but not classified as a central heating system.

Fuels

Principal fuels used in heating systems are gas (natural and propane), oil, electricity, and coal. The selection of a fuel depends largely on its availability and its cost compared to other kinds.

Fuel oils No. 1 and No. 2 are generally used in residential oil-fired heating systems. While No. 2 oil is slightly more expensive, it gives more heat per gallon. It is used in both atomizing and rotary oil burners. No. 1 oil is used more in vaporizing or pot-type burners. The manufacturers of oil burners usually stipulate the grade of fuel oil to be used. The Underwriters Laboratories, Inc. and the Underwriters Laboratories of Canada also stipulate the proper grade of fuel oil. The BTU of No. 1 fuel oil is approximately 136,000 BTU per gallon. The BTU content of No. 2 fuel oil is approximately 140,000 BTU per gallon.

Gas-type fuels are natural gas, manufactured gas, and liquefied petroleum. Each has different combustion characteristics and different heat values when burned.

Natural gases are those found in or near crude petroleum deposits. It is piped under pressure from gas fields to consumer centers that range over a large part of the country. The heating value of natural gas varies from about 1,000 BTU to 1,100 BTU per cubic foot.

Manufactured gases are gases produced from such raw materials as coal, oil, and coke. Most manufactured gases have low calorific values ranging between 500 BTU and 1,000 BTU per cubic foot.

Liquefied petroleum gas, extracted from "wet" natural gas, is sold commercially as propane, butane, or bottled gas (LP). Propane gas is used extensively for domestic heating. It contains 2,516 BTU per cubic foot. It is delivered in bulk and stored in large holding tanks.

Electricity is used as a fuel or source of heat in central heating systems that employ electric-fired boilers for hot water systems; electric-fired furnaces for forced warm air systems; and in heat pumps. Some of the advantages of electric fuel are safety, economy of space, quiet operation, and a more uniform temperature. Disadvantages may be higher installation costs and operation. The heat value of electricity is approximately 3,413 BTU per kilowatt hour (kwh).

Coal fuel is commonly classified as anthracite, a clean hard coal that burns with little or no flame or smoke, and *bituminus,* which burns with a smoky yellow flame. Anthracite contains about 14,400 BTU per lb.; bituminus contains about 11,000 BTU to 14,400 BTU per lb.

Warm Air Heating Systems

Warm air heating systems are classified as *gravity* or *forced warm air.* In a gravity-type system the warm air is delivered to the rooms above through a system of ducts extending from the furnace in the basement. Registers are located in the floors or walls of the rooms. The cooler air in the rooms returns back to the furnace through return grills and ducts. It is reheated and recirculated.

A gravity system may lose considerable heat during the slow transfer of air through the ducts. A more rapid movement minimizes the heat loss. Installation of a blower in the furnace plenum to speed up air transfer changes the system to a *forced warm air system.* The return-air ducts are joined before reaching the furnace, which distinguishes this system from the gravity type.

The plenum is the air chamber to collect warm air and is usually located on the top of the furnace. However, some furnaces have a *reversed flow* (down-flow) with the plenum located on the bottom. Buildings built on slabs (no basement), with the furnace on the same level, require this type. The furnace may be located in a closet or similar confined area, and the ducts are located in the floor or walls of the rooms being served.

In the increasingly popular *perimeter-loop heating system*, ducts are imbedded in the concrete slab or suspended below in the crawl space. A main, round, continuous duct extends along the perimeter of the building.

Hot Water and Steam Heating Systems

Furnaces that supply hot water and steam heat are similar. The fuel is usually oil or gas, although electric-fired furnaces are available. Steam heat is more often used in large residential and commercial buildings, while circulating hot water is used both in homes and large commercial and industrial buildings. Circulation of the hot water to and from radiators and convectors may be by gravity or forced through the pipes by an electric pump.

There are four basic piping arrangements in hot water heating systems: one-pipe system; direct return two-pipe system; reverse-return two-pipe system; and series-loop system in which the radiators and convectors are a part of the piping circuit.

Steam heating systems are essentially the same as hot water systems.

Electric Radiant Heat

Electric radiant heat consists of electric heating cable imbedded in floors, walls, or ceilings. Ceiling radiant heating is the most common type. The heat is controlled in each room by wall thermostats. It is not recommended for rooms with ceiling heights under 7 feet.

Electric Heat Pumps

An electric heat pump is basically a reversible air-conditioning system. In summer the heat is extracted from the inside air and expelled outdoors, cooling the building. In winter, heat is extracted from outside air, ground, or well water and transferred to the interior of the building. In either case the warm or the cool air is distributed by means of a duct system.

The size of the heat pump is very important and should be estimated by a trained and experienced individual. Too large or too small a unit will not perform satisfactorily. The heat gain and the heat loss are essential factors in computing the size of the heat pump.

CENTRAL AIR CONDITIONING

Central air conditioning can be installed in all types of forced-air heating systems. By installing a refrigeration-type water chiller in steam or hot water heating systems, cold water is circulated through the heating convectors to cool the rooms.

The subject of central air conditioning, as with central heating systems, is far too complex to attempt an in-depth presentation in this chapter. There are excellent texts available for reference and study including Audel's three-volume *Heating, Ventilating and Air Conditioning Library;* and the *Pocket Handbook for Air Conditioning, Heating, Ventilation and Refrigeration,* of ASHRAE.

SELECTED REFERENCE

Refrigeration and Air Conditioning, 3rd, Langley, B., Prentice-Hall, Englewood Cliffs, NJ, 1997.

CHAPTER 29

MATHEMATICS FOR CALCULATING MATERIAL AND LABOR COSTS

Battery- and solar-powered pocket calculators make it possible to rapidly and accurately calculate areas and volumes in the field, shortening many operations formerly done with pencil and paper. Square roots, decimal equivalents of fractions, percentages, and calculations of feet and inches are done in seconds.

SURFACE AREA MEASUREMENT

Polygons

A polygon is a closed plane figure consisting of three or more sides. A *regular* polygon has all equal sides and equal angles; an *irregular* polygon has unequal sides or angles. Names are given to polygons according to the number of sides.

- *Triangles* have three sides.
- *Quadrilaterals* have four sides.
- *Pentagons* have five sides.
- *Hexagons* have six sides.
- *Heptagons* have seven sides.
- *Octagons* have eight sides.

- *Nonagons* have nine sides.
- *Decagons* have ten sides.
- *Dodecagons* have twelve sides.

The area of any regular polygon is one-half the sum of its sides multiplied by the length of a perpendicular from its center to one of its sides. Table 29–3 (on pages 552–553) may be used to obtain *areas, sides,* and *radiuses* of regular polygons when the length of one side (s); the radius of a circumscribed circle (R); or the radius of an inscribed circle (r) is known.

Triangles

The area of a triangle is one-half of the base times the height (altitude).

EXAMPLES

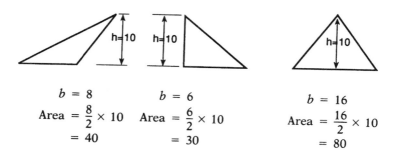

$$b = 8$$
$$\text{Area} = \frac{8}{2} \times 10$$
$$= 40$$

$$b = 6$$
$$\text{Area} = \frac{6}{2} \times 10$$
$$= 30$$

$$b = 16$$
$$\text{Area} = \frac{16}{2} \times 10$$
$$= 80$$

Hero's Formula for the Area of a Triangle

The area of any triangle can be determined by the following formula, where *s* equals one-half the sum of the sides of the triangle.

$$\text{Area} = \sqrt{s(s - a)(s - b)(s - c)}$$

EXAMPLE

$$\text{Area} = \sqrt{12(12 - 10)(12 - 6)(12 - 8)}$$
$$= \sqrt{12(2)(6)(4)}$$
$$= \sqrt{576}$$
$$= 24$$

$b = 6$ $c = 8$

$a = 10$

$$s = \frac{(10 + 6 + 8)}{2} = 12$$

Rule of Pythagoras for Right Triangles

In any right triangle the square of the hypotenuse is equal to the sum of the squares of the other two sides. By this formula, if any two sides of a right triangle are known, the third side can be computed.

$$c^2 = a^2 + b^2$$
$$a^2 = c^2 - b^2$$
$$b^2 = c^2 - a^2$$

b c a

EXAMPLE

$$a = 6 \qquad b = 8 \qquad c = 10$$

$$c = \sqrt{6^2 + 8^2} \qquad a = \sqrt{10^2 - 8^2} \qquad b = \sqrt{10^2 - 6^2}$$
$$= \sqrt{36 + 64} \qquad\; = \sqrt{100 - 64} \qquad = \sqrt{100 - 36}$$
$$= \sqrt{100} \qquad\qquad = \sqrt{36} \qquad\qquad = \sqrt{64}$$
$$= 10 \qquad\qquad\;\; = 6 \qquad\qquad\;\; = 8$$

Parallelograms

A parallelogram is a four-sided polygon with parallel opposite sides. A *rectangle* is a parallelogram with four right angles. A *square* is a rectangle with all four of its sides equal.

The area of a rectangle is its length times its width when measured in a horizontal plane, such as the floor of a room. The area is its length times its height when measured in a vertical plane, such as the wall of a room.

EXAMPLES

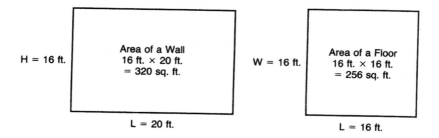

H = 16 ft.

Area of a Wall
16 ft. × 20 ft.
= 320 sq. ft.

L = 20 ft.

W = 16 ft.

Area of a Floor
16 ft. × 16 ft.
= 256 sq. ft.

L = 16 ft.

Rhomboids

While rectangles and squares are parallelograms, distinguished by having four right angles, another type of parallelogram is the *rhomboid*. It has equal opposite sides and no right angles.

The area of a rhomboid is the height (altitude) times the length (base).

EXAMPLE

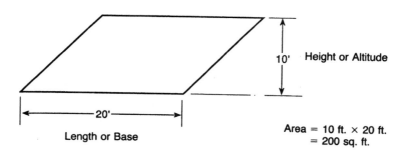

10'   Height or Altitude

20'

Length or Base

Area = 10 ft. × 20 ft.
= 200 sq. ft.

Trapezoids

A *trapezoid* is a quadrilateral with only two of its sides parallel. The area of a trapezoid is one-half the sum of the parallel sides times the height or altitude.

EXAMPLES

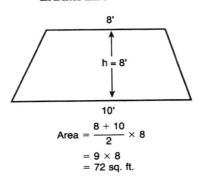

$$\text{Area} = \frac{8 + 10}{2} \times 8$$
$$= 9 \times 8$$
$$= 72 \text{ sq. ft.}$$

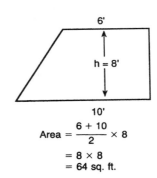

$$\text{Area} = \frac{6 + 10}{2} \times 8$$
$$= 8 \times 8$$
$$= 64 \text{ sq. ft.}$$

Circles

The *circumference* of a circle is its perimeter, the distance around the exterior.

The *diameter* is the length of a straight line from one side of the circumference to the opposite side and passing through the center of the circle.

The *radius* is one-half of the diameter of a circle.

$$\frac{\text{Circumference}}{\text{Diameter}} = \pi \text{ (pi)} = 3.1416 \text{ or } 3\frac{1}{7}$$

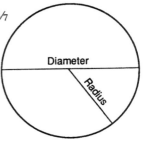

- Circumference = Diameter × 3.1416
- Circumference = Radius × 6.2832
- Diameter = Circumference × .3183
- Radius = Circumference × .159155
- Area = Diameter^2 × .785398
- Area = Radius^2 × 3.1416
- Area = $\dfrac{\text{Diameter} \times \text{Circumference}}{2}$

RECTANGULAR VOLUME MEASUREMENT

The formula for the volume of rectangular solids is $V = l \times w \times h$.

EXAMPLES

A concrete wall is 20 ft. long, 12 in. thick, and 8 ft. high. How many cubic feet of concrete does it contain?

Volume = 20 ft. × 1 ft. × 8 ft.
= 160 cu. ft.

The basement of a building is 24 ft. × 20 ft. × 7 ft. How many cubic yards of soil must be removed?

Volume = 24 ft. × 20 ft. × 7 ft.
= 3,360 cu. ft.
= $\dfrac{3,360}{27}$
= 124.44 cu. yds.

PROPERTIES OF CYLINDERS AND SPHERES

Cylinders

- The *surface area* of a cylinder is its circumference times its height. The formula is:

πDh, where
π = 3.1416
D = Diameter
h = height or length

EXAMPLE

A cylinder or a tank with a diameter of 18 ft. and a height of 40 ft. has a surface area of:

3.1416 × 18 × 40 = 2,261.95 cu. ft.

The *cubic volume* of a cylinder is the area of the base times its height or length. The formula is:

$$\pi\, r^2\, h, \text{ where}$$
$$\pi = 3.1416$$
$$r = \text{radius}$$
$$h = \text{height or length}$$

EXAMPLE
Using the cylinder or tank in the previous example, the volume is:

$$3.1416 \times 9^2 \times 40 \text{ ft.}$$
$$= 3.1416 \times 81 \times 40 \text{ ft.}$$
$$= 10{,}178.78 \text{ cu. ft.}$$

Spheres

The *surface area* of a sphere is the diameter times the circumference. The formula is:

$$\text{Area} = 4\,\pi\, r^2$$

EXAMPLE
What is the volume of a sphere with a radius of 10 feet?

$$\text{Area} = 4 \times 3.1416 \times 10^2$$
$$= 1{,}256.64 \times 100$$
$$= 125{,}664 \text{ sq. ft.}$$

The *volume* of a sphere is $4/3 \times \pi\, r^3$, or $4.1888\, r^3$

EXAMPLE
The sphere in the preceding example has a radius of 10 ft.

$$\text{Volume} = 4.1888 \times 10^3$$
$$= 4.1888 \times 1{,}000$$
$$= 4{,}189 \text{ cu. ft.}$$

TABLE 29-1

Table to Convert Inches and Fractions of Inches to Decimals of a Foot

INCHES	0'	1/16"	1/8"	3/16"	1/4"	5/16"	3/8"	7/16"	1/2"	9/16"	5/8"	11/16"	3/4"	13/16"	7/8"	15/16"
0	.00	.0052	.0104	.0156	.0208	.0260	.0312	.0365	.0417	.0469	.0521	.0573	.0625	.0677	.0729	.0781
1	.0833	.0885	.0938	.0990	.1042	.1094	.1146	.1198	.1250	.1302	.1354	.1406	.1458	.1510	.1562	.1615
2	.1667	.1719	.1771	.1823	.1875	.1927	.1979	.2031	.2083	.2135	.2188	.2240	.2292	.2344	.2396	.2448
3	.25	.2552	.2604	.2656	.2708	.2760	.2812	.2865	.2917	.2969	.3021	.3073	.3125	.3177	.3229	.3281
4	.3333	.3385	.3437	.3489	.3542	.3594	.3646	.3698	.375	.3802	.3854	.3906	.3958	.4010	.4062	.4115
5	.4167	.4219	.4271	.4323	.4375	.4427	.4479	.4531	.4583	.4635	.4688	.474	.4792	.4844	.4896	.4948
6	.50	.5052	.5104	.5156	.5208	.5260	.5312	.5365	.5417	.5469	.5521	.5573	.5625	.5677	.5729	.5780
7	.5833	.5885	.5837	.5990	.6042	.6094	.6146	.6198	.625	.6302	.6354	.6406	.6458	.6510	.6562	.6615
8	.6667	.6719	.6771	.6823	.6875	.6927	.6979	.7031	.7083	.7135	.7188	.724	.7292	.7344	.7396	.7448
9	.75	.7552	.7604	.7656	.7708	.7760	.7812	.7865	.7917	.7969	.8021	.8073	.8125	.8177	.8229	.8281
10	.8333	.8385	.8437	.8490	.8542	.8594	.8646	.8698	.8750	.8802	.8854	.8906	.8958	.9010	.9062	.9115
11	.9167	.9219	.9271	.9323	.9375	.9427	.9479	.9531	.9583	.9635	.9688	.974	.9792	.9844	.9896	.9948
12	1.000	1.005	1.010	1.016	1.210	1.026	1.031	1.0365	1.0417	1.047	1.052	1.0573	1.0625	1.0677	1.073	1.0781

TABLE 29–2

Common Fractions and Their Decimal Equivalents

4ths	8ths	16ths	32nds	EXACT DECIMAL EQUIVALENT	4ths	8ths	16ths	32nds	EXACT DECIMAL EQUIVALENT
			1	.03 125				17	.53 125
		1	2	.06 25			9	18	.56 25
			3	.09 375				19	.59 375
	1	2	4	.12 5		5	10	20	.625
			5	.15 625				21	.65 625
		3	6	.18 75			11	22	.68 75
			7	.21 875				23	.71 875
1	2	4	8	.25	3	6	12	24	.75
			9	.28 125				25	.78 125
		5	10	.31 25			13	26	.81 25
			11	.34 375				27	.84 375
	3	6	12	.375		7	14	28	.875
			13	.40 625				29	.90 625
		7	14	.43 75			15	30	.93 75
			15	.46 875				31	.96 875
2	4	8	16	.50	4	8	16	32	1.00

TABLE 29–3

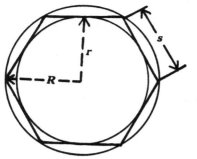

Table to Determine Areas, Sides, and Radiuses of Regular Polygons When the Length of One Side (s); the radius of a *circumscribed* circle (R); or the radius of an *inscribed* circle (r) is known.

NUMBER OF SIDES (n)	AREA $s = 2R \sin \dfrac{180°}{n}$	AREA $R = \tfrac{1}{2}s \csc \dfrac{180°}{n}$	AREA $r = \tfrac{1}{2}s \cot \dfrac{180°}{n}$	R $s = 2R \sin \dfrac{180°}{n}$
3	$s^2 \times 0.4330$	$R^2 \times 1.299$	$r^2 \times 5.196$	$s \times 0.5774$
4	$s^2 \times 1.000$	$R^2 \times 2.000$	$r^2 \times 4.000$	$s \times 0.7071$
5	$s^2 \times 1.721$	$R^2 \times 2.378$	$r^2 \times 3.633$	$s \times 0.8507$
6	$s^2 \times 2.598$	$R^2 \times 2.598$	$r^2 \times 3.464$	$s \times 1.000$
7	$s^2 \times 3.634$	$R^2 \times 2.736$	$r^2 \times 3.371$	$s \times 1.152$
8	$s^2 \times 4.828$	$R^2 \times 2.828$	$r^2 \times 3.314$	$s \times 1.307$
9	$s^2 \times 6,182$	$R^2 \times 2.893$	$r^2 \times 3.276$	$s \times 1.462$
10	$s^2 \times 7.694$	$R^2 \times 2.939$	$r^2 \times 3.249$	$s \times 1.618$
12	$s^2 \times 11.20$	$R^2 \times 3.000$	$r^2 \times 3.215$	$s \times 1.932$
15	$s^2 \times 17.64$	$R^2 \times 3.051$	$r^2 \times 3.188$	$s \times 2.405$
16	$s^2 \times 20.11$	$R^2 \times 3.062$	$r^2 \times 3.183$	$s \times 2.563$
20	$s^2 \times 31.57$	$R^2 \times 3.090$	$r^2 \times 3.168$	$s \times 3.196$

TABLE 29-3 (continued)

R	s	s	r	r
$r = \frac{1}{2}s \cot \frac{180°}{n}$	$R = \frac{1}{2}s \csc \frac{180°}{n}$	$r = \frac{1}{2}s \cot \frac{180°}{n}$	$R = \frac{1}{2}s \csc \frac{180°}{n}$	$s = 2R \sin \frac{180°}{n}$
$r \times 2.000$	$R \times 1.732$	$r \times 3.464$	$R \times 0.5000$	$s \times 0.2887$
$r \times 1.414$	$R \times 1.414$	$r \times 2.000$	$R \times 0.7071$	$s \times 0.5000$
$r \times 1.236$	$R \times 1.176$	$r \times 1.453$	$R \times 0.8090$	$s \times 0.6882$
$r \times 1.155$	$R \times 1.000$	$r \times 1.155$	$R \times 0.866$	$s \times 0.8660$
$r \times 1.110$	$R \times 0.8678$	$r \times 0.9631$	$R \times 0.9010$	$s \times 1.038$
$r \times 1.082$	$R \times 0.7654$	$r \times 0.8284$	$R \times 0.9239$	$s \times 1.207$
$r \times 1.064$	$R \times 0.6840$	$r \times 0.7279$	$R \times 0.9397$	$s \times 1.374$
$r \times 1.052$	$R \times 0.6180$	$r \times 0.6498$	$R \times 0.9511$	$s \times 1.539$
$r \times 1.035$	$R \times 0.5176$	$r \times 0.5359$	$R \times 0.9659$	$s \times 1.866$
$r \times 1.022$	$R \times 0.4158$	$r \times 0.4251$	$R \times 0.9781$	$s \times 2.352$
$r \times 1.020$	$R \times 0.3902$	$r \times 0.3978$	$R \times 0.9808$	$s \times 2.514$
$r \times 1.013$	$R \times 0.3129$	$r \times 0.3168$	$R \times 0.9877$	$s \times 3.157$

TABLE 29–4

Square Root of Numbers

Number	Square Root	Number	Square Root	Number	Square Root	Number	Square Root	Number	Square Root
1	1.00	48	6.93	95	9.75	142	11.92	189	13.75
2	1.41	49	7.00	96	9.80	143	11.96	190	13.78
3	1.73	50	7.07	97	9.85	144	12.00	191	13.82
4	2.00	51	7.14	98	9.90	145	12.04	192	13.86
5	2.24	52	7.21	99	9.95	146	12.08	193	13.89
6	2.45	53	7.28	100	10.00	147	12.12	194	13.93
7	2.65	54	7.35	101	10.05	148	12.17	195	13.96
8	2.83	55	7.42	102	10.10	149	12.21	196	14.00
9	3.00	56	7.48	103	10.15	150	12.25	197	14.04
10	3.16	57	7.55	104	10.20	151	12.29	198	14.07
11	3.32	58	7.62	105	10.25	152	12.33	199	14.11
12	3.46	59	7.68	106	10.30	153	12.37	200	14.14
13	3.61	60	7.75	107	10.34	154	12.41	201	14.18
14	3.74	61	7.81	108	10.39	155	12.45	202	14.21
15	3.87	62	7.87	109	10.44	156	12.49	203	14.25
16	4.00	63	7.94	110	10.49	157	12.53	204	14.28
17	4.12	64	8.00	111	10.54	158	12.57	205	14.32
18	4.24	65	8.06	112	10.58	159	12.61	206	14.35
19	4.36	66	8.12	113	10.63	160	12.65	207	14.39
20	4.47	67	8.19	114	10.68	161	12.69	208	14.42
21	4.58	68	8.25	115	10.72	162	12.73	209	14.46
22	4.69	69	8.31	116	10.77	163	12.77	210	14.49
23	4.80	70	8.37	117	10.82	164	12.81	211	14.53
24	4.90	71	8.43	118	10.86	165	12.85	212	14.56
25	5.00	72	8.49	119	10.91	166	12.88	213	14.59
26	5.10	73	8.54	120	10.95	167	12.92	214	14.63
27	5.20	74	8.60	121	11.00	168	12.96	215	14.66
28	5.29	75	8.66	122	11.05	169	13.00	216	14.70
29	5.39	76	8.72	123	11.09	170	13.04	217	14.73
30	5.48	77	8.78	124	11.14	171	13.08	218	14.76
31	5.57	78	8.83	125	11.18	172	13.11	219	14.80
32	5.66	79	8.89	126	11.23	173	13.15	220	14.83
33	5.74	80	8.94	127	11.27	174	13.19	221	14.87
34	5.83	81	9.00	128	11.31	175	13.23	222	14.90
35	5.92	82	9.06	129	11.36	176	13.27	223	14.93
36	6.00	83	9.11	130	11.40	177	13.30	224	14.97
37	6.08	84	9.17	131	11.45	178	13.34	225	15.00
38	6.16	85	9.22	132	11.49	179	13.38	226	15.03
39	6.25	86	9.27	133	11.53	180	13.42	227	15.07
40	6.32	87	9.33	134	11.58	181	13.45	228	15.10
41	6.40	88	9.38	135	11.62	182	13.49	229	15.13
42	6.48	89	9.43	136	11.66	183	13.53	230	15.17
43	6.56	90	9.49	137	11.70	184	13.56	231	15.20
44	6.63	91	9.54	138	11.75	185	13.60	232	15.23
45	6.71	92	9.59	139	11.79	186	13.64	233	15.26
46	6.78	93	9.64	140	11.83	187	13.67	234	15.30
47	6.86	94	9.70	141	11.87	188	13.71	235	15.33

TABLE 29–4 (continued)

Number	Square Root	Number	Square Root	Number	Square Root	Number	Square Root	Number	Square Root
236	15.36	289	17.00	342	18.49	395	19.87	448	21.17
237	15.39	290	17.03	343	18.52	396	19.90	449	21.19
238	15.43	291	17.06	344	18.55	397	19.92	450	21.21
239	15.46	292	17.09	345	18.57	398	19.95	451	21.24
240	15.49	293	17.12	346	18.60	399	19.97	452	21.26
241	15.52	294	17.15	347	18.63	400	20.00	453	21.28
242	15.56	295	17.18	348	18.65	401	20.02	454	21.31
243	15.59	296	17.20	349	18.68	402	20.05	455	21.33
244	15.62	297	17.23	350	18.71	403	20.07	456	21.35
245	15.65	298	17.26	351	18.73	404	20.10	457	21.38
246	15.68	299	17.29	352	18.76	405	20.12	458	21.40
247	15.72	300	17.32	353	18.79	406	20.15	459	21.42
248	15.75	301	17.35	354	18.81	407	20.17	460	21.45
249	15.78	302	17.38	355	18.84	408	20.20	461	21.47
250	15.81	303	17.41	356	18.87	409	20.22	462	21.49
251	15.84	304	17.44	357	18.89	410	20.25	463	21.52
252	15.87	305	17.46	358	18.92	411	20.27	464	21.54
253	15.91	306	17.49	359	18.95	412	20.30	465	21.56
254	15.94	307	17.52	360	18.97	413	20.32	466	21.59
255	15.97	308	17.55	361	19.00	414	20.35	467	21.61
256	16.00	309	17.58	362	19.03	415	20.37	468	21.63
257	16.03	310	17.61	363	19.05	416	20.40	469	21.66
258	16.06	311	17.64	364	19.08	417	20.42	470	21.68
259	16.09	312	17.66	365	19.10	418	20.45	471	21.70
260	16.12	313	17.69	366	19.13	419	20.47	472	21.73
261	16.16	314	17.72	367	19.16	420	20.49	473	21.75
262	16.19	315	17.75	368	19.18	421	20.52	474	21.77
263	16.22	316	17.78	369	19.21	422	20.54	475	21.79
264	16.25	317	17.80	370	19.24	423	20.57	476	21.82
265	16.28	318	17.83	371	19.26	424	20.59	477	21.84
266	16.31	319	17.86	372	19.29	425	20.62	478	21.86
267	16.34	320	17.89	373	19.31	426	20.64	479	21.89
268	16.37	321	17.92	374	19.34	427	20.66	480	21.91
269	16.40	322	17.94	375	19.36	428	20.69	481	21.93
270	16.43	323	17.97	376	19.39	429	20.71	482	21.95
271	16.46	324	18.00	377	19.42	430	20.74	483	21.98
272	16.49	325	18.03	378	19.44	431	20.76	484	22.00
273	16.52	326	18.06	379	19.47	432	20.78	485	22.02
274	16.55	327	18.08	380	19.49	433	20.81	486	22.05
275	16.58	328	18.11	381	19.52	434	20.83	487	22.07
276	16.61	329	18.14	382	19.54	435	20.86	488	22.09
277	16.64	330	18.17	383	19.57	436	20.88	489	22.11
278	16.67	331	18.19	384	19.60	437	20.90	490	22.14
279	16.70	332	18.22	385	19.62	438	20.93	491	22.16
280	16.73	333	18.25	386	19.65	439	20.95	492	22.18
281	16.76	334	18.28	387	19.67	440	20.98	493	22.20
282	16.79	335	18.30	388	19.70	441	21.00	494	22.23
283	16.82	336	18.33	389	19.72	442	21.02	495	22.25
284	16.85	337	18.36	390	19.75	443	21.05	496	22.27
285	16.88	338	18.38	391	19.77	444	21.07	497	22.29
286	16.91	339	18.41	392	19.80	445	21.10	498	22.32
287	16.94	340	18.44	393	19.82	446	21.12	499	22.34
288	16.97	341	18.47	394	19.85	447	21.14	500	22.36

TABLE 29–5

American System of Weights and Measures

Linear (length) Measure

12 inches	= 1 foot
3 feet	= 1 yard
5½ yards (16½ feet)	= 1 rod
40 rods (220 yards)	= 1 furlong
8 furlongs	= 1 mile
. 320 rods	= 1 mile
5,280 feet	= 1 mile
6,080 feet	= 1 nautical mile
3 miles	= 1 league
6 feet	= 1 fathom

Square (area) Measure

144 square inches	= 1 square foot
9 square feet	= 1 square yard
100 square feet	= 1 square
30¼ square yards	= 1 square rod
160 square rods	= 1 acre
4,840 square yards	= 1 acre
43,560 square feet	= 1 acre
640 acres	= 1 square mile
" "	= 1 section

Cubic (volume) Measure

1,728 cubic inches	= 1 cubic foot
27 cubic feet	= 1 cubic yard
231 cubic inches	= 1 U.S. gallon
277.27 cubic inches	= 1 Imperial gallon

TABLE 29-5

(continued)

Avoirdupois (weight) Measure

437½ grains	= 1 ounce
16 drams	= 1 ounce
16 ounces	= 1 pound
100 pounds	= 1 hundredweight
2,000 pounds	= 1 short ton
2,240 pounds	= 1 long ton

Liquid Measure

4 fluid ounces	= 1 gill
4 gills	= 1 pint
16 fluid ounces	= 1 pint
2 pints	= 1 quart
32 ounces	= 1 quart
4 quarts	= 1 gallon
31½ gallons	= 1 barrel
2 barrels (63 gallons)	= 1 hogshead
7.4805 gallons	= 1 cubic foot

Dry Measure

2 pints	= 1 quart
8 quarts	= 1 peck
4 pecks	= 1 bushel

Surveyor's Measure

7.92 inches	= 1 link
25 links (16½ feet)	= 1 rod
4 rods (66 feet)	= 1 chain
10 square chains	= 1 acre
640 acres	= 1 square mile

TABLE 29-6

American Measure Conversion Factors

To Convert From	To	Multiply By
acres	square feet	43,560.0
acres	square yards	4,047.0
acres	square miles	0.0015625
bushels	dry pints	64.0
bushels	dry quarts	32.0
bushels	pecks	4.0
cubic feet	cubic inches	1,728.0
cubic feet	cubic yards	0.037037
cubic feet	gallons (U.S.A.)	7.481
cubic feet	gallons (Imperial)	6.22905
cubic feet of water	pounds	62.37
cubic inches	cubic feet	0.0005787
cubic inches	gallons	0.004329
cubic inches	cubic yards	0.0002143
cubic yards	cubic inches	46,656.0
cubic yards	cubic feet	27.0
fathoms	feet	6.0
feet	fathoms	0.16667
feet	inches	12.0
feet	yards	0.3333
feet	miles	0.00018939
inches	feet	0.083333
inches	yards	0.027778
gallons (U.S.A.)	cubic feet	0.13368
gallons (U.S.A.)	cubic inches	231.0
gallons (U.S.A.)	ounces	128.0
gallons (U.S.A.)	gallons (Imperial)	0.832702
gallons (Imperial)	gallons (U.S.A.)	1.20091
miles (statute)	feet	5,280.0
miles (statute)	yards	1,760.0
miles (statute)	miles (nautical)	0.8684
miles (nautical)	miles (statute)	1.1516
miles (nautical)	feet	6,080.204

TABLE 29–6

(continued)

yards	inches	36.0
yards	feet	3.0
yards	miles	0.00056818
square inches	square feet	0.00694444
square inches	square yards	0.000771605
square feet	square inches	144.0
square feet	square yards	0.111111
square feet	acres	0.0000229
square yards	square inches	1,296.0
square yards	square feet	9.0
square miles	square feet	27,878,400.0
square miles	square yards	3,097,600.0
square miles	acres	640.0
square yards	square inches	1,296.0
square yards	square feet	9.0
tons (long)	pounds	2,240.0
tons (short)	pounds	2,000.0

TABLE 29–7

The Metric System

Linear (length) Measure				
10 millimeters	= 1 centimeter	=	0.3937	inches
10 centimeters	= 1 decimeter	=	3.9370	inches
10 decimeters	= 1 meter	=	39.370	inches
100 centimeters		=	3.280	feet
10 meters	= 1 dekameter	= 393.70		inches
10 dekameters	= 1 hectometer	= 328.08		feet
10 hectometers	= 1 kilometer	=	0.62137	miles
10 kilometers	= 1 myriameter	=	6.2137	miles

TABLE 29-8

American Measure and Metric
Equivalents

Linear (length) Measure			
1 millimeter	=	0.03937	inches
1 centimeter	=	0.3937	inches
1 meter	=	39.37	inches
1 meter	=	3.2808	feet
1 kilometer	=	0.062137	miles
1 kilometer	=	1,093.63	yards
1 inch	=	2.54	centimeters
1 inch	=	25.40	milimeters
1 foot	=	0.3048	meters
1 yard	=	0.9144	meters
1 mile	=	1.609344	kilometers
1 nautical mile	=	1.852	kilometers
Square (area) Measure			
1 square centimeter	=	0.1550	square inches
1 square meter	=	1.196	square yards
1 square meter	=	10.7639	square feet
1 square kilometer	=	0.3861	square miles
1 hectare	=	2.4711	acres
1 square inch	=	6.4516	square centimeters
1 square foot	=	0.0929	square meters
1 square yard	=	0.836	square meters
1 square mile	=	2.590	square kilometers
1 acre	=	0.4047	hectare
Cubic (volume) Measure			
1 cubic centimeter	=	0.06102	cubic inches
1 cubic meter	=	1.308	cubic yards
1 cubic meter	=	35.316	cubic feet
1 cubic inch	=	16.3871	cubic centimeters
1 cubic foot	=	0.028	cubic meters
1 cubic yard	=	0.76455	cubic meters

TABLE 29–8

(continued)

Avoirdupois (weight) Measure

1 grain	=	64.7989	milligram
1 milligram	=	0.0154	grains
1 gram	=	0.0353	ounces
1 gram	=	15.432	grains
1 kilogram	=	2.2046	pounds
1 metric ton	=	1.1	short tons
1 ounce	=	28.3495	grams
1 pound	=	0.4536	kilograms
1 hundredweight	=	45.359	kilograms
1 ton	=	909.1	kilograms
1 ton	=	0.0907	metric tor

Liquid Measure

1 liter	=	1.0567	quarts
1 ounce	=	29.5727	millimete
1 pint	=	0.473	liters
1 quart	=	0.9463	liters
1 gallon	=	3.785	liters

Dry Measure

1 pint	=	0.05506	liters
1 peck	=	8.8095	liters
1 liter	=	1.8162	pints

INDEX